Lecture Notes in Mathematics

T0238608

Editors:
J.-M. Morel, Cachan
F. Takens, Groningen
B. Teissier, Paris

Alexander Y. Khapalov

Controllability of Partial Differential Equations Governed by Multiplicative Controls

 Springer

Alexander Y. Khapalov
Washington State University
Department of Mathematics
Pullman, WA 99163
USA
khapala@math.wsu.edu

ISBN: 978-3-642-12412-9   e-ISBN: 978-3-642-12413-6
DOI: 10.1007/978-3-642-12413-6
Springer Heidelberg Dordrecht London New York

Lecture Notes in Mathematics ISSN print edition: 0075-8434
                              ISSN electronic edition: 1617-9692

Library of Congress Control Number: 2010927757

Mathematics Subject Classification (2000): 35, 93, 76, 49, 92

Cover design: SPi Publisher Services

Printed on acid-free paper

springer.com

To Irina, Elena and Dasha

Foreword

In a typical mathematical model of a controlled distributed parameter process one usually finds either boundary or internal locally distributed controls to serve as the means to describe the effect of external actuators on the process at hand. However, these classical controls, entering the model equations as *additive* terms, are not suitable to deal with a vast array of processes that can change their principal intrinsic properties due to the control actions. Important examples here include (but not limited to) the *chain reaction*-type processes in biomedical, nuclear, chemical and financial applications, which can change their (reaction) rate when certain "catalysts" are applied, and the so-called "smart materials", which can, for instance, alter their frequency response.

The goal of this monograph is to address the issue of global controllability of partial differential equations in the context of *multiplicative (or bilinear) controls*, which enter the model equations as coefficients. The mathematical models of our interest include the linear and nonlinear parabolic and hyperbolic PDE's, the Schrödinger equation, and coupled hybrid nonlinear distributed parameter systems associated with the swimming phenomenon.

Pullman, WA, USA   *Alexander Khapalov*
January 2010

vii

Preface

This monograph developed from the research conducted in 2001–2009 in the area of controllability theory of partial differential equations. The concept of controllability is a principal component of Control Theory which was brought to life in the 1950's by numerous applications in engineering, and has received the most significant attention both from the engineering and the mathematical communities since then.

A typical control problem deals with an evolution process which can be affected by a certain parameter, called control. Normally, the goal of a control problem is to steer this process from the given initial state to the desirable target state by selecting a suitable control among available options. If this is indeed possible, then one usually desires to achieve this steering while optimizing a certain criterion, solving what is called an optimal control problem.

Controllability theory studies the first part of the above-described control process. Namely, given any initial state, it studies the richness of the range of the mapping: *control* → *state of the process* (at some moment of time).

Controllability theory was originally developed in the 1960's for the linear ordinary differential equations, governed by the additive controls. Later, since the 1970's it became the subject of keen interest for the researchers working in the area of partial differential equations as well. As a result, nowadays there exists a quite comprehensive controllability theory for the linear pde's governed by the additive controls which can act inside the system's space domain (locally distributed or point controls) or on its boundary (boundary controls). In such context, the respective mathematical methods are essentially the methods of the theory of linear operators, particularly, of the duality theory.

In this monograph, however, the subjects of interest are the multiplicative controls that enter the system equations as coefficients. Therefore, the aforementioned *control* → *state of the process* mapping becomes highly nonlinear, even if the original pde is linear. This gives rise to the necessity of developing a different methodology for this type of controllability problems.

In this monograph we address this issue in the context of linear and semilinear parabolic and hyperbolic equations, as well as the Schrödinger equation. Particular attention is given to nonlinear swimming models. In the introduction we discuss the motivation for the use of multiplicative controls as opposed to the classical additive ones, and compare the mathematical methods involved in the respective studies.

Acknowledgments:

The author's research presented in this monograph was supported in part by the NSF Grants DMS-0204037 and DMS-0504093.

The author wishes to express his gratitude to Ron Mohler who introduced him to the fascinating area of multiplicative controls.

The author wishes to express his gratitude to the Department of Mathematics of the University of Rome II "Tor Vergata" and to the Instituto Nazionale di Alta Matematica (Italy) for hospitality throughout his sabbatical in the Spring semester of 2008 during which parts of this monograph were written.

The author also wishes to thank the reviewers for their comments and suggestions which made a significant impact on the content of this monograph.

Pullman, WA, USA   *Alexander Khapalov*
January 2010

Contents

Chapter 1
Introduction

1.1 Controlling PDE's: Why Multiplicative Controls?

In the mathematical models associated with controlled distributed parameter systems evolving in bounded domains two types of controls – boundary and internal locally distributed – are typically used. These controls enter the model as *additive* terms (having in mind that the boundary controls can be modeled by making use of suitable additive Dirac's functions) and have localized support. The latter is either a part of the boundary or a set within the system's space domain. Such control can, for example, be a source in a heat/mass-transfer process or a piezoceramic actuator placed on a beam. Publications in this area are so numerous that it is simply impossible to mention them all here – allow us just to refer the reader to our very limited bibliography below, associated mostly with the immediate content of this monograph and to the references therein.

In terms of applications it appears that the above-mentioned classical additive controls can adequately model only those controlled processes which do not change their principal physical characteristics due to the control actions. They rather describe the affect of various externally added "alien" sources or forces on the process at hand. This limitation, however, excludes a vast array of new and not quite new technologies, such as, for example, "smart materials" and numerous biomedical, chemical and nuclear chain reactions, which are able to change their principal parameters (e.g., the frequency response or the reaction rate) under certain purposefully induced conditions ("catalysts").

The intent of this monograph is to address the just-outlined issues in the context of global controllability of partial differential equations through the introduction and study of *multiplicative* (also known as *bilinear*) controls. These controls enter the system equations as *coefficients*. Accordingly they can change at least some of the principal parameters of the process at hand, such as, for example, a natural frequency response of a beam or the rate of a chemical reaction. In the former case this can be caused, e.g., by embedded "smart" alloys and in the latter case by various catalysts and/or by the speed at which the reaction ingredients are mechanically mixed. Our *main goal in this monograph is to introduce a new controllability methodology suitable for the study of linear and nonlinear pde's in the framework of multiplicative controls.*

A.Y. Khapalov, *Controllability of Partial Differential Equations Governed by Multiplicative Controls*, Lecture Notes in Mathematics 1995, DOI 10.1007/978-3-642-12413-6_1, © Springer-Verlag Berlin Heidelberg 2010

It is also important to notice that currently there are only very few publications available in the area of controllability of distributed parameter systems by means of multiplicative controls. This is in a very sharp contrast with the corresponding research in the framework of ordinary differential equations where many interesting results were obtained for the period of several decades now (see, e.g., the survey [5] and remarks on bibliography in the end of this introduction).

Let us give now several examples of important applications motivating the use of multiplicative controls in the framework of pde's. As the reader will see later, different types of pde's give rise to different concepts of controllability. (In other words: "What is the "right question" to ask here?)

Example 1.1 (The nuclear chain reaction). This chain reaction is characterized by the fact that the number of particles of the diffusing material increases by the reaction with the surrounding medium. For example, a nuclear fission results from the collision of neutrons with active uranium nuclei, which leads to the occurrence of new neutrons whose number is greater than one. These neutrons, in turn, react with active nuclei in the same way and hence the number of neutrons increases. If this process is treated approximately as a linear diffusion process, we arrive to the following (simplified) equation:

$$u_t = a^2 \Delta u + v(x,t)u, \qquad (1.1)$$

where $u(x,t) \geq 0$ is the neutron density at point x at time t and $v > 0$, since the chain reaction is equivalent to the existence of sources of diffusing materials (neutrons) proportional to their concentration (neutron density).

In a nuclear plant the chain reaction is typically controlled by means of so-called "control rods," which in turn can absorb neutrons. In equation (1.1) this can be associated with the change of the value and sign of the coefficient v, which thus can be regarded as a multiplicative control.

Note that, if one wants to use the traditional additive control to describe the above fission model, then this would lead one to an equation like

$$u_t = a^2 \Delta u + v(x,t),$$

where $v(x,t)$ is the additive locally distributed control. In terms of applications, this type of control would amount to controlling the chain reaction by somehow adding into or withdrawing out of the chamber at will a certain amount of neutrons, which is not realistic.

A similar modeling approach applies to numerous biomedical, chemical and heat- and mass-transfer reactions and other processes (e.g., arising in the population dynamics and financial mathematics), involving various types of "catalysts." The corresponding models can also be nonlinear.

Example 1.2 (Biomedical applications). The following system of nonlinear equations models the interaction of leukocytes and "invading" bacteria in a cell (see [3] and, e.g., [60], p. 499).

Denote by u, y and z respectively the leukocyte, bacterial, and attractant concentrations. Then we have the following:

1. Bacteria diffuse, reproduce, and are destroyed when they come in contact with leukocytes:

$$y_t = \mu y_{xx} + (k_1 - k_2 u)y. \tag{1.2}$$

2. The chemoattractant is produced by bacterial metabolism and diffuses:

$$z_t = D z_{xx} + k_3 y. \tag{1.3}$$

3. The leukocytes are chemotactically attracted to the attractant and they die as they digest the bacteria, so that

$$u_t = J_x + (k_4 - k_5 y)u, \tag{1.4}$$

where $J = k_6 u + k_7 u z_x$ is the leukocyte flux.

In the above $k_i, i = 1, \ldots, 7$ and D are various coefficients, and equations (1.2)–(1.4) are complemented by a set of boundary conditions – we refer to [60] for details. The question of interest here is – *if and when the leukocytes can successfully fight against a bacterial invasion* (again see [60] and the references therein for different ways to approach this issue). One can try to analyze this very challenging nonlinear problem as a controllability one with the goal to achieve the steering to a suitable equilibrium by means of (some of) the coefficients k_i's treated as bilinear controls. The corresponding control actions can be interpreted, e.g., as the use of a drug to create the conditions (that is, to change the reaction rate) such that bacteria will die. (To the contrary, the use of traditional additive controls would mean that one has an option to add into or withdraw out of the given cell some leukocytes or bacteria, which does not seem realistic.)

Example 1.3 (Non-homogeneous bilinear system). Denote by $u(x,t)$ the temperature of a rod of unit length at point x at time t. Then the following model describes the heat-transfer in this rod in accordance with Newton's Law (e.g., [142]):

$$u_t = u_{xx} + v(u - \theta(x,t)) \text{ in } Q_T = (0,1) \times (0,T), \tag{1.5}$$
$$u = 0 \text{ in } \Sigma_T, \quad u \mid_{t=0} = u_0 \in L^2(\Omega).$$

Here the term $v(u - \theta(x,t))$ describes the heat exchange between the rod and the surrounding medium of temperature $\theta(x,t)$. We can regard v as a bilinear control. It is known that v is proportional to the heat-transfer coefficient, which depends on the environment, the substance at hand, and its surface area. Note that in this example the "mathematical boundary" in the corresponding initial and boundary value problem is not the same as the actual physical boundary of the body.

Non-homogeneous models as in (1.5) arise also in many other applications dealing with the heat- and mass-transfer. If the heat (mass)-transfer involves fluids (air), the corresponding bilinear control v also depends on the speed of the fluid. The latter can be controlled in some applications by the induced magnetic field. Alternatively, the surface area can be changed when the substance at hand is a polymer (e.g., a planar array of gel fibers can be controlled to maximize the surface area exposed to the surrounding fluid). Also, we refer to the so-called "extended" surface applications ("smartly" added/controlled fins, pins, studs, etc.) when one wishes to increase/decrease the exchange between the source and ambient fluid.

Example 1.4 (Variable vibration response). An important practical example here is the SMA-composite beam containing NiTi fibres that can change its vibration response when heated by an electrical current (this can be interpreted as a variable load).

We found only two early references related to the area of bilinear controllability for the linear wave and beam equations. Namely, in the pioneering work [8] by J.M. Ball, J.E. Marsden, and M. Slemrod the approximate controllability of the rod equation

$$u_{tt} + u_{xxxx} + v(t)u_{xx} = 0$$

with hinged ends and of the wave equation

$$u_{tt} - u_{xx} + v(t)u = 0$$

with Dirichlet boundary conditions, where v is control (the axial load), was shown making use of the nonharmonic Fourier series approach under the additional "non-traditional" assumption that all the modes in the initial data are active. The results of [8] are discussed in detail in Chapter 9 below.

We also refer to [90] further exploring the ideas of [8] in the context of simultaneous control of the rod equation and Schrödinger equation.

Example 1.5 (Multiplicative controllability of the Schrödinger equation). This equation arises in such modern technologies as nuclear magnetic resonance, laser spectroscopy and quantum information science.

Let us give an example of a possible setup of multiplicative controllability problem for the Schrödinger equation due to Rouchon [131], Beauchard [11], and Beauchard and Coron [16].

Consider a quantum particle of mass m with potential V in a non-Galilean frame of absolute position $D(t)$ in R^1. It can be represented by a complex-valued wave function $\phi(t,z)$ which solves the following Schrödinger equation:

$$i\hbar\frac{\partial\phi}{\partial t}(t,z) = -\frac{\hbar^2}{2m}\frac{\partial^2\phi}{\partial z^2}(t,z) + V(z-D(t))\phi(t,z). \qquad (1.6)$$

With a suitable change of variables (see [16]) and assuming that $m = 1$ and $\hbar = 1$, equation (1.6) can be re-written as follows:

$$i\frac{\partial \psi}{\partial t}(t,x) = -\frac{1}{2}\frac{\partial^2 \psi}{\partial x^2}(t,x) + (V(x) - u(t)x)\psi(t,x), \qquad (1.7)$$

where $u(t) = -\ddot{D}(t)$. Equation (1.7) describes the nonrelativistic motion of particle with potential V in a uniform electrical field $t \to u(t) \in R$, which can be viewed as a *multiplicative control*.

Recently, a substantial progress has been made in the study of controllability properties of (1.7), see [11, 12, 14, 16, 18, 26, 35, 124] and the references therein. We discuss these results in Part IV below.

Example 1.6 (Swimming phenomenon). The swimming phenomenon is undoubtedly among the most interesting mathematical problems arising in the fluid mechanics. We discuss this very intriguing phenomenon from the multiplicative controllability viewpoint in Part III.

1.2 Additive Controls vs Multiplicative Controls: Methodology

Let us try to highlight, in an informal setting, some of the principal differences between additive and multiplicative controls in terms of approach to the concept of controllability itself and of the mathematical methods which are typically used to study the respective properties.

1.2.1 Additive Controls

Consider the following abstract evolution equation governed by an *additive control*:

$$\frac{dy(t)}{dt} = Ay(t) + Bv(t), \quad t \in (0, T), \ T > 0, \qquad (1.8)$$

$$y(0) = y_0 \in H.$$

Here A is a self-adjoint operator with dense domain in the real Hilbert space H, generating a C^o semigroup of bounded linear operators on H, $v \in L^2((0,T);V)$ is control with values in the Hilbert space V and $B : V \to H$ is a linear bounded operator. Let us assume that we are interested in the study of controllability properties of solutions to (1.8) in $C([0,t_0];H)$.

Example 1.7. An example of this setup can be the following mixed problem for the heat equation governed by the locally-distributed control:

$$u_t = \Delta u + \kappa_\omega v \text{ in } \Omega \times (0,T) = Q_T, \qquad (1.9)$$

$$u = 0 \text{ in } \partial\Omega \times (0,T),$$

$$u \mid_{t=0} = u_0 \in L^2(\Omega) = H.$$

Here Ω is a bounded domain in R^n with boundary $\partial\Omega$, $v \in L^2(Q_T)$ is control supported in the given subdomain ω of Ω, and $\kappa(x)$ is the characteristic function of ω: $\kappa(x) = 1$ for $x \in \omega$ and $\kappa(x) = 0$ elsewhere in Ω.

Other typical examples, motivating a system like (1.8), are the linear parabolic and hyperbolic equations governed either by the locally distributed controls or by controls supported on (a part of) the boundary $\partial\Omega$ or by the point controls (in the last two cases one needs to use a suitable δ-function for a setup like (1.8)).

Denote the set of all states of system (1.8) that can be achieved at time $T > 0$ from the given initial state y_0 by $\mathscr{Y}(T, y_0)$. Then, if this set is dense in H for any $y_0 \in H$ it is said that (1.8) is *approximately controllable in H at time T*. If $\mathscr{Y}(T, y_0) = H$ for any $y_0 \in H$, then it is said that system (1.8) is *controllable or exactly controllable in H at time T*. Due to the affine nature of control term Bv, to establish the above properties, it suffices to study the properties of the set $\mathscr{Y}(T, y_0 = 0)$ only.

It is well-known that the issue of controllability of (1.8) is linked to the observability properties of its so-called *dual* system. The duality theory was first developed in the 1960's to study controllability/observability properties of the linear ODE's. In the context of infinite dimensional systems it was formalized by S. Dolecki and D. Russell in [34]. J.-L. Lions introduced in [111, 112] a very close theory, known as the *Hilbert Uniqueness Method* which employs uniqueness theorems to construct Hilbert spaces in which exact controllability holds for the given control operator.

The dual system for (1.8) is as follows:

$$\frac{d\varphi(t)}{dt} = -A\varphi(t), \quad t \in (0,T),$$

$$\varphi(T) = \varphi_T \in H.$$

Introduce the following, so-called *observation* operator:

$$\mathbf{G} : H \to L^2((0,T);V), \quad \mathbf{G}\varphi_T = B^*\varphi,$$

where B^* is dual of the control operator B. Then, assuming that $w_0 = 0$ in (1.8), we have the *duality relation*:

$$< y(T), \varphi_T >_H = < v, \mathbf{G}\varphi_T >_{L^2((0,T);V)},$$

where $< \cdot, \cdot >$ stands for the inner product in the respective Hilbert space. This relation yields that *system (1.8) is approximately controllable in H at time T if the operator G has an inverse, and it is exactly controllable if G^{-1} is bounded.*

Thus, in the case of linear system like (1.8) the solution to the controllability problem can be reduced to the analysis of operator G:

- To show the approximate controllability of (1.8) in H at time $T > 0$, one needs to establish the respective *unique continuation property*, namely, that the equality $G\varphi_T = 0$ yields $\varphi_T = 0$.
- To show the exact controllability of (1.8) in H at time $T > 0$, one needs to establish an *observability* estimate like

$$\| G\varphi_T \|_{L^2((0,T);V)} \geq \gamma \| \varphi_T \|_H \quad \forall \varphi_T \in H, \ \gamma > 0.$$

Example 1.8. The following mixed problem describes the dual system for (1.9):

$$\varphi_t = -\Delta\varphi \quad \text{in } Q, \tag{1.10}$$
$$\varphi = 0 \quad \text{in } \partial\Omega \times (0,T),$$
$$\varphi \mid_{t=T} = \varphi_T \in L^2(\Omega).$$

The observation operator $G : L^2(\Omega) \rightarrow L^2(0,T;L^2(\omega))$ is defined by the equality $G\varphi_T = \varphi$ in $\omega \times (0,T)$. To show that system (1.9) is approximately controllable in $L^2(\Omega)$ at time $T > 0$, one needs to verify that, if φ vanishes in $\omega \times (0,T)$, then φ_T vanishes in Ω. This can be done by making use of the respective unique continuation result for solutions to (1.10) (namely, from $\omega \times (0,T)$ to $\Omega \times (0,T)$), see, e.g., [10]. One does not have the exact controllability in $L^2(\Omega)$ due to the smoothing property of solutions to the heat equation.

The controllability theory of linear pde's with additive controls has received a significant attention in the literature in the last four decades. Many powerful mathematical methods were introduced and adopted to tackle a variety of controllability problems along the above-outlined duality approach, including harmonic and nonharmonic analysis, the multiplier method, unique continuation, Carleman's estimates and microlocal analysis. It is impossible to mention here all the fundamental results obtained within this area. Therefore let us refer the reader only to some of the respective publications, including several excellent monograph on this subject (alphabetically): C. Bardos, G. Lebeau, J. Rauch [20], A.G. Butkovskii [22], R. Dáger and E. Zuazua [33], H.O. Fattorini and D.L. Russell [38], A. Fursikov and O. Imanuvilov [50], A. Khapalov [69], V. Komornik [93], V. Komornik and P. Loreti [94], J. Lagnese [98], J. Lagnese and J.-L. Lions [99], J. Lagnese, G. Leugering, and E.J.P.G. Schmidt [100], I. Lasiecka [101], I. Lasiecka and R. Triggiani [102–104], J.L. Lions [111–113], W.A.J. Luxemburg, and J. Korevaar [115], D. Russell [132], D. Tataru [138], E. Zuazua [150–152], and the references therein.

The methods developed for the linear pde's with additive controls were also successfully extended to the semilinear versions of these equations, making use of the inverse mapping or implicit function theorems, see, e.g., in addition to the above references (alphabetically): V. Barbu [9], P. Cannarsa, V. Komornik, and P. Loretti [24], C. Fabre, J.-P. Puel and E. Zuazua [36], H.O. Fattorini [39], E. Fernández-Cara [43], E. Fernández-Cara and E. Zuazua [46], J. Henry [57], A.Y. Khapalov [61,63,64,66], I. Lasiecka and R. Triggiani [102], E. Zuazua [146,147,149] and the references therein.

1.2.2 Multiplicative Controls

A representative, matching (1.8), example of an abstract evolution equation, governed by a multiplicative control, can be written as follows:

$$\frac{dw(t)}{dt} = Aw(t) + p\mathscr{B}(w(t)), \quad t > 0, \tag{1.11}$$
$$w(0) = w_0,$$

where p is *multiplicative* control from a suitable Banach space P and $\mathscr{B} : H \to H$ is a bounded operator. To (informally) compare systems (1.8) and (1.11), we can assume, without loss of generality, that A is the same self-adjoint operator as in (1.8) and that system (1.11) has a unique generalized solution in $C([0,t_0];H)$.

Example 1.9. An example of this setup, matching to system (1.9) with additive control, can be written as follows:

$$u_t = \Delta u + p(x,t)u, \quad \text{in } \Omega \times (0,T) = Q_T, \tag{1.12}$$
$$u = 0 \quad \text{in } \partial\Omega \times (0,T),$$
$$u\,|_{t=0} = u_0 \in L^2(\Omega),$$

where $p \in L^\infty(Q)$ is control (see Part I).

An intrinsic feature of the linear system (1.8) is that its solution mapping $v \to y(t)$ is linear for $y_0 = 0$. In comparison, the bilinear nature of control term in equation (1.11) implies that the solution mapping $p \to w(t)$ is *highly nonlinear* for any initial datum $w_0 \neq 0$. Furthermore, $w = 0$ is the fixed point for system (1.11), which means that it cannot be steered anywhere from $w_0 = 0$. Therefore, the duality approach developed to study controllability properties of the linear system (1.8) does not apply to system (1.11). This gives rise to the need for a new methodology to study controllability of pde's, governed by multiplicative controls.

In their turn, the classical definitions of approximate and exact controllability, developed for additive controls, will also require modification. Let us highlight some of such possible modifications:

- First of all, instead of studying the properties of the set $\mathscr{Y}(T,y_0)$ of all states of the *linear* system (1.8) reachable from the given initial state at a *fixed time* T, in the case of the *bilinear* system (1.11) we will study the set of all its states that are reachable at *any finite time*. Denote this set for the given w_0 by $\mathscr{W}(w_0)$:

$$\mathscr{W}(w_0) = \{w(t) \mid w \text{ is the solution to } (1.11) \text{ for some } p \in P,\, t > 0\}.$$

Respectively, we will say that (1.11) is *globally approximately controllable from* w_0 in H if $\mathscr{W}(w_0)$ is dense in H and that it is *locally approximately controllable from* w_0 in H if $\mathscr{W}(w_0)$ is dense in some neighborhood of w_0 in H. These two controllability properties, i.e., *with no preassigned duration of control time*, are also often regarded as the *reachability* or *attainability* properties of the controlled system at hand.

- In the case of system (1.8), the local approximate (or exact) controllability implies the global one. This is not the case of the multiplicative controls.
- Typically, one cannot expect the global or local exact controllability of (1.11) in H, see Theorem 9.2 in Chapter 9 (due to [8]).
- In general, we also do not have approximate controllability in H for a bilinear equation like (1.11). For example:

  - In the case of *the heat equation* with the homogeneous Dirichlet boundary conditions, governed via the coefficient in the reaction term as given in (1.12) (see Part I), the maximum principle implies that if the initial condition is non-negative, then its solution will remain nonnegative, regardless of the choice of control. Therefore, a natural controllability question would rather be: *Can we steer this equation from any nonzero nonnegative initial state to any non-negative state in $H = L^2(\Omega)$ as close as we wish?* We will call the respective property – the *nonnegative* (or, alternatively, *nonpositive*) approximate controllability in $L^2(\Omega)$.
  - In the case of multiplicative controls the choice of initial state can be critical. For example, one can have the global approximate controllability only from certain nonzero initial states. In the case of hyperbolic systems, governed by the time-dependent controls only, these could be the ones *with all modes active*, see Chapter 9 in Part II.
  - One can also study the (exact) controllability in subspaces of H (see the same chapter).
  - In the case of the Schrödinger equation as in Example 1.5 above, the L^2-norm of its solutions remains unchanged, regardless of the choice of the real-valued control applied. Therefore, one can only study both local and global controllability properties of this equation on the L^2-sphere of the radius which is equal to the L^2-norm of the initial state (see Part IV).

Multiplicative methodology. The methods of the pioneering paper [8] (we discuss them in detail in Chapter 9) make use of the inverse function theorem and of the concept of Riesz bases under the aforementioned *additional assumption* that all the modes in the initial state are present.

The general methodological strategy in Parts I–III can be characterized as a *constructive qualitative approach*. Its main idea is to:

- distinguish a desirable intrinsic dynamics of the system at hand and then
- use a carefully selected multiplicative control which will make it dominating on the time-interval (asymptotically short or very long) when this control is applied.
- One can "switch" between such dynamics to steer the system at hand towards the desirable target state.

In Parts I and II, where we study controllability properties of linear and semilinear parabolic and hyperbolic pde's, the *critical first step is the selection of a static (time-independent) multiplicative control in such a way that the desirable target state becomes one of the eigenfunctions* of the resulting pde at hand (or as "close" to it as we wish). Then we apply a sequence of constant multiplicative controls which will change the values of the eigenvalues on the respective time-intervals when these controls are active. Suitable choice of these controls and the duration of their actions will allow us to switch between the intrinsic dynamics of the system at hand to steer it to the target state.

For example, in the case of the semilinear parabolic equation (see Chapters 2 and 3 in Part I) the following three dynamics will be distinguished:

- *Pure diffusion dynamics* – associated with the Laplace operator.
- *Pure reaction dynamics* – due to the controlled reaction term, which can dominate on some time-interval when a suitable multiplicative control is active.
- *Nonlinear "disturbance"* – the dynamics caused by the (generic) nonlinear term. It can be suppressed for some period of time by the actions of "large" multiplicative controls.

Respectively, one can try to select a multiplicative control in such a way that the corresponding trajectories of the parabolic system at hand can be approximated by those associated with the first two dynamics, while suppressing the effect of nonlinearity.

In Part II (Chapters 6–7), the key qualitative observation is the following. Let $\{c_k(t), k = 1, \ldots\}$ be the sequence of coefficients in the Fourier series representation of the solution to the wave equation (with no nonlinear term) at time $t > 0$, which is acted upon by a constant multiplicative control. Then the available motions of the point $(c_k(t), \dot{c}_k(t))$ in the two dimensional space, describing the dynamics of the kth mode of the solution, include: (i) the *ellipsoidal motions* around the origin and (ii) the *horizontal motions* towards and away from the origin. We will show that one can find a sequence of constant controls (i.e., the resulting control will be piecewise constant in time) in such a way that the pairs $\{c_k(t), \dot{c}_k(t)\}, k = 1, \ldots$ can be controlled in a desirable way, one at a time, *independently* from the subsequent control actions with respect to other modes.

In Part III, where we study controllability properties of swimming models, such a key qualitative observation is the fact that *a body in a fluid will move in the direction where it meets the least resistance.* The main idea here is to distinguish a principal nonlinear dynamics of the model at hand, which corresponds to this phenomenon, and use it to derive a constructive formula for micromotions of the swimmer it describes.

In Part IV we discuss the recent results on controllability properties of the Schrödinger equation on a unit L^2-sphere. The dominating method to study the local exact controllability properties is the classical idea to use the implicit function or inverse mapping theorems. For nonlocal results, this approach is complemented by a compactness argument. For the approximate controllability two approaches will be discussed. The first is based on the use of bilinear ode's techniques, applied to the Galerkin approximations. The established finite dimensional controllability result is then extended to the original pde. The second approach makes use of the stabilization technique, combined with the complex-conjugate time-reversibility of the Schrödinger equation and the fact that the distance between any two of its solutions with the same real-valued control remains constant in the L^2-norm.

Remarks on early publication in the area of multiplicative controllability:

- To our knowledge, there were no published works on bilinear controllability of reaction-diffusion-convection equations or swimming models prior to author's recent works [67, 70–72], providing the basis for Part I of this monograph.
- Among earlier works on the controllability of other types of (linear) pde's by means of bilinear controls we can only name already mentioned [8] and [90] (further exploring the ideas of [8]), and recent [13] and [15].
- In contrast to the above-described situation, an extensive and thorough bibliography on controllability of bilinear ode's is available. Let us just mention (alphabetically) along these lines the works by A. Agrachev, B. Bonnard, W.M. Boothby, R.W. Brockett, J.P. Gauthier, V. Judrjevic, D.E. Koditschek, I. Kupka, K.S. Narendra, R.E. Rink, G. Sallet, H.S. Sussman, E.N. Wilson, to name just a few – see, e.g., the survey [5]. Research in this area was seemingly originated in the 60s, on the one hand, by the works of J. Kucera [95], who linked this area to the Lie Algebra approach, and, on the other hand, by the works of R.R. Mohler and R. Rink [120,130] (see also [83–89] and the references therein), who pursued the qualitative approach and numerous applications.
- **Optimal control of bilinear pde's.** We also refer to the works by S. Lenhart and others [21, 105–107] (see also the references therein) on the issue of *optimal* bilinear control for various pde's.
- Aside from controllability, a closely related issue is *stabilization by means of bilinear controls.* For example, this is the way how a controlled chain reaction can be viewed. Again, we can point out only at a very limited number of publications in this area in terms of pde's, see [6, 7, 25, 122].

The monograph is organized as follows. In Part I we discuss the issues of global multiplicative controllability in the context of both linear and semilinear parabolic equations, which may also include superlinear terms. Part II deals with the same

issues for both linear and semilinear hyperbolic equations. In Part III we introduce
a series of simplified abstract swimming models and discuss their well-posedness
and both local and global controllability. In Part IV we give an account of some
recent results on controllability properties of the Schrödinger equation governed by
the real-valued multiplicative control.

Part I
Multiplicative Controllability
of Parabolic Equations

Chapter 2
Global Nonnegative Controllability of the 1-*D* Semilinear Parabolic Equation

Abstract In this chapter we study the global approximate controllability of the one dimensional semilinear convection-diffusion-reaction equation governed in a bounded domain via a coefficient in the reaction term. Even in the linear case, due to the maximum principle, such system is not globally or locally controllable in any reasonable linear space. It is also well known that for the superlinear terms admitting a power growth at infinity the global approximate controllability by traditional additive controls of localized support is out of question. However, we will show that a system like that can be steered in $L^2(0,1)$ from any non-negative nonzero initial state into any neighborhood of any desirable non-negative target state by at most three static (*x*-dependent only) multiplicative controls, applied subsequently in time, while only one such control is needed in the linear case.

2.1 Introduction

We begin Part I by considering in Chapter 2 the one dimensional semilinear convection-diffusion-reaction equation governed via the coefficient in the additive reaction term. Besides the actual controllability results, one of our main goals here is to illustrate the conceptual difference between a controllability problem by means of *classical additive controls* and a controllability problem by means of *multiplicative controls*, which, in particular, gives rise to a new definition for controllability.

In the end of Chapter 2 we discuss the further layout of the remainder of Part I, explicitly focusing on the methodological differences between the problems we consider in Part I.

2.1.1 Problem Formulation and a Concept of Controllability

In this section we consider the following Dirichlet boundary problem:

$$u_t = u_{xx} + \alpha u - f(x,t,u,u_x) \quad \text{in } Q_T = \Omega \times (0,T), \ \Omega = (0,1), \quad (2.1)$$
$$u(0,t) = u(1,t) = 0, \quad t \in (0,T), \quad u\mid_{t=0} = u_0 \in L^2(\Omega),$$

A.Y. Khapalov, *Controllability of Partial Differential Equations Governed by Multiplicative Controls*, Lecture Notes in Mathematics 1995,
DOI 10.1007/978-3-642-12413-6_2, © Springer-Verlag Berlin Heidelberg 2010

where $\alpha \in L^\infty(Q_T)$ is a control function. Given $T > 0$, we further assume that $f(x, t, u, p)$ is Lebesgue's measurable in x, t, u, p, and continuous in u, p for almost all $(x, t) \in Q_T$, and is such that

$$| f(x, t, u, p) | \le \beta | u |^{r_1} + \beta | p |^{r_2} \quad \text{a.e. in } Q_T \text{ for } u, p \in R, \quad (2.2)$$

$$\int_\Omega f(x, t, \phi, \phi_x) \phi \, dx \ge 0 \quad \forall \phi \in H_0^1(0, 1), \quad (2.3)$$

where $\beta > 0$, and

$$r_1 \in (1, 5), \quad r_2 \in (1, 5/3). \quad (2.4)$$

A simple example of a function f satisfying conditions (2.2)–(2.4) is $f(u) = u^3$. Note that (2.2) and (2.4) mean that $| f |$ is bounded above by a strictly superlinear function in u and p near the origin as well (see also Remark 2.5 below in this respect).

Here and below we use the standard notations for Sobolev spaces such as $H_0^{1,0}(Q_T) = L^2(0, T; H_0^1(\Omega)) = \{\phi \mid \phi, \phi_x \in L^2(Q_T), \phi(\cdot, t) \mid_{\partial\Omega} = 0\}$ and $H_0^1(\Omega) = \{\phi \mid \phi, \phi_x \in L^2(\Omega), \phi \mid_{\partial\Omega} = 0\}$. We refer, e.g., to [97] (p. 466), where it was shown that system (2.1), (2.2)–(2.4) admits at least one generalized solution in $C([0, T]; L^2(\Omega)) \cap H_0^{1,0}(Q_T) \cap L^6(Q_T)$, but its uniqueness is not guaranteed.

Our goal in this section is to study the global approximate controllability properties of system (2.1), (2.2)–(2.4).

Let us remind the reader that, in its classical form, it is said that the system at hand is globally *approximately controllable* in the given (*linear* phase-) space H at time $T > 0$ if it can be steered in H from any initial state into any neighborhood of any desirable target state at time T, by selecting a suitable available (traditionally, additive) control. In turn, the system at hand is globally *exactly null-controllable* in H if it can be steered in H from any initial state to the zero-state exactly within the given time-interval $[0, T]$. However, it is not unnatural to expect that *the use of multiplicative controls will "demand" some principal modifications of the aforementioned concepts*.

Note, first of all, that, even in the case when $f = 0$, that is, when (2.1) becomes the following linear boundary problem:

$$y_t = y_{xx} + \alpha y \quad \text{in } Q_T = (0, 1) \times (0, T), \quad (2.5)$$
$$y(0, t) = y(1, t) = 0, \quad t \in (0, T), \quad y \mid_{t=0} = y_0 \in L^2(0, 1),$$

its solution *still* depends highly nonlinearly on α, which makes the associated controllability problem *nonlinear* as well.

Secondly, whenever the zero state is a fixed point for the system at hand, it is immediately clear that such a system is not globally controllable from $y_0 = 0$ and its exact null controllability is out of question as well.

Thirdly, even if $y_0 \ne 0$ and $f = 0$ the resulting seemingly "simpler" linear system (2.5) still cannot be approximately controllable by bilinear control α in any "reasonable" linear space. Indeed, e.g., if $y_0(x)$ is non-negative, then, the maximum principle implies that $y(x,t)$ must remain non-negative for all $t > 0$, regardless of the choice of $\alpha \in L^\infty(Q_T)$ (for more general regularity assumptions on system (2.5) for which the maximum principle still holds we refer the reader, e.g., to [97], p. 181). In other words, one is unable to reach any of the "negative" (or, alternatively, posi- tive) target states from a non-negative (or, alternatively, from non-positive) initial state. Accordingly, it seems of interest to study the approximate controllability of (2.1)/(2.5) while dealing exclusively with the initial and target states of the same sign.

In terms of applications, such modification is very natural. For example, if (2.5) describes a diffusion process with $y(x,t)$ being the concentration of a substance at point x at time t, then, of course, it cannot accept the negative values. Alternatively, if (2.5) describes a heat-transfer process, with $y(x,t)$ being the temperature at point x at time t, then it is also natural to assume that the temperature cannot fall below certain level, determined by the initial temperature distribution and the boundary conditions (not to mention the physical concept of the absolute zero).

For certainty, and without loss of generality, *everywhere below in this section we will deal with non-negative initial and target states only (unless we specifically say otherwise)*.

Also, the fact that the bilinear control α in (2.1)/(2.5) is a coefficient opens, in turn, certain possibilities to steer the system at hand to the desirable target state by creating suitable "drift" motions towards it. Namely, by making this target state a new equilibrium. This can be achieved by static (time-independent) bilinear con- trols, i.e., of a much simpler structure to implement, compared to, e.g., the traditional x- and t-dependent locally distributed $L^2(Q_T)$-controls. On the other hand, such static controls will not allow us to comply with the "traditional approximate con- trollability" requirement that the control time should be the same for any pair of the initial and target states (in which case, of course, the magnitudes of respective "traditional" controls must generally increase when control time decreases and/or the distance between the initial and target states increases).

Finally, the above-cited classical definition of approximate controllability be- comes ill-posed (and rather "questionable" in terms of applications) when the system at hand admits multiple solutions.

We summarize the above discussion by introducing the following definition.

Definition 2.1. We will say that system (2.1)–(2.4), generally admitting multiple solutions, is "non-negatively" globally approximately controllable in $L^2(\Omega)$ if for every $\varepsilon > 0$ and non-negative $u_0, u_d \in L^2(\Omega), u_0 \neq 0$ there is a $T = T(\varepsilon, u_0, u_d)$ and multiplicative/bilinear control $\alpha \in L^\infty(Q_T)$ such that for all (i.e., possibly multiple) solutions of (2.1)–(2.4), corresponding to it

$$\| u(\cdot, T) - u_d \|_{L^2(\Omega)} \le \varepsilon. \tag{2.6}$$

Remark 2.1. One can, of course, also study the multiplicative controllability of the mixed problem like (2.1)/(2.5) in the case when the initial datum $u_0(x)$ can change sign in $(0, 1)$. In particular, in the recent paper [23] we investigated the case when the initial and target states admit no more than finitely many changes of sign. The method of [23] is different from what we will apply below in this chapter, but it is also of qualitative nature.

2.1.2 Main Results

We formulate the results of this section separately for the linear and semilinear versions of the mixed boundary problem (2.1), (2.2)–(2.4). They are as follows.

Theorem 2.1 (The case of the linear equation). *System (2.5) is "non-negatively" approximately controllable in $L^2(0,1)$ in the sense of Definition 2.1 by means of static controls $\alpha = \alpha(x), \alpha \in L^\infty(0,1)$ only. Moreover, the corresponding solution to (2.5) remains non-negative at all times. (No multiple solutions in this case.)*

Theorem 2.2 (The semilinear case). *System (2.1), (2.2)–(2.4) is "non-negatively" approximately controllable in $L^2(\Omega)$ in the sense of Definition 2.1. The corresponding steering can be achieved by subsequent applying of three suitable static bilinear controls.*

Corollary 2.1. *The condition that the initial states y_0 and u_0 in Theorems 2.1 and 2.2 are nonzero and non-negative can be replaced with the following more general assumptions:*

$$\int_0^1 y_0 y_d dx > 0, \quad \int_0^1 u_0 u_d dx > 0. \tag{2.7}$$

(However, $y(x,t)$ may now accept both negative and positive values during the steering process.)

Corollary 2.1 extends the results of Theorems 2.1 and 2.2 to a wider set of initial states (i.e., not necessarily non-negative), but, formally, it excludes the zero target states $y_d = u_d = 0$. However, we show below in the first step of the proof of Theorem 2.2 that the zero target state can approximately be reached arbitrarily fast from any initial state in $L^2(0,1)$.

The remainder of this chapter is organized as follows. Theorem 2.1 is proven in section 2.2. Section 2.3 deals with some preliminary estimates, further used in section 2.4 to prove Theorem 2.2. In section 2.5 we further discuss our main results and the layout of the remainder of Part I.

Remark 2.2. It is noteworthy that the global approximate controllability of the semilinear heat equation with superlinear power-like terms as in (2.1), (2.2)–(2.4) is not possible by means of additive locally distributed controls [50, 57]. More precisely, solutions to such an equation remain uniformly bounded outside the control support, regardless of the magnitude of control applied. (Hence, the non-negative global controllability in the sense of Definition 2.1 is out of question as well.) On the positive global controllability results for semilinear parabolic equations with superlinear terms and *additive* controls of localized support we refer to [4, 9, 43, 46, 64–66].

2.2 Proof of Theorem 2.1: The Case of the Linear Heat Equation

Our central idea here is to try to select $\alpha = \alpha(x)$ in such a way that the target state y_d (or its "close" approximation) becomes co-linear to the first (non-negative) eigenfunction associated with the diffusion-reaction term $y_{xx} + \alpha(x)y$ in (2.5), which is then approached by the corresponding trajectory of (2.5) in $L^2(\Omega)$ as t increases.

2.2.1 Preliminaries

Denote by λ_k and $\omega_k(x)$, $k = 1,\ldots$ respectively the eigenvalues and orthonormalized in $L^2(0,1)$ eigenfunctions of the spectral problem $\omega_{xx} + \alpha(x)\omega = \lambda\omega$, $\omega \in H_0^1(0,1)$ (which, in fact, is a linear ode). It is known that

$$\| \alpha \|_{L^\infty(0,1)} \geq \lambda_1 > \lambda_2 > \ldots, \tag{2.8}$$

and $\lambda_k \to -\infty$ as k increases. The unique solution to (2.5) in $C([0,T];L^2(0,1)) \cap H_0^{1,0}(Q_T)$ admits the following representation:

$$y(x,t) = \sum_{k=1}^{\infty} e^{\lambda_k t} \left(\int_0^1 y_0(r)\omega_k(r)dr \right) \omega_k(x). \tag{2.9}$$

Note that, for the given α, we can endow the space $H_0^1(0,1)$ with the norm

$$\| \omega \|_{H_0^1(0,1)} = \left(\int_0^1 (\omega_x^2 + (-\alpha(x) + c)\omega^2(x))dx \right)^{1/2},$$

where c is any positive constants exceeding $\| \alpha \|_{L^\infty(0,1)}$. Then,

$$\| \omega_k \|_{H_0^1(0,1)} = \left((c - \lambda_k) \int_0^1 \omega_k^2(x)dx \right)^{1/2},$$

and

$$\| y(\cdot,t) \|_{L^2(0,1)} \leq e^{\lambda_1 t} \| y_0 \|_{L^2(0,1)}, \quad t \geq 0, \tag{2.10}$$

$$\| y(\cdot,t) \|_{C[0,1]} \leq C_* \| y(\cdot,t) \|_{H_0^1(0,1)} \leq C(t) \| y_0 \|_{L^2(0,1)}, \quad t > 0, \tag{2.11}$$

where C_* is a positive constant associated with the continuous embedding $H_0^1(0,1) \subset C[0,1]$ and the function $C(t), t > 0$ is nondecreasing. (Both C_* and $C(t)$ depend on α.)

2.2.2 Proof of Theorem 2.1

Our plan of the proof of Theorem 2.1 is as follows:

(i) We intend to show that for "almost any" given non-negative target state $y_d \in L^2(0,1)$ we can select an $\alpha_*(x)$ such that $y_d \parallel y_d \parallel_{L^2(0,1)}$ becomes the first eigenfunction for (2.5), associated with the largest eigenvalue in the representation (2.9). In other words, in (2.9),

$$\omega_1(x) = \frac{y_d(x)}{\parallel y_d \parallel_{L^2(0,1)}}. \tag{2.12}$$

(ii) Then we will show that the actual control can be selected as

$$\alpha(x) = \alpha_*(x) + a,$$

where a is a constant chosen so that the first term in the corresponding representation (2.9) converges to y_d, while the remainder of the series converges to zero as t increases. Note that adding a in the above does shift the eigenvalues corresponding to α_* (denote them by $\{\lambda_k\}_{k=1}^{\infty}$ now) from λ_k to $\lambda_k + a$, but the eigenfunctions remain the same for α_* and $\alpha_* + a$.

Step 1. Maximum principle. Let us recall first that according to the generalized maximum principle

$$0 \le y(x,t) \le e^{bt} \parallel y_0 \parallel_{L^\infty(0,1)}, \quad b = \parallel \alpha \parallel_{L^\infty(0,1)}, \quad t > 0, \tag{2.13}$$

whenever $y_0 \in L^\infty(0,1)$ and is a.e. non-negative. Indeed, if $\alpha(x) \le 0$ this is true with $b = 0$ (in this case (2.13) becomes the standard generalized maximum principle.) Otherwise, (2.13) follows from the maximum principle applied to the function $z(x,t) = e^{-bt}y(x,t)$, satisfying

$$z_t = z_{xx} + (\alpha(x) - b)z \text{ in } Q_T,$$

$$z(0,t) = z(1,t) = 0, \quad t \in (0,T), \quad z \mid_{t=0} = y_0 \in L^\infty(0,1)$$

with $\alpha(x) - b \le 0$.

Note now that any non-negative $y_0 \in L^2(0,1)$ can be approximated in $L^2(0,1)$ by a sequence of non-negative $y_{0k} \in C[0,1]$. Due to (2.11) the solutions to (2.5) corresponding to y_0 and y_{0k} converge to each other in $C[0,1]$ for every positive t. Hence, in view of (2.13), we also have

$$y(x,t) \geq 0 \text{ in } (0,1) \times (0,\infty), \tag{2.14}$$

whenever y_0 is non-negative element of $L^2(0,1)$.

Step 2. To prove Theorem 2.1, it is sufficient to consider any set of non-negative target states y_d which is dense in the set of all non-negative elements of $L^2(0,1)$. To this end, we will consider only (a) nonzero non-negative continuously differentiable functions $y_d = y_d(x), x \in [0,1]$ that (b) vanish at $x = 0, 1$ and (c) whose second derivatives are piecewise continuous with finitely many discontinuities of the first kind (hence, $y_d \in H^2(0,1) \cap H_0^1(0,1)$) such that (d)

$$y_d(x) > 0 \text{ in } (0,1) \quad \text{and} \quad \begin{cases} \frac{y_{dxx}}{y_d}, & \text{where } y_d(x) \neq 0, \\ 0, & \text{where } y_d(x) = 0, \end{cases} \in L^\infty(0,1). \tag{2.15}$$

To ensure the last condition in (2.15) it is sufficient to select y_d that, in addition to the above, is linear near the endpoints $x = 0, 1$. This would guarantee that $y_{dxx} = 0$ near $x = 0, 1$, while elsewhere, due to the first condition in (2.15), the denominator in (2.15) is strictly separated from 0.

Let us show that any non-negative element $g \in L^2(0,1)$ can indeed be approximated in this space by a sequence of functions described in the above.

Firstly, note that without loss of generality, we can assume that $g(x) \geq c > 0$ for almost all $x \in (0,1)$ (since any non-negative function g can be approximated, e.g., by a sequence of positive functions $g(x) + 1/k$, where $k \to \infty$).

Secondly, recall that any $g \in L^2(0,1)$, positive a.e. in $(0,1)$ as described in the above, can be approximated in this space by a sequence of piecewise constant positive functions $g_k(x)$ with possible jumps at $x_j = j/k$, $j = 1, \ldots, k$, $x_0 = 0$, $x_k = 1$, where

$$g_k(x) = \frac{1}{k} \int_{x_{j-1}}^{x_j} g(x)dx, \quad x \in [x_{j-1}, x_j), \quad j = 1, \ldots, k.$$

Note that $g_k(x)$ are strictly separated from zero in $(0,1)$.

Thirdly, each of such piecewise constant functions can in turn be approximated in $L^2(0,1)$ by continuous piecewise linear functions that vanish at $x = 0, 1$ and everywhere else are strictly positive, whose graphs, accordingly, do not have vertical pieces. (Namely, for that one just needs, e.g., to connect the graphs of the former functions by the pieces of straight lines near the discontinuity points.)

Finally, each of these broken lines can be "smoothened" at the corners, e.g., by using pieces of circles of "sufficiently small" radia with centers located on the bisectors of the angles generated by the corresponding adjacent straight lines of the graphs (so that these lines are tangent to the associated circles). These "smoothened" lines are the graphs of the functions satisfying all the conditions (a)–(d) described in the above and approximate in $L^2(0, 1)$ the above-constructed broken lines (and, hence, eventually an arbitrarily selected non-negative $g \in L^2(0, 1)$) as the above-mentioned radia tend to zero.

Step 3. Select any non-negative nonzero $y_0 \in L^2(0, 1)$ and any y_d as described in the above. Set

$$\alpha_*(x) = -\frac{y_{dxx}(x)}{y_d(x)}, \quad x \in (0, 1). \tag{2.16}$$

Note that $\alpha_*(x)$ is not identically zero in $L^\infty(0, 1)$ (otherwise, $y_d \equiv 0$.) The eigenvalues associated with this α_* we further denote by $\{\lambda_k\}_{k=1}^\infty$.

(2.16) means that the function

$$\frac{y_d(x)}{\| y_d \|_{L^2(0,1)}}$$

is an eigenfunction for (2.5) with $\alpha = \alpha_*$, say

$$\frac{y_d(x)}{\| y_d \|_{L^2(0,1)}} = \omega_{k*}(x), \tag{2.17}$$

associated with the zero eigenvalue $\lambda_{k*} = 0$.

Note now that, since all the eigenfunctions are orthogonal in $L^2(0, 1)$, that is,

$$\int_0^1 \omega_k \omega_m dx = 0 \quad \text{for } k \neq m,$$

and $\omega_{k*} > 0$ in $(0, 1)$, the function (2.17) is *the only non-negative eigenfunction* for (2.5) with $\alpha = \alpha_*$ as in (2.16) (more precisely, it is the only eigenfunction that does not change sign in $(0, 1)$).

Also, since $\omega_{k*}(x) > 0$, whenever $y_0(x) \geq 0, y_0(x) \neq 0$ in $(0, 1)$,

$$\int_0^1 y_0 \omega_{k*} dx > 0 \tag{2.18}$$

(see also Remark 2.3 below).

Step 4. Moreover,

$$k* = 1 \quad \text{(that is, } \lambda_1 = 0\text{)}. \tag{2.19}$$

Indeed, otherwise (i.e., if $k* > 1$ and $\omega_1(x) \in H_0^1(0, 1) \subset C[0, 1]$ changes sign in $(0, 1)$) we can select a nonzero non-negative y_0 such that

$$\int_0^1 y_0 \omega_1 dx \neq 0.$$

Therefore, the solution to (2.5) with $\alpha = \alpha_* - \lambda_1 = \alpha_*$, according to the (generic) formula (2.9), is represented by the series

$$\int_0^1 y_0(r)\omega_1(r)dr\omega_1(x) + \sum_{k=2}^{\infty} e^{\lambda_k t} \left(\int_0^1 y_0(r)\omega_k(r)dr \right) \omega_k(x),$$

converging as t increases in $C[0,1]$ to the function (recall that $\lambda_k < \lambda_1 = 0$ for $k = 2,\ldots$)

$$\int_0^1 y_0(r)\omega_1(r)dr\omega_1(x)$$

excepting (as we assumed it arguing by contradiction) negative values somewhere in $(0,1)$. This contradicts the maximum principle (2.14) and hence (2.19) holds.

Step 5. Select now the bilinear control α of the type

$$\alpha = \alpha_* + a, \quad a \in R.$$

Then, the corresponding solution to (2.5) is as follows:

$$y(x,t) = e^{at} \int_0^1 y_0(r)\omega_1(r)dr\omega_1(x) + r(x,t)$$

$$= e^{at} \int_0^1 y_0(r)\omega_1(r)dr\omega_1(x)$$

$$+ \sum_{k>1} e^{(\lambda_k + a)t} \left(\int_0^1 y_0(r)\omega_k(r)dr \right) \omega_k(x), \qquad (2.20)$$

where

$$\lambda_k < 0 \text{ for } k > 1.$$

Accordingly, making use of (2.10) and (2.17)–(2.19), we have the following estimate:

$$\| y(\cdot,t) - y_d \|_{L^2(0,1)} \leq | e^{at} \int_0^1 y_0\omega_1 dx - \| y_d \|_{L^2(0,1)} | + \| r \|_{L^2(0,1)}$$

$$\leq | e^{at} \int_0^1 y_0\omega_1 dx - \| y_d \|_{L^2(0,1)} |$$

$$+ e^{(\lambda_2 + a)t} \| y_0 \|_{L^2(0,1)} . \qquad (2.21)$$

Select now a and $T > 0$ such that

$$e^{aT} \int_0^1 y_0 \omega_1 dx - \| y_d \|_{L^2(0,1)},$$

that is,

$$a = \frac{1}{T} \ln \left(\frac{\| y_d \|_{L^2(0,1)}}{\int_0^1 y_0 \omega_1 dx} \right).$$

Then, it follows from (2.21) that

$$\| y(\cdot, T) - y_d \|_{L^2(0,1)} \leq e^{\lambda_2 T} \frac{\| y_d \|_{L^2(0,1)}}{\int_0^1 y_0 \omega_1 dx} \| y_0 \|_{L^2(0,1)} \rightarrow 0 \qquad (2.22)$$

as T increases. This ensures (2.6) for some pair $\{a, T\}$ and ends the proof of Theorem 2.1. □

Remark 2.3. Given a nonzero non-negative y_T, in the above argument we used the condition that y_0 is a nonzero non-negative function only to ensure that the first Fourier coefficient in the solution representation (2.20) is positive. This proves Corollary 2.1 in respect of Theorem 2.1.

2.3   The Semilinear Case: A Priori Estimates

To prove Theorem 2.2 we will need the following estimates.
   Denote $\mathscr{B}(0, T) = C([0, T]; L^2(0, 1)) \cap H_0^{1,0}(Q_T)$ and put

$$\| q \|_{\mathscr{B}(0,T)} = \left(\max_{t \in [0,T]} \| q(\cdot, t) \|_{L^2(0,1)}^2 + 2 \int_0^T \int_0^1 q_x^2 dx ds \right)^{1/2}.$$

We have the following a priori estimate.

Lemma 2.1. *Given $T > 0$ and $\alpha(x) \leq 0$, any solution to (2.1)–(2.4) (if there are multiple solutions) satisfies the following two estimates:*

$$\| u \|_{\mathscr{B}(0,T)}, \ \| u \|_{L^6(Q_T)} \leq C \| u_0 \|_{L^2(0,1)} . \qquad (2.23)$$

The difference $z = u - y$ between any solution u to (2.1), (2.2)–(2.4) (if there are multiple ones) and the unique solution to (2.5) with $y_0 = u_0$ satisfies the following two estimates:

$$\| z \|_{\mathscr{B}(0,T)}, \ \| z \|_{L^6(Q_T)} \leq C \left(T^{\frac{5}{6}(1-\frac{r_1}{5})} \| u_0 \|_{L^2(0,1)}^{r_1} + T^{\frac{5}{6}(1-\frac{3r_2}{5})} \| u_0 \|_{L^2(0,1)}^{r_2} \right).$$
$$(2.24)$$

Here and below we routinely use symbols c and C to denote (different) *generic* positive constants.

Proof. Recall [97] that $f(\cdot,\cdot,w,w_x) \in L^{6/5}(Q_T)$ and that the following energy equality holds for (2.1) treated as a linear equation with the source term $f(x,t,u,u_x)$, e.g., [97] (p. 142):

$$\frac{1}{2} \| u \|^2_{L^2(0,1)} \Big|_0^t + \int_0^t \int_0^1 (u_x^2 - \alpha u^2 + f(x,s,u,u_x)u)\,dxds = 0 \quad \forall t \in [0,T]. \quad (2.25)$$

Here and everywhere below, if there exist several solutions to (2.1), we always deal separately with one of them at a time, while noticing that all the estimates involved hold uniformly over all possible solutions.

Combining (2.25) and (2.3) and the assumption that $\alpha(x) \le 0$ immediately yields:

$$\| u(\cdot,t) \|^2_{L^2(0,1)} + 2 \int_0^t \int_0^1 u_x^2(x,s)\,dxds \le \| u_0 \|^2_{L^2(0,1)} \quad \forall t \in [0,T]. \quad (2.26)$$

This provides the first estimate in (2.23) with respect to the $\mathscr{B}(0,T)$-norm. The second estimate (with properly arranged generic constant) follows by the continuity of the embedding $\mathscr{B}(0,T)$ into $L^6(Q_T)$ (e.g., [97], pp. 467, 75):

$$\| u \|_{L^6(Q_T)} \le c \| u \|_{\mathscr{B}(0,T)}. \quad (2.27)$$

We now intend to evaluate the difference between the solution u to (2.1) with $\alpha(x) \le 0$ and that to its truncated version (2.5). If $z = u - y$, then

$$z_t = z_{xx} + \alpha z - f(x,t,u,u_x) \quad \text{in} \quad Q_T,$$
$$z \mid_{x=0,1} = 0, \quad z \mid_{t=0} = 0.$$

Similar to (2.25) and (2.26) we have for any $\delta > 0$,

$$\| z(\cdot,t) \|^2_{L^2(0,1)} + 2 \int_0^t \int_0^1 z_x^2(x,s)\,dxds$$

$$\le -2 \int_0^t \int_0^1 zf(x,s,u,u_x)dxds$$

$$\le 2 \| z \|_{L^6(Q_t)} \| f(\cdot,\cdot,u,u_x) \|_{L^{6/5}(Q_t)}$$

$$\le 2c \| z \|_{\mathscr{B}(0,t)} \| f(\cdot,\cdot,u,u_x) \|_{L^{6/5}(Q_T)}$$

$$\le \delta \| z \|^2_{\mathscr{B}(0,t)} + \frac{c^2}{\delta} \| f(\cdot,\cdot,u,u_x) \|^2_{L^{6/5}(Q_T)} \quad \forall t \in [0,T], \quad (2.28)$$

where we have used Hölder's and Young's inequalities and, again, (2.27).

From (2.28), we have

$$\max_{\tau \in (0,t)} \| z(\cdot,\tau) \|^2_{L^2(0,1)} \leq \delta \| z \|^2_{\mathscr{B}(0,t)} + \frac{c^2}{\delta} \| f(\cdot,\cdot,u,u_x) \|^2_{L^{6/5}(Q_t)} \quad \forall t \in [0,T].$$

Hence, again from (2.28),

$$\| z \|^2_{\mathscr{B}(0,t)} \leq 2\delta \| z \|^2_{\mathscr{B}(0,t)} + \frac{2c^2}{\delta} \| f(\cdot,\cdot,u,u_x) \|^2_{L^{6/5}(Q_T)} \quad \forall t \in [0,T]$$

and

$$(1 - 2\delta) \| z \|^2_{\mathscr{B}(0,t)} \leq \frac{2c^2}{\delta} \| f(\cdot,\cdot,u,u_x) \|^2_{L^{6/5}(Q_T)} \tag{2.29}$$

or

$$\| z \|_{\mathscr{B}(0,T)} \leq \frac{\sqrt{2}c}{\sqrt{\delta(1-2\delta)}} \| f(\cdot,\cdot,u,u_x) \|_{L^{6/5}(Q_T)}, \tag{2.30}$$

provided that

$$0 < \delta < \frac{1}{2}.$$

Now, using (2.2) and Hölder's inequality (as in [97], p. 469; [61], p. 863), we obtain:

$$\| f(\cdot,\cdot,u,u_x) \|_{L^{6/5}(Q_T)} \leq \beta T^{\frac{5}{6}(1-\frac{r_1}{5})} \| u \|^{r_1}_{L^6(Q_T)}$$

$$+ \beta T^{\frac{5}{6}(1-\frac{3r_2}{5})} \| u_x \|^{r_2}_{L^2(Q_T)}. \tag{2.31}$$

Combining (2.30), (2.31) with (2.23) and, again, with (2.27) yields (2.24). This completes the proof of Lemma 2.1.   □

Remark 2.4. We can derive the estimates similar to (2.23) and (2.24), for the case when α is a positive constant (see Chapter 5 for more general situation in this respect). They are as follows:

$$\| u \|_{\mathscr{B}(0,T)}, \| u \|_{L^6(Q_T)} \leq Ce^{\alpha T} \| u_0 \|_{L^2(0,1)}, \tag{2.32}$$

$$\| z \|_{\mathscr{B}(0,T)}, \| z \|_{L^6(Q_T)} \leq Ce^{\frac{2\alpha T}{1-\delta}} \frac{1}{\sqrt{\delta}} (T^{\frac{5}{6}(1-\frac{r_1}{5})} \| u \|^{r_1}_{L^6(Q_T)}$$

$$+ (T^{\frac{5}{6}(1-\frac{3r_2}{5})} \| u_x \|^{r_2}_{L^2(Q_T)}) \quad \forall \delta \in (0,1/2), \tag{2.33}$$

where C does not depend on α. We will also use them in the proof of Theorem 2.2 below.

2.4   The Semilinear Case: Proof of Theorem 2.2

Our plan to prove Theorem 2.2 is as follows:

- Given the initial and target states u_0 and u_d, first we will steer system (2.1), (2.2)–(2.4) in $L^2(\Omega)$ from u_0 to a state

$$u_* = s_* u_d + o(s_*) \tag{2.34}$$

  for some small parameter $s_* > 0$. To this end, we will make use of the "superlinearity" assumptions (2.2)–(2.4), which ensure that the system at hand behaves "almost" like the linear one near the origin.
- Then, we "stretch" the state u_* to approximate the desirable target state u_d.

Proof of Theorem 2.2. Fix any non-negative $u_0 \neq 0$ in $L^2(0,1)$ and u_d satisfying the assumptions on y_d described in (2.15).

Step 1. Select any $t_* > 0$. On the interval $(0, t_*)$ we intend to apply a negative constant control $\alpha(x) = \lambda$ (its value will be chosen later). Then for the corresponding solution y to (2.5) with $y_0 = u_0$ we have (using the generic representation (2.9)):

$$y(x,t_*) = e^{\lambda t_*} \sum_{k=1}^{\infty} e^{\lambda_k t_*} \left(\int_0^1 y_0(r) \omega_k(r) dr \right) \omega_k(x), \tag{2.35}$$

where in this case $\lambda_k = -(\pi k)^2$, $\omega_k(x) = \sqrt{2} \sin \pi kx$.

Fix any $s \in (0,1)$ and select now (a constant on $(0,t_*)$ bilinear control) $\lambda = \lambda(t_*, s) < 0$ such that

$$e^{\lambda t_*} = s.$$

Then, it follows from (2.35) that

$$y(\cdot, t_*) \longrightarrow sy_0(\cdot) = su_0 \text{ as } t_* \to 0+.$$

Furthermore, using the estimates (2.24) we also obtain the same convergence for the corresponding solution to (2.1) (uniform over all the possible multiple solutions). In other words, we can steer in $L^2(\Omega)$, *arbitrarily fast*, (2.1) from any initial state $u_0 = y_0$ to a state as follows (compare it to (2.34)):

$$u(\cdot, t_*) = sy_0 + o(s) = su_0 + o(s) \text{ as } s \to 0+ \tag{2.36}$$

for some $t_* = t_*(s)$ ($t_*(s) \to 0$ as $s \to 0+$), where

$$\| o(s)/s \|_{L^2(0,1)} \longrightarrow 0 \text{ as } s \to 0. \tag{2.37}$$

(For example, to ensure (2.37), we can select λ of the form

$$\lambda = \frac{\ln t_*^\kappa}{t_*}$$

for some κ, in which case $s = t_*^\kappa$.)

We further assume that s is "small" (to ensure (2.36) and (2.37) whenever it will be necessary).

Step 2. Select

$$\alpha_*(x) = -\frac{u_{dxx}(x)}{u_d(x)}$$

as in (2.16). The corresponding first eigenfunction for the parabolic operator in (2.5) (see (2.17)–(2.19)) will be

$$\frac{u_d(x)}{\| u_d \|_{L^2(0,1)}} = \omega_1(x).$$

We now apply Theorem 2.1 for the linear system (2.5) on some interval (t_*,t^*), where t^* will be selected later, with

$$su_0 + o(s) \qquad\qquad (2.38)$$

from (2.36) in place of the initial state (note that for small s Remark 2.3 applies – this will eventually yield Corollary 2.1 for Theorem 2.2) and with

$$s^{1+\xi}u_d, \text{ where } \xi \in (0, \min\{r_1, r_2\} - 1) \qquad\qquad (2.39)$$

(r_1 and r_2 are from (2.2)–(2.4)), in place of the target state. Accordingly (along the lines (2.20)–(2.22) and using (2.38) and (2.39)), we will have the estimate as in (2.22):

$$\| y(\cdot,t^*) - s^{1+\xi}u_d \|_{L^2(0,1)} \le e^{\lambda_2(t^*-t_*)} \frac{s^{1+\xi} \| u_d \|_{L^2(0,1)}}{\int_0^1 (sy_0 + o(s))\omega_1 dx} \| sy_0 + o(s) \|_{L^2(0,1)}$$

$$(2.40)$$

for some $t^* > t_*$ (where λ_k's are the eigenvalues associated with α_*, $\lambda_1 = 0$). Here, as in Step 5 of the proof of Theorem 2.1, y is the solution to (2.5) on (t_*,t^*) with bilinear control $\alpha(x) = \alpha_*(x) + a$ such that

$$e^{a(t^*-t_*)} \int_0^1 (sy_0 + o(s))\omega_1 dx = s^{1+\xi} \| u_d \|_{L^2(0,1)}$$

or, whenever $a \neq 0$,

$$t^* = t_* + \frac{1}{a} \ln \left(\frac{s^{1+\xi} \| u_d \|_{L^2(0,1)}}{\int_0^1 (s y_0 + o(s)) \omega_1 dx} \right), \tag{2.41}$$

and λ_1 and λ_2 are the first two eigenvalues for (2.5) with $\alpha = \alpha_*$.

Select now

$$a = - \| \alpha_* \|_{L^\infty(0,1)} - \rho, \tag{2.42}$$

where $\rho > 0$ is some (fixed) constant. Since $\lambda_1 = 0$,

$$a < 0 \quad \text{and} \quad \alpha(x) = \alpha_*(x) + a < 0, \ x \in [0,1].$$

Hence, by (2.41),

$$t^* - t_* = t^*(s) - t_*(s) \ \to \ \infty \quad \text{as } s \to 0+$$

and also the estimate (2.24) applies on the interval (t_*, t^*):

$$\| u(\cdot, t^*) - s^{1+\xi} u_d \|_{L^2(0,1)} \le \| u(\cdot, t^*) - y(\cdot, t^*) \|_{L^2(0,1)} + \| y(\cdot, t^*) - s^{1+\xi} u_d \|_{L^2(0,1)}$$

$$\le C(t^* - t_*)^{\max\{\frac{5}{6}(1-\frac{r_1}{3}), \frac{5}{6}(1-\frac{3r_2}{5})\}} s^{\min\{r_1, r_2\}}$$

$$+ \left(C s^{\xi \lambda_2 / a} \right) s^{1+\xi} \| u_d \|_{L^2(0,1)}$$

$$= o(s^{1+\xi}) \quad \text{as } s \to 0+$$

(we remind the reader that C denotes a generic positive constant). Here we also used (2.40) and (2.42) to show that as $s \to 0+$:

$$e^{\lambda_2(t^* - t_*)} = \left(\frac{s^{1+\xi} \| u_d \|_{L^2(0,1)}}{\int_0^1 (s y_0 + o(s)) \omega_1 dx} \right)^{\frac{\lambda_2}{a}} \le C s^{\xi \lambda_2 / a}$$

(with $\lambda_2 < \lambda_1 = 0$ and a as in (2.42)) and that

$$(t^* - t_*)^{\max\{\frac{5}{6}(1-\frac{r_1}{3}), \frac{5}{6}(1-\frac{3r_2}{5})\}} s^{\min\{r_1, r_2\}}$$

$$= \left(\frac{1}{a} \ln \left(\frac{s^{1+\xi} \| u_d \|_{L^2(0,1)}}{\int_0^1 (s y_0 + o(s)) \omega_1 dx} \right) \right)^{\max\{\frac{5}{6}(1-\frac{r_1}{3}), \frac{5}{6}(1-\frac{3r_2}{5})\}} s^{\min\{r_1, r_2\}}$$

$$\le \left(C \, | \ln s \, |^{\max\{\frac{5}{6}(1-\frac{r_1}{3}), \frac{5}{6}(1-\frac{3r_2}{5})\}} s^{\min\{r_1, r_2\} - 1 - \xi} \right) s^{1+\xi},$$

also recalling that $r_1, r_2 > 1$ in (2.2)–(2.4) and $\xi \in (0, \min\{r_1, r_2\} - 1)$.

Thus, we showed that

$$u(\cdot,t^*) = s^{1+\xi}u_d + o(s^{1+\xi}) \text{ as } s \to 0+.$$

Remark 2.5. We would like to point out here that the "superlinearity" condition (2.4) is essential in the above to ensure that $\min\{r_1, r_2\} > 1$.

Step 3. Now we apply the last step of our plan of the proof as outlined in the beginning of this section with $s_* = s^{1+\xi}$. Namely, on some interval (t^*, T) (with T to be selected later) we apply positive constant control

$$\alpha(x) \equiv \alpha > 0, \quad t \in (t^*, T).$$

In the fashion of Steps 1 and 2 in the above, but based on the estimates (2.32) and (2.33) in Remark 2.4 in place of Lemma 2.1, it can be shown that with parameters $\varepsilon = T - t^* = T - t^*(s)$, $s_* = s^{1+\xi}$, and $\alpha = \alpha(\varepsilon, s_*)$, selected as follows:

(A) $s_* \to 0+$;

(B) $e^{\alpha\varepsilon} = \dfrac{1}{s_*}$ or $\alpha = \dfrac{1}{\varepsilon}\ln\dfrac{1}{s_*}$;

(C) $\varepsilon \to 0$ (i.e., $T - t^*(s) \to 0$) in such a way that also

$$e^{\frac{2\alpha\varepsilon}{1-\delta}}\varepsilon^{\min\{\frac{5}{6}(1-\frac{r_1}{5});\frac{5}{6}(1-\frac{3r_2}{5})\}} = \varepsilon^{\min\{\frac{5}{6}(1-\frac{r_1}{5});\frac{5}{6}(1-\frac{3r_2}{5})\}}s_*^{-\frac{2}{1-\delta}} \to 0,$$

we will have for some $T > 0$ that

$$\| u(\cdot,T) - u_d \|_{L^2(0,1)} \to 0. \tag{2.43}$$

This completes the proof of Theorem 2.2. $\square$

2.5 Concluding Remarks and the Layout of the Remainder of Part I

It is worth noticing, that in terms of applications, the method used to prove our main linear result - Theorem 2.1 - allows one to deal with relatively small and simple (and hence "practical") static controls, which, however, may need to be applied for a relatively long time. (These controls were also used in the crucial part of the proof of Theorem 2.2.) This method requires the first eigenvalues be simple, which is an obvious property when $n = 1$.

With the changes necessary to take care about the regularity of solutions to (2.5)/(2.1) and the approximation procedure described in Step 2 of the proof of Theorem 2.1, the arguments of Theorems 2.1 and 2.2 can be extended to the general n-dimensional case as well, provided that the first eigenvalue of (2.5)/(2.1) associated with α_* as in (2.16) is *simple*, which in turn appears to be the case, see, e.g., [129], pp. 202-203, [128], pp. 276-278.

Plan for the remainder of Part I:

In Chapter 3 we will discuss a different non-negative controllability result again for a system like in (2.1) but *now in several space dimensions* with the terms f which can again be superlinear at infinity but *now they do not have to be superlinear near the origin*. Contrary to the above, the main result of Chapter 3 always requires that at least three "large" static bilinear controls (whose magnitudes increase as the precision of steering increases) be applied subsequently for *very short times*. Unlike the method of the present section, based on the use of the dynamics imposed by the diffusion-reaction term like $y_{xx} + \alpha(x)y$, the method of Chapter 3 focuses on the "suppression" of the effect of diffusion. It makes use of the part of the dynamics of (2.1) which can be approximated by the trajectories of the ordinary differential equation $dz/dt = \alpha(x)z$ in $L^2(\Omega)$. Accordingly, the method of Chapter 3 does not apply to obtain the results of Theorem 2.1 (and of Corollary 2.1) of this chapter, allowing the use of *"small" single* static controls.

Chapter 4 deals with a different approach to the bilinear controllability, which is as follows. It is known that the heat equation is approximately controllable in $L^2(\Omega)$ by the additive static controls $v = v(x)$ with support everywhere in Ω. Then a suitable bilinear control for (2.5) can be sought as some "well-posed modification" of the expression $\alpha(x,t) = v(x)/y(x,t)$, in which case the "original" additive static control $v(x)$ is artificially "transformed" into the bilinear term $\alpha(x,t)y(x,t)$. Note however, that this approach deals with essentially more "complex" controls (e.g., in terms of practical implementation), namely, the functions of both x and t. In Chapter 4 it is used to investigate the exact null-controllability of a semilinear nonhomogeneous version of the equation (2.5) with the bilinear term like $\alpha(x,t)(f(x,t,y) - \theta(x))$, where f is sublinear and θ is given. Note that the exact null-controllability discussed in Chapter 4 is out of question for the homogeneous bilinear system like (2.1) (or (2.5)) of the present section, since the origin is always a fixed point for such a system for any control α. (We also want to refer along these lines to a conference talk [47] where a similar method was discussed in the context of the non-negative approximate controllability for a special class of the semilinear parabolic equations whose solutions satisfy the maximum principle, see also [109] and [110] in this respect.)

In *Chapter 5* we consider the semilinear heat equation with superlinear term, governed by a *pair of controls*: (I) a traditional internal either locally distributed or lumped control <u>and</u> (II) a single multiplicative control. The introduction of the latter is motivated by the lack of global controllability properties for this class of pde's when they are steered solely by the additive controls. Based on an asymptotic technique allowing us to *separate and combine* the impacts generated by the above-mentioned two types of controls, we manage to establish the global approximate controllability for such superlinear pde in the classical sense.

Chapter 3
Multiplicative Controllability of the Semilinear Parabolic Equation: A Qualitative Approach

Abstract In this chapter we establish the global non-negative approximate controllability property for a rather general semilinear heat equation with superlinear term, governed in a bounded domain $\Omega \subset R^n$ by a multiplicative control in the reaction term like $vu(x,t)$, where v is the control. We show that any non-negative target state in $L^2(\Omega)$ can approximately be reached from any non-negative, nonzero initial state by applying at most three static bilinear $L^\infty(\Omega)$-controls subsequently in time. This result is further applied to discuss the controllability properties of the nonhomogeneous version of this problem with bilinear term like $v(u(x,t) - \theta(x))$, where θ is given. Our approach is based on an asymptotic technique allowing us to distinguish and make use of the pure diffusion and/or pure reaction parts of the dynamics of the system at hand, while suppressing the effect of a nonlinear term.

3.1 Introduction

In this chapter we consider the following Dirichlet boundary problem, governed in a bounded domain $\Omega \subset R^n$ by a multiplicative control $v \in L^\infty(Q_T)$ in the reaction term:

$$\frac{\partial u}{\partial t} = \Delta u + vu - f(x,t,u,\nabla u) \quad \text{in } Q_T = \Omega \times (0,T), \tag{3.1}$$

$$u = 0 \quad \text{in } \Sigma_T = \partial\Omega \times (0,T), \quad u \mid_{t=0} = u_0 \in L^2(\Omega).$$

We assume that f is the given function satisfying the following conditions:

- $f(x,t,q,p)$ is Lebesgue's measurable in x,t,q,p, is continuous in q,p for almost all $(x,t) \in Q_T$;
- there exist a non-negative function ψ in $L^{1+n/(n+4)}(\Omega)$, $\beta > 0$, and

$$r_1 \in [0, 1 + \frac{4}{n}), \quad r_2 \in [0, 1 + \frac{2}{n+2}) \tag{3.2}$$

A.Y. Khapalov, *Controllability of Partial Differential Equations Governed by Multiplicative Controls*, Lecture Notes in Mathematics 1995, DOI 10.1007/978-3-642-12413-6_3, © Springer-Verlag Berlin Heidelberg 2010

such that

$$| f(x,t,q,p) | \le \psi(x,t) + \beta \, | \, q \, |^{r_1} + \beta \, \| \, p \, \|_{R^n}^{r_2} \quad \text{a.e. in } Q_T \text{ for } q \in R, \, p \in R^n;$$
(3.3)

- there exist $\rho > 0$ and $\nu > 0$ such that

$$\int_\Omega f(x,t,\phi,\nabla\phi) \, \phi \, dx \ge (\nu - 1) \int_\Omega \| \, \nabla\phi \, \|_{R^n}^2 \, dx - \rho \int_\Omega (1 + \phi^2) dx \quad \forall \phi \in H_0^1(\Omega).$$
(3.4)

For $n = 1$ a simple example of a function f satisfying conditions (3.2)–(3.4) is $f(u) = u^3$.

We refer, e.g., to [97] (p. 466), where it was shown that system (3.1), (3.2)–(3.4) admits at least one generalized solution in the space $C([0,T]; L^2(\Omega)) \cap H_0^{1,0}(Q_T) \cap L^{2+4/n}(Q_T)$, while its uniqueness is not guaranteed.

We intend to analyze the global controllability properties of the homogeneous bilinear system (3.1) and of its nonhomogeneous version.

3.2   Main Results

In Chapter 2, where we studied the 1-D version of model (3.1), the central idea was to select the bilinear control $v = v(x)$ in such a way that the target state (or its "close" approximation) becomes co-linear to the first (non-negative) eigenfunction for the truncated linear version of (3.1). The selected target state is then approached by the system at hand as t increases. On the one hand, this method allows one to deal with "practical" relatively small and simple static controls, but, on the other hand, the control time can be relatively large (that is, we have a "trade-off"). However, such approach requires the first eigenvalues be simple for any a control $v(x)$ – hence one needs to assume that $n = 1$.

Another principal limitation of the method of Chapter 2 is the assumption that the nonlinear term must be superlinear near the origin as well (i.e., in particular, it must vanish at the origin, which is an equilibrium in this case). In this way it can be assured that the system at hand behaves "almost" like a linear one near the origin, which enabled us to make use of the bilinear controllability properties of the latter. This limitation also did not allow us to extend the methods of Chapter 2 to a non-homogeneous bilinear system like

$$\frac{\partial z}{\partial t} = \Delta z + v(z - \theta(x)) - f(x,t,z,\nabla z) \quad \text{in } Q_T,$$
(3.5)

$$z = 0 \quad \text{in } \Sigma_T, \quad z \, |_{t=0} = z_0 \in L^2(\Omega),$$

where f does not necessarily vanish at the origin and $\theta \ne 0$ is given.

We now intend to prove the same as in Chapter 2 non-negative controllability result, <u>but now</u>:

- in *any* space dimension and
- within *an arbitrarily small time-interval* $(0, T)$ given in advance.
- We employ a different qualitative approach which allows us to *eliminate the assumptions of Chapter 2 on the one dimensionality* of the system at hand and
- to *get rid of the superlinearity assumption on the nonlinear term near the origin.*

Our central idea below is to view the evolution of system (3.1) as an interaction of the following three dynamics associated with the respective three terms in the right-hand side of (3.1):

- *Pure diffusion dynamics* - when $v = 0$ and $f = 0$;
- *Pure reaction dynamics* - caused by the reaction term only. We further associate it with the system

$$\frac{\partial y}{\partial t} = vy \quad \text{in } Q_T, \tag{3.6}$$

$$y \mid_{t=0} = y_0.$$

- *Nonlinear "disturbance"* - the dynamics caused by the nonlinear term of class (3.2)–(3.4).

Accordingly, our strategy to achieve the desirable controllability result will be to try to select the multiplicative control in such a way that the corresponding trajectories of (3.1) can be approximated by those associated with the pure diffusion and/or the pure reaction like in (3.6), while the effect of nonlinearity is to be suppressed. The latter appears to be unavoidable when dealing with a general class of highly nonlinear terms like in (3.2)–(3.4).

Our main results are as follows.

Theorem 3.1. *Let $T > 0$ be given and the pair of the initial and target states $u_0 \in H_0^1(\Omega)$ and $u_d \in L^2(\Omega)$ be such that*

$$\frac{u_d}{u_0} \in H^2(\Omega), \quad \nabla\left(\frac{u_d}{u_0}\right) \in [L^\infty(\Omega)]^n, \quad \Delta\left(\frac{u_d}{u_0}\right) \in L^\infty(\Omega) \tag{3.7}$$

and

$$0 < c_1 \le \frac{u_d(x)}{u_0(x)} \le c_2 < 1 \text{ a.e. in } \Omega, \tag{3.8}$$

where c_1 and c_2 are some positive constants. Then for every $\varepsilon > 0$ there is a $T_* \in (0, T)$ such that for all, i.e., possibly multiple solutions to (3.1), (3.2)–(3.4) generated by control

$$v(x) = \frac{1}{T_*} \ln\left(\frac{u_d(x)}{u_0(x)}\right), \tag{3.9}$$

we have the same uniform estimate

$$\| u(\cdot, T_*) - u_d \|_{L^2(\Omega)} \leq \varepsilon. \tag{3.10}$$

Theorem 3.1 can be reformulated as follows.

Theorem 3.2. *Given $T > 0$, let $v_*(x)$ be any function such that*

$$v_* \in L^\infty(\Omega) \bigcap H^2(\Omega), \quad \nabla v_* \in [L^\infty(\Omega)]^n, \quad \Delta v_* \in L^\infty(\Omega), \quad v_*(x) \leq L < 0 \text{ a.e.. in } \Omega,$$

where L is some negative constant. Then for any $u_0 \in H_0^1(\Omega)$ and every $\varepsilon > 0$ there is a $T_* \in (0, T)$ such that for all, i.e., possibly multiple solutions to (3.1)–(3.4) with control

$$v(x) = \frac{1}{T_*} v_*(x),$$

we have the same estimate

$$\| u(\cdot, T_*) - e^{v_*(\cdot)} u_0 \|_{L^2(\Omega)} \leq \varepsilon.$$

Theorems 3.1 and 3.2 provide the basis for the following non-negative controllability result within any time-interval $(0, T)$ given in advance.

Theorem 3.3 (Non-negative controllability of (3.1)). *Given $T > 0$, assume that the boundary $\partial\Omega$ of domain Ω is of class $C^{3+[n/2]}$ (where $[n/2]$ denotes the largest non-negative integer which does not exceed $n/2$). Then for every $\varepsilon > 0$ and non-negative $u_0, u_d \in L^2(\Omega), u_0 \neq 0$ there exist a $T_* = T_*(\varepsilon, u_0, u_d) \in (0, T)$ and a control $v \in L^\infty(Q_{T_*})$ such that for all (i.e., possibly multiple) solutions of (3.1)–(3.4), corresponding to the latter, (3.10) holds. Suitable v can be selected as a combination of at most three static controls applied subsequently in time.*

An immediate consequence of Theorem 3.3 for the non-homogeneous system (3.5) is as follows.

Theorem 3.4 (Non-homogeneous case). *Given $T > 0$, let $\theta \in H^2(\Omega) \bigcap H_0^1(\Omega)$ and the boundary $\partial\Omega$ of domain Ω be of class $C^{3+[n/2]}$. For every $\varepsilon > 0$ and $z_0, z_d \in L^2(\Omega), z_0 \neq \theta, z_0(x) \geq \theta(x), z_d(x) \geq \theta(x)$ a.e. in Ω, there exist a $T_* = T_*(\varepsilon, z_0, z_d) \in (0, T)$ and a bilinear control $v \in L^\infty(Q_{T_*})$ such that for all (i.e., possibly multiple) solutions of (3.5), (3.2)–(3.4), corresponding to the latter, (3.10) holds with z_0, z_d in place of u_0, u_d. Again, suitable v can be selected as a combination of at most three static controls applied subsequently in time.*

Indeed, to prove Theorem 3.4 it is sufficient to notice that the substitution $w = z - \theta$ transforms (3.5) into

$$\frac{\partial w}{\partial t} = \Delta w + vw - f_*(x,t,w,\nabla w) \quad \text{in } Q_T, \tag{3.11}$$

$$w = 0 \quad \text{in } \Sigma_T, \quad w\mid_{t=0} = z_0 - \theta \in L^2(\Omega),$$

where $f_*(x,t,w,\nabla w) = -\Delta\theta + f(x,t,w+\theta,\nabla w+\theta)$. Making use of Young's inequality and the inequality $(a+b)^\gamma \leq C(a^\gamma + b^\gamma)(a,b,\gamma \geq 0, C = C(\gamma) > 0)$, one can check that conditions (3.2)–(3.4) hold for this function as well (in general, with a different set of parameters β, v, ρ). Then Theorem 3.4 follows immediately from Theorem 3.3 applied to (3.11).

The remainder of Chapter 3 is organized as follows. In section 3.3 we introduce several auxiliary technical estimates. In section 3.4 we prove Theorems 3.1 and 3.2, while Theorem 3.3 is proven in section 3.5.

3.3 Auxiliary Estimates

In this section we prove the following three lemmas containing several estimates, which are heavily used in the proofs of Theorems 3.1–3.3.

Denote $\mathscr{B}(0,T) = C([0,T];L^2(\Omega))\cap H_0^{1,0}(Q_T)$ and

$$\| \phi \|_{\mathscr{B}(0,T)} = \left(\max_{t\in[0,T]} \| \phi(\cdot,t) \|_{L^2(\Omega)}^2 + 2v \int_0^T \int_\Omega \| \nabla\phi \|_{R^n}^2 \, dxds \right)^{1/2},$$

where $v > 0$ is from (3.4).

Lemma 3.1. *Given $T > 0$ and $v \in L^\infty(\Omega)$, any solution to system (3.1)–(3.4) (if there are multiple solutions) satisfies the following two estimates:*

$$\| u \|_{\mathscr{B}(0,T)}, \ \| u \|_{L^{2+4/n}(Q_T)} \leq Ce^{(\alpha+\rho)T} \left(\| u_0 \|_{L^2(\Omega)}^2 + 2\rho T \right)^{1/2}, \tag{3.12}$$

where $\alpha = \| v \|_{L^\infty(\Omega)}$ and C is a generic positive constant (it does not depend on v).

Consider now the truncated version of system (3.1) as follows:

$$h_t = \Delta h + vh \quad \text{in } Q_T, \tag{3.13}$$

$$h = 0 \text{ in } \Sigma_T, \quad h\mid_{t=0} = h_0 \in L^2(\Omega).$$

Lemma 3.2. *Given $T > 0$, $v \in L^\infty(\Omega)$, $\delta \in (0,1/4)$, we have the following two estimates for the difference $\xi = u - h$ between any corresponding solution u to*

(3.1)–(3.4) (if there are multiple ones) and the unique corresponding solution to (3.13):

$$\| \xi \|_{\mathcal{B}(0,T)}, \| \xi \|_{L^{2+4/n}(Q_T)} \leq Ce^{2\sqrt{2}\alpha T}\{\| u_0 - h_0 \|^2_{L^2(\Omega)}$$

$$+ \frac{1}{\sqrt{\delta}} T^{\frac{n+4}{2(n+2)}(1-\frac{r_1 n}{n+4})} \| u \|^{r_1}_{L^{2+4/n}(Q_T)}$$

$$+ T^{\frac{n+4}{2(n+2)}(1-\frac{(n+2)r_2}{(n+4)})} \| \nabla u \|^{r_2}_{[L^2(Q_T)]^n}$$

$$+ \| \psi \|_{L^{1+n/(n+4)}(Q_T)}\} \tag{3.14}$$

where the (generic) constant C does not depend on $\alpha = \| v \|_{L^\infty(\Omega)}.$

Next we have the following result, based on Lemma 3.1.

Lemma 3.3. *Given* $T > 0$, *the following estimate holds for solutions to (3.13) with* $h_0 \in H^1_0(\Omega)$, $v \in L^\infty(\Omega) \cap H^2(\Omega)$, $\nabla v \in [L^\infty(\Omega)]^n$, $\Delta v \in L^\infty(\Omega)$, $v(x) \leq 0$:

$$\int\int_{Q_T} (\Delta h)^2 dx dt + \frac{1}{2}\int_\Omega \| \nabla h(x,T) \|^2_{R^n} dx$$

$$\leq \frac{1}{2}\int_\Omega \| \nabla h_0 \|^2_{R^n} dx + \frac{1}{2} \| \Delta v \|_{L^\infty(\Omega)} TC^2 e^{2\alpha T} \| h_0 \|^2_{L^2(\Omega)}, \tag{3.15}$$

where C is from Lemma 3.1 and $\alpha = \| v \|_{L^\infty(\Omega)}.$

Proof of Lemma 3.1. Recall (see, e.g., [97]) that $f(\cdot,\cdot,u,\nabla u) \in L^{1+n/(n+4)}(Q_T)$ and that the following energy equality holds for (3.1) treated as a linear equation with the source term $f(x,t,u,\nabla u)$, e.g., [97] (p. 142):

$$\frac{1}{2} \| u \|^2_{L^2(\Omega)}|^t_0 + \int_0^t \int_\Omega (\| \nabla u \|^2_{R^n} - vu^2 + f(x,s,u,\nabla u)u) dx ds = 0 \ \forall t \in [0,T].$$

$$\tag{3.16}$$

Here and everywhere below, if there exist several solutions to (3.1), we always deal separately with a selected one, while noticing that all the estimates hold uniformly.

Combining (3.16) and (3.4) yields for $t \in [0,T]$:

$$\| u(\cdot,t) \|_{L^2(\Omega)}^2 + 2v \int_0^t \int_\Omega \| \nabla u \|_{R^n}^2 \, dxds$$

$$\leq \| u_0 \|_{L^2(\Omega)}^2 + 2(\alpha + \rho) \int_0^t \int_\Omega u^2 dxds + 2\rho T$$

$$\leq \left(\| u_0 \|_{L^2(\Omega)}^2 + 2\rho T \right) + 2(\alpha + \rho)$$

$$\times \int_0^t \left(\| u(\cdot,\tau) \|_{L^2(\Omega)}^2 + 2v \int_0^\tau \int_0^1 \| \nabla u \|_{R^n}^2 \, dxds \right) d\tau. \quad (3.17)$$

Applying Gronwall-Bellman inequality to (3.17) yields the first estimate in (3.12) with respect to the $\mathscr{B}(0,T)$-norm with $C = \sqrt{2}$. The second estimate follows by the continuity of the embedding of $\mathscr{B}(0,T)$ into $L^{2+4/n}(Q_T)$ (e.g., [97], pp. 467, 75), due to which,

$$\| \xi \|_{L^{2+4/n}(Q_T)} \leq c \| \xi \|_{\mathscr{B}(0,T)} \quad (3.18)$$

for some constant $c > 0$ independent of T. In this case $C = c\sqrt{2}$ in (3.12). This ends the proof of Lemma 3.1.   □

Proof of Lemma 3.2. We now intend to evaluate the difference between any possible (multiple) solution u to (3.1) and that to its truncated version (3.13).

Denote $\xi = u - h$, then

$$\xi_t = \Delta \xi + v\xi - f(x,t,u,\nabla u) \quad \text{in} \quad Q_T, \quad (3.19)$$
$$\xi|_{\Sigma_T} = 0, \quad \xi|_{t=0} = u_0 - h_0.$$

Multiplying (3.19) by ξ and integrating it by parts in $Q_t = \Omega \times (0,t)$, we obtain the following chain of estimates for all $t \in [0,T]$:

$$\| \xi(\cdot,t) \|_{L^2(\Omega)}^2 + 2v \int_0^t \int_\Omega \| \nabla \xi \|_{R^n}^2 (x,s) \, dxds$$

$$= \| \xi(\cdot,0) \|_{L^2(\Omega)}^2 + 2 \int_0^t \int_\Omega v\xi^2 dxds - 2 \int_0^t \int_\Omega \xi f(x,s,u,\nabla u) dxds$$

$$\leq \| \xi(\cdot,0) \|_{L^2(\Omega)}^2 + 2\alpha \int_0^t \int_0^1 \xi^2 dxds + 2 \| \xi \|_{L^{2+4/n}(Q_t)} \| f(x,t,u,\nabla u) \|_{L^{1+n/(n+4)}(Q_t)}$$

$$\leq \parallel \xi(\cdot,0) \parallel^2_{L^2(\Omega)} +2\alpha \int\limits_0^t \int\limits_\Omega \xi^2 dx ds + 2c \parallel \xi \parallel_{\mathscr{B}(0,t)} \parallel f(x,t,u,\nabla u) \parallel_{L^{1+n/(n+4)}(Q_T)}$$

$$\leq \parallel \xi(\cdot,0) \parallel^2_{L^2(\Omega)} +2\alpha \int\limits_0^t \parallel \xi \parallel^2_{\mathscr{B}(0,s)} ds + \delta \parallel \xi \parallel^2_{\mathscr{B}(0,t)}$$

$$+\frac{c^2}{\delta} \parallel f(\cdot,\cdot,u,\nabla u) \parallel^2_{L^{1+n/(n+4)}(Q_T)}, \tag{3.20}$$

where we made use of (3.18) and Hölder's and Young's inequalities.

It follows from (3.20) that we have

$$\parallel \xi \parallel^2_{\mathscr{B}(0,t)} \leq 2 \parallel \xi(\cdot,0) \parallel^2_{L^2(\Omega)} +4\alpha \int\limits_0^t \parallel \xi \parallel^2_{\mathscr{B}(0,s)} ds + 2\delta \parallel \xi \parallel^2_{\mathscr{B}(0,t)}$$

$$+\frac{2c^2}{\delta} \parallel f(\cdot,\cdot,u,\nabla u) \parallel^2_{L^{1+n/(n+4)}(Q_T)} \quad \forall t \in [0,T].$$

Thus,

$$\parallel \xi \parallel^2_{\mathscr{B}(0,t)} \leq 4 \parallel \xi(\cdot,0) \parallel^2_{L^2(\Omega)} + 8\alpha \int\limits_0^t \parallel \xi \parallel^2_{\mathscr{B}(0,s)} ds$$

$$+\frac{4c^2}{\delta} \parallel f(\cdot,\cdot,u,\nabla u) \parallel^2_{L^{1+n/(n+4)}(Q_T)}, \tag{3.21}$$

provided that $0 < \delta < \frac{1}{4}$.

Making use of Gronwall-Bellman inequality, we derive from (3.21) that

$$\parallel \xi \parallel_{\mathscr{B}(0,T)} \leq e^{2\sqrt{2}\alpha T} \left(2 \parallel \xi(\cdot,0) \parallel_{L^2(\Omega)} + \frac{2c}{\sqrt{\delta}} \parallel f(\cdot,\cdot,u,\nabla u) \parallel_{L^{1+n/(n+4)}(Q_T)} \right).$$
$$\tag{3.22}$$

Now, using (3.3) and Hölder's inequality (as in [97], p. 469; [61], p. 863), we obtain:

$$\parallel f(\cdot,\cdot,u,\nabla u) \parallel_{L^{1+n/(n+4)}(Q_T)} \leq \beta T^{\frac{n+4}{2(n+2)}(1-\frac{r_1 n}{n+4})} \parallel u \parallel^{r_1}_{L^{2+4/n}(Q_T)}$$

$$+\beta T^{\frac{n+4}{2(n+2)}(1-\frac{(n+2)r_2}{(n+4)})} \parallel \nabla u \parallel^{r_2}_{[L^2(Q_T)]^n}$$

$$+ \parallel \psi \parallel_{L^{1+n/(n+4)}(Q_T)}. \tag{3.23}$$

Combining (3.22) and (3.23) yields the result of Lemma 3.2. □

Proof of Lemma 3.3. We need to evaluate Δh in $L^2(Q_T)$, where h satisfies (3.13) with $h_0 \in H_0^1(\Omega)$.

Consider any

$$v \in L^\infty(\Omega)\bigcap H^2(\Omega), \quad \nabla v \in [L^\infty(\Omega)]^n, \quad \Delta v \in L^\infty(\Omega), \quad v(x) \le 0 \quad \text{in } \Omega. \quad (3.24)$$

Multiplying the equation (3.13) by Δh and further integrating it over Q_T yield:

$$\int\int_{Q_T} \left(-h_t \Delta h + (\Delta h)^2 \right) dxdt = -\int\int_{Q_T} v h \Delta h dxdt. \quad (3.25)$$

In turn,

$$-\int\int_{Q_T} h_t \Delta h dxdt = \frac{1}{2}\int_\Omega \sum_{k=1}^n h_{x_k}^2(x,T)dx - \frac{1}{2}\int_\Omega \sum_{k=1}^n h_{x_k}^2(x,0)dx,$$

while, having in mind (3.24),

$$-\int\int_{Q_T} v h \Delta h dxdt = -\int\int_{Q_T} \sum_{k=1}^n v h h_{x_k x_k} dxdt$$

$$= \int\int_{Q_T} \sum_{k=1}^n v h_{x_k}^2 dxdt + \frac{1}{2}\int\int_{Q_T} \sum_{k=1}^n v_{x_k}(h^2)_{x_k} dxdt$$

$$\le \frac{1}{2}\| \Delta v \|_{L^\infty(\Omega)} \int\int_{Q_T} h^2 dxdt.$$

Combining all the above yields:

$$\int\int_{Q_T} (\Delta h)^2 dxdt + \frac{1}{2}\int_\Omega \| \nabla h(x,T) \|_{R^n}^2 dx$$

$$\le \frac{1}{2}\int_\Omega \| \nabla h_0 \|_{R^n}^2 dx + \frac{1}{2}\| \Delta v \|_{L^\infty(\Omega)} \int\int_{Q_T} h^2 dxdt$$

$$\le \frac{1}{2}\int_\Omega \| \nabla h_0 \|_{R^n}^2 dx$$

$$+ \frac{1}{2}\| \Delta v \|_{L^\infty(\Omega)} T \max_{t\in[0,T]} \int_\Omega h^2(x,t)dx. \quad (3.26)$$

From (3.26) and estimate (3.12) (applied to system (3.13)) we obtain the estimate (3.15). This ends the proof of Lemma 3.3. □

3.4 Proofs of Theorems 3.1 and 3.2

3.4.1 Proof of Theorem 3.1: Controllability to "Smaller" Target States of the Truncated System

In this subsection we consider the truncated version (3.13) of system (3.1)

Step 1. Consider h_0 and h_d satisfying the assumptions of Theorem 3.1 in place of u_0 and u_d.

Denote

$$v_*(x) = \ln\left(\frac{h_d(x)}{h_0(x)}\right). \tag{3.27}$$

Then

$$h_d(x) = e^{v_*(x)} h_0(x). \tag{3.28}$$

Remark 3.1. Note that, since (3.6) is, in fact, a linear ordinary differential equation in $L^2(\Omega)$, in view of (3.27) and (3.28), its solution y satisfies the following property:

$$y(x, 1/s) = h_d(x) \quad \text{when } y_0 = h_0,\ v(x) = sv_*(x) \tag{3.29}$$

for any number $s > 0$.

Step 2. To prove Theorem 3.1 in the case when $f = 0$, we need to show that h_d can be approximated by a suitable solution to (3.13). To this end we intend to study the difference between the solutions to (3.13) and to (3.6) on $(0, T)$.

Consider any $v \in L^\infty(\Omega)$ and denote $g = h - y$. Then, assuming that $y_0 = h_0$ in (3.6) and (3.13), we obtain:

$$g_t = vg + \Delta h \text{ in } Q_T,$$
$$g\,|_{t=0} = 0.$$

Thus

$$g(x, t) = \int_0^t e^{v(x)(t-\tau)} \Delta h(x, \tau) d\tau,\ t \in [0, T]. \tag{3.30}$$

Let now v be of the form as in (3.29), namely,

$$v(x) = sv_*(x),$$

where we now treat the positive number s as a parameter (its value will be selected later in Step 4).

Then, since (see (3.8))

$$\ln c_1 \leq v_* \leq \ln c_2 < 0,$$

formula (3.30) yields:

$$\| g(\cdot,t) \|_{L^2(\Omega)}^2 \le \left(\frac{e^{2st \ln c_2} - 1}{s \ln c_2} \right) \| \Delta h \|_{L^2(Q_t)}^2, \quad Q_t = \Omega \times (0,t), \ t \in [0,T]. \quad (3.31)$$

Step 3. Making use of the estimate (3.15) from Lemma 3.3 applied with $v = sv_*$, we derive from (3.31) that

$$\| g(\cdot,t) \|_{L^2(\Omega)}^2 \le \left(\frac{e^{2st \ln c_2} - 1}{s \ln c_2} \right)$$
$$\times \left(\frac{1}{2} \int_\Omega \| \nabla h_0 \|_{\mathbb{R}^n}^2 \, dx + \frac{1}{2} s \| \Delta v_* \|_{L^\infty(\Omega)} \, t C^2 e^{2\alpha st} \| h_0 \|_{L^2(\Omega)}^2 \right), \quad (3.32)$$

where $\alpha = \| v_* \|_{L^\infty(\Omega)}$.

Step 4. Select now $s > 0$ and $T_* \in (0,T)$ such that

$$T_* = \frac{1}{s} \ \text{ or } \ T_* s = 1. \quad (3.33)$$

Then we obtain from (3.32), (3.33) and the property (3.29) that

$$\| g(\cdot,T_*) \|_{L^2(\Omega)} = \| h(\cdot,T_*) - y(\cdot,T_*) \|_{L^2(\Omega)} = \| h(\cdot,T_*) - h_d \|_{L^2(\Omega)} \to 0 \quad (3.34)$$

as $s \to \infty$ (or $T_* \to 0+$), which ensures (3.10) for any given in advance ε for some pair (s, T_*) as in (3.33). This ends the proof of Theorem 3.1 when $f = 0$. □

3.4.2 Proof of Theorem 3.1: The General Case

It follows from the above argument by making use of Lemma 3.2, in which we evaluated the difference between the (possible multiple) solutions u to (3.1) and h to (3.13) in a uniform way as given in (3.14). Indeed, it follows from (3.14) that under condition (3.33) and with $h_0 = u_0$:

$$\| u(\cdot,T_*) - h(\cdot,T_*) \|_{L^2(\Omega)} \to 0$$

as $s = 1/T_* \to \infty$, which ensures (3.10) whenever it holds for h as in (3.34). This ends the proof of Theorem 3.1 in the general case. □

3.4.3 Proof of Theorem 3.2: Controllability Properties with Negative Multiplicative Controls

It is immediate from the proof of Theorem 3.1, since the assumptions on h_0 and h_d in the latter are used specifically to ensure the properties of v_* required in Theorem 3.2. $\qquad\qquad\qquad\qquad\qquad\qquad\qquad\qquad\qquad\qquad\qquad\qquad\qquad$ □

3.5 Proof of Theorem 3.3: Nonnegative Controllability

3.5.1 The Truncated Linear Equation

Again we study first the truncated problem (3.13).

Consider any pair of initial and target states $h_0, h_d \in L^2(\Omega)$, which are nonnegative (almost everywhere) in Ω and $h_0 \neq 0$. Since we study the issue of approximate controllability and because the set of infinitely differentiable functions with compact support (denote it by $C_0^\infty(\Omega)$) is dense in $L^2(\Omega)$, without loss of generality we can further assume that

$$h_d \in C_0^\infty(\Omega), \quad h_d \neq 0, \quad h_d(x) \geq 0 \ \forall x \in \Omega. \tag{3.35}$$

We plan to approximate h_d by using three static bilinear controls, applied subsequently in time:

(a) Firstly, we will use $v = 0$ on some time-interval $(0, t_1)$ to steer our system to a state $h(\cdot, t_1)$ which is strictly positive in the interior of Ω.

(b) Secondly, we will use a relatively "large" positive constant control v on some time-interval (t_1, t_2) to steer our system to a state which is "larger" than the given h_d in (3.35).

(c) Finally, we will use a static control as described in Theorem 3.1 on some time-interval (t_2, t_3), to steer our system further to a desirable neighborhood of h_d.

Step 1. Pick any $t_1 > 0$ and apply in (3.13) the zero bilinear control $v = 0$ on $(0, t_1)$. Then, at time t_1 system (3.13) reaches the state

$$h(\cdot, t_1) \in H_0^1(\Omega) \bigcap H^{3+[n/2]}(\Omega) \subset C^2(\bar{\Omega}). \tag{3.36}$$

(We refer, e.g., to [119] for the corresponding regularity and embedding results.)

Note also that, due to the smoothing effect, the solution h to (3.13) is classical in $\bar{\Omega} \times [\beta, t_1]$, for any $\beta \in (0, t_1]$ [119]. Furthermore, due to the strong maximum principle, see, e.g., [49],

$$h(x, t_1) > 0 \text{ in the interior of } \Omega, \quad h(x, t_1)\,|_{\partial\Omega} = 0. \tag{3.37}$$

Step 2. Consider any $t_2 > t_1$. On the interval (t_1, t_2) we apply a positive constant control $v(x) = v$ (its value will be chosen later). Then for the corresponding solution h to (3.13) on (t_1, t_2) we have

$$h(x, t_2) = e^{v(t_2 - t_1)} \sum_{k=1}^{\infty} e^{\lambda_k(t_2 - t_1)} \left(\int_{\Omega} h(r, t_1) \omega_k(r) dr \right) \omega_k(x), \qquad (3.38)$$

where λ_k ($\lambda_k \to -\infty$ as $k \to \infty$) and $\omega_k(x)$ ($\| \omega_k \|_{L^2(\Omega)} = 1$), $k = 1, \dots$ are respectively the eigenvalues and eigenfunctions associated with the spectral problem $\Delta \omega = \lambda \omega$, $\omega |_{\partial \Omega} = 0$ in $H_0^1(\Omega)$.

Consider any number $\gamma > 1$ (its value will be chosen more precisely a little bit later) and select a constant (in t and x) control $v > 0$ such that

$$e^{v(t_2 - t_1)} = \gamma, \quad \text{namely,} \quad v = \frac{\ln \gamma}{t_2 - t_1}. \qquad (3.39)$$

(Thus, v depends on γ and $t_2 - t_1$.)

Then, it follows from (3.38) that with this control

$$h(\cdot, t_2) \to \gamma h(\cdot, t_1) \text{ as } t_2 \to t_1 + \text{ in } C(\bar{\Omega}), \qquad (3.40)$$

as implied by the estimate

$$\| h(\cdot, t_2) - \gamma h(\cdot, t_1) \|_{C(\bar{\Omega})} \leq C \| h(\cdot, t_2) - \gamma h(\cdot, t_1) \|_{H^{1 + [n/2]}(\Omega)}$$

$$\leq C \left(\sum_{k=1}^{\infty} \lambda_k^{1 + [n/2]} \left(e^{\lambda_k(t_2 - t_1)} - 1 \right)^2 \right.$$

$$\times \left. \left(\int_{\Omega} h(r, t_1) \omega_k(r) dr \right)^2 \right)^{1/2},$$

where $C > 0$ is a (generic) constant associated with the continuous embedding $H^{1 + [n/2]}(\Omega) \subset C(\bar{\Omega})$ (see, e.g., [119]).

Select now the value of $\gamma > 1$ in such a way that

$$\gamma h(x, t_1) \geq h_d(x) + 1 \quad \forall x \in \text{supp } h_d,$$

which is possible due to (3.37) and (3.35), where supp h_d stands for the set of all x where $h_d(x) \neq 0$ (i.e., where $h_d(x) > 0$).

For this γ and any given $\sigma \in (0,1)$ (to be selected more precisely later) select any positive number $t_2 > t_1, t_2 = t_2(\gamma,\sigma)$ and v as in (3.39) such that

$$\gamma h(x,t_1) + \sigma/2 \geq h(x,t_2) \geq \gamma h(x,t_1) - \sigma/2 \geq -\sigma/2 \quad \forall x \in \Omega, \quad (3.41)$$

$$h(x,t_2) \geq h_d(x) \quad \forall x \in \text{supp } h_d. \quad (3.42)$$

This is possible due to (3.35), (3.37) and (3.40).

Step 3. We will now apply Theorem 3.2 to the system (3.13) on some interval (t_2,t_3) with the initial state $h(x,t_2)$ and the static control

$$v(x) = \frac{1}{t_3 - t_2} v_\sigma(x),$$

where

$$v_\sigma = \ln\left(\frac{h_d + \sigma^2/2}{h(\cdot,t_2) + \sigma}\right) \in C^2(\bar{\Omega}).$$

(Note that the additional "regularizing" terms $\sigma^2/2$ and σ ensure that the argument of the logarithmic function in the above is positive everywhere in $\bar{\Omega}$.)

Since, in view of (3.41)–(3.42),

$$\frac{\sigma^2/2}{\max_{x\in\bar{\Omega}} h(x,t_2) + \sigma} \leq \frac{h_d(x) + \sigma^2/2}{h(x,t_2) + \sigma}$$

$$\leq \begin{cases} \frac{h(x,t_2)+\sigma-\sigma+\sigma^2/2}{h(x,t_2)+\sigma} \leq 1 - \frac{\sigma-\sigma^2/2}{\max_{x\in\bar{\Omega}} h(x,t_2)+\sigma} & \text{for } x \in \text{supp } h_d, \\ \frac{\sigma^2/2}{-\sigma/2+\sigma} = \sigma & \text{for } x \in \Omega\backslash\text{supp } h_d, \end{cases}$$

we have in Ω:

$$\ln\left(\frac{\sigma^2/2}{\max_{x\in\bar{\Omega}} h(x,t_2) + \sigma}\right) < v_\sigma(x)$$

$$\leq \ln\left(\max\left\{\sigma, 1 - \frac{\sigma-\sigma^2/2}{\max_{x\in\bar{\Omega}} h(x,t_2) + \sigma}\right\}\right) < 0.$$

According to Theorem 3.2, this v will steer (3.13) in $L^2(\Omega)$ at some time t_3 from $h(\cdot,t_2) \in H_0^1(\Omega)$ as close as we wish to a state

$$e^{v_\sigma(x)} h(x,t_2) = h(x,t_2)\left(\frac{h_d(x) + \sigma^2/2}{h(x,t_2) + \sigma}\right), \quad (3.43)$$

provided that t_3 is sufficiently close to t_2 from the right. For example, there is a $t_3 > t_2$ $(t_3 = t_3(\sigma))$ such that

$$\| h(\cdot,t_3) - h(\cdot,t_2) \left(\frac{h_d + \sigma^2/2}{h(\cdot,t_2) + \sigma} \right) \|_{L^2(\Omega)} \leq \sigma. \tag{3.44}$$

To finish the proof of Theorem 3.3 for the case $f = 0$, it remains to notice that, in view of (3.41)–(3.42), the expression in (3.43) converges in $L^2(\Omega)$ to the desirable target state h_d as $\sigma \to 0+$.

Indeed, we have

$$| h(x,t_2) \left(\frac{h_d(x) + \sigma^2/2}{h(x,t_2) + \sigma} \right) - h_d(x) | = | \frac{h(x,t_2)\sigma^2/2 - h_d(x)\sigma}{h(x,t_2) + \sigma} |$$

$$\leq \sigma^2/2 \frac{h(x,t_2)}{h(x,t_2) + \sigma} + \sigma \frac{h_d(x)}{h(x,t_2) + \sigma}$$

$$\leq \sigma^2/2 + \sigma \tag{3.45}$$

for all $x \in$ supp h_d, where, in view of (3.42), $h(x,t_2) \geq h_d(x) > 0$ and hence

$$0 \leq \frac{h(x,t_2)}{h(x,t_2) + \sigma} \leq 1, \tag{3.46}$$

and

$$0 \leq \frac{h_d(x)}{h(x,t_2) + \sigma} \leq \frac{h(x,t_2) + \sigma}{h(x,t_2) + \sigma} = 1. \tag{3.47}$$

In turn, since h_d vanishes elsewhere, for $x \in \Omega \backslash$ supp h_d we have:

$$h(x,t_2) \left(\frac{h_d(x) + \sigma^2/2}{h(x,t_2) + \sigma} \right) - h_d(x) = h(x,t_2) \left(\frac{\sigma^2/2}{h(x,t_2) + \sigma} \right) \tag{3.48}$$

where, due to (3.41),

$$| h(x,t_2) \left(\frac{\sigma^2/2}{h(x,t_2) + \sigma} \right) | \leq | h(x,t_2) | \left(\frac{\sigma^2/2}{-\sigma/2 + \sigma} \right) |$$

$$\leq (\gamma \| h(\cdot,t_1) \|_{C(\bar{\Omega})} + \sigma/2)\sigma. \tag{3.49}$$

Thus, combining (3.45)–(3.49), we obtain that

$$\| h(\cdot,t_2) \left(\frac{h_d + \sigma^2/2}{h(\cdot,t_2) + \sigma} \right) - h_d \|_{C(\bar{\Omega})} \leq \sigma^2 + \sigma + \sigma\gamma \| h(\cdot,t_1) \|_{C(\bar{\Omega})} \to 0$$

as $\sigma \to 0+$, which, in view of (3.34), completes the proof of Theorem 3.3 in the case when $f = 0$. □

3.5.2 The General Semilinear Case

As in the case of Theorem 3.1, it follows from estimate (3.14) evaluating the difference between the solutions u to (3.1) and h to (3.13) (uniformly with respect to possible multiple solutions to (3.1)).

Indeed, (3.14) implies that whenever the product of the $L^\infty(\Omega)$-norm of the static bilinear control $v(x)$, applied, say, on the time-interval (a,b), and its duration $(b-a)$, namely, $\|v\|_{L^\infty(\Omega)}(b-a)$ remains bounded, we have

$$\| u(\cdot,b) - h(\cdot,b) \|_{L^2(\Omega)} \to 0,$$

provided that $b \to a+$ and $\|u(\cdot,a) - h(\cdot,a)\|_{L^2(\Omega)} \to 0$.

Indeed, in the above:

- On the interval $(0,t_1)$ we used $v = 0$ and can select t_1 as small as we wish.
- On the interval (t_1,t_2) we applied a constant v as in (3.39) such that for any (fixed) $\gamma > 1$ we have $v(t_2 - t_1) = \ln \gamma$.
- On the interval (t_2,t_3) we used the condition like in (3.33), exactly as it is described in the proof of Theorem 3.1 in the general case. Again, $(t_3 - t_2)$ can be arbitrarily small.

Thus applying (3.14) subsequently three times on the aforementioned intervals, while selecting sufficiently small $t_i, i = 1,2,3$ and the bilinear controls as described in the above, we can ensure (3.10). This completes the proof of Theorem 3.3 in the general case. □

Chapter 4
The Case of the Reaction-Diffusion Term Satisfying Newton's Law

Abstract We discuss both the approximate and exact null-controllability of the diffusion-reaction equation governed via a coefficient in the reaction term, modeled according to Newton's Law. Both linear and semilinear versions of this term are considered.

4.1 Introduction

4.1.1 Problem Formulation and Motivation

In this chapter we consider the following Dirichlet boundary problem in a bounded domain $\Omega \subset R^n$:

$$\frac{\partial u}{\partial t} = \Delta u + \alpha(x,t)(u - \theta(x,t)) \quad \text{in} \quad Q_T = \Omega \times (0,T), \tag{4.1}$$

$$u = 0 \quad \text{in} \quad \Sigma_T = \partial\Omega \times (0,T), \quad u\mid_{t=0} = u_0 \in L^2(\Omega), \quad \alpha \in L^\infty(Q_T), \quad \theta \in L^2(Q_T).$$

Here θ is the given function and α is the control. It is well-known (e.g., [97]) that for any $T > 0$ system (4.1) admits a unique solution in the space $C([0,T];L^2(\Omega))$ $\bigcap L^2(0,T;H_0^1(\Omega))$. For $n \leq 2$ the same result holds for $\alpha \in L^2(Q_T)$. We further assume that the boundary $\partial\Omega$ of domain Ω is of class $C^{1+[n/2]}$.

In Chapter 3 we already obtained a global controllability result in a "half-space" (more precisely, for $u_0(x), u_d(x) \geq \theta(x)$) for a nonhomogeneous bilinear mixed problem similar to (4.1). It was derived from the respective non-negative controllability result for the associated homogeneous bilinear problem by a simple substitution (see Theorem 3.4 there). In this chapter we focus primarily on the global *exact null*-controllability problem, which requires a completely different method.

Let us recall that (4.1) is globally *exactly null-controllable* in H if it can be steered in H from any initial state to the zero-state exactly. (Note that the latter is an equilibrium for (4.1) when $\alpha = 0$.)

A.Y. Khapalov, *Controllability of Partial Differential Equations Governed by Multiplicative Controls*, Lecture Notes in Mathematics 1995, DOI 10.1007/978-3-642-12413-6_4, © Springer-Verlag Berlin Heidelberg 2010

4.1.2 Main Results

Exact null-controllability. Our main results here are as follows.

Theorem 4.1. *Assume that*

$$\theta \in L^\infty(Q_\infty), \ |\theta(x,t)| \geq v_0 > 0 \text{ a.e. in } Q_\infty = \Omega \times (0,\infty) \qquad (4.2)$$

for some positive constant $v_0 > 0$. Then there is a $T = T(\theta) > 0$ such that for any $u_0 \in L^2(\Omega)$ there exists an $\alpha \in L^\infty(Q_T)$ for which the corresponding solution to (4.1) vanishes at time T:

$$u(\cdot, T) = 0. \qquad (4.3)$$

Without much of extra cost, Theorem 4.1 can be extended to the semilinear equation like

$$\frac{\partial y}{\partial t} = \Delta y + \alpha(x,t)(f(x,t,y) - \theta(x,t)) \quad \text{in } Q_T, \qquad (4.4)$$

$$y = 0 \quad \text{in } \Sigma_T, \quad y\mid_{t=0} = y_0 \in L^2(\Omega),$$

where f satisfies two conditions:

- It is sublinear:

$$|f(x,t,p)| \leq M|p| \quad \forall p \in R, \text{ a.a. } (x,t) \in Q_\infty. \qquad (4.5)$$

- There is an $\alpha_* \in L^\infty(\Omega)$, $\alpha_* \neq 0$ and $\rho > 1/2$ such that

$$\alpha_*(x)f(x,t,p)p \leq -\rho p^2 \|\alpha_*\|_{L^\infty(\Omega)} \quad \forall p \in R, \text{ a.e. in } Q_\infty. \qquad (4.6)$$

Note that system (4.1) is a particular case of (4.4), namely, when $f(p) = p$ (in which case one can select, e.g., $\alpha_* \equiv -1$, $\rho = 3/4$ in (4.6)). The above two conditions are further explained in Remark 4.4 below. It is well known that for any $T > 0$, $\alpha \in L^\infty(Q_T)$, and $\theta \in L^2(Q_T)$ system (4.4) admits a unique solution in $C([0,T];L^2(0,1)) \cap L^2(0,T;H_0^1(\Omega))$ (e.g., [97]).

Corollary 4.1. *The result of Theorem 4.1 holds for the semilinear mixed problem (4.4)–(4.6).*

Remark 4.1. In Theorem 4.1 and Corollary 4.1 in conditions (4.2), (4.5) and (4.6) the set Q_∞ can, of course, be replaced with $Q_{T(\theta)}$ after $T(\theta)$ is found (see the proof of Theorem 4.1).

Our next result deals with the case when $n = 1$ and $\alpha \in L^2(Q_T)$ vanishing outside of the given strict subdomain of Ω.

Assume that condition (4.2) holds locally in Ω. Namely, there is an open subset ω of Ω such that

$$\theta \in L^\infty(Q_\infty), \quad |\theta(x,t)| \geq v_0 > 0 \text{ a.e. in } \omega \times \infty. \tag{4.7}$$

We begin with the local exact null-controllability result.

Theorem 4.2. *Let $n = 1$, $\Omega = (0,1)$, and $T > 0$ be given. There exists a $\sigma > 0$ such that for any $u_0 \in L^2(\Omega)$, $\| u_0 \|_{L^2(\Omega)} \leq \sigma$ there is a bilinear control*

$$\alpha \in L^2(\omega \times (0,T)), \quad \alpha = 0 \text{ in } Q_T \backslash (\omega \times (0,T))$$

for which the corresponding solution to (4.1) vanishes at time T, provided that condition (4.7) holds as well.

Corollary 4.2. *Assume that θ is static, i.e., $\theta = \theta(x)$, and condition (4.7) holds. Then the result of Theorem 4.2 holds for any $u_0 \in L^2(\Omega)$ with one correction - the corresponding control time required to steer the system at hand to the zero-state depends on the choice of u_0 as well.*

Remark 4.2. The restriction on the space dimension in Theorem 4.2 and Corollary 4.2 is due to the following reasons. Firstly, we make use of the known existence and uniqueness results for system (4.1) with the $L^2(Q_T)$-coefficient in the reaction term (see [97]) which hold only for the space dimension not exceeding 2. In turn, the necessity for $L^2(Q_T)$-controls arises due to the fact that our method makes use of the known results on the exact null-controllability of the linear parabolic pde's governed by *additive* locally distributed controls which require such $L^2(Q_T)$-regularity. Secondly, our method also uses the space $C(\bar{\Omega})$ (or $L^\infty(\Omega)$) for solutions to (4.1), which for $L^2(Q_T)$-multiplicative controls is the case when $n = 1$ (due to the embedding $H^1_0(0,1) \subset C[0,1]$, if, e.g., $\Omega = (0,1)$, see also [97], p. 181 for the generalized maximum principle). Note that at the first glance, it may seem that one can use the smoothing effect to overcome the above-mentioned difficulty. However, in the context of the exact null-controllability there is "no action" *after* (i.e., when the smoothing effect could possibly "help") the system reached the equilibrium exactly.

Approximate controllability. We have the following result, which requires that $\theta(x,t)$ vanishes beyond the time-interval $(0, T(\theta))$ described in Theorem 4.1.

Theorem 4.3. *Let $T = T(\theta)$ be as found in Theorem 4.1. Assume now that condition (4.2) holds only on $(0,T)$ and $\theta(x,t) = 0$ for $t > T$. Then for any $\bar{T} > T(\theta)$, any $\varepsilon > 0, u_0, u_d \in L^2(\Omega)$ there exist a $\hat{T} \in (T, \bar{T})$ and an $\alpha \in L^\infty(Q_{\hat{T}})$ for which*

$$\| u(\cdot, \hat{T}) - u_d \|_{L^2(\Omega)} \leq \varepsilon.$$

Remark 4.3. • The assumption on θ in Theorem 4.3 may, in fact, be very reasonable in some applications. Indeed, if $\theta(x,t)$ models the temperature of the surrounding medium, this assumption means that one is able (a) to select two settings for it – the first is, say, static $\theta \neq 0$ (in the sense of (4.2)) and the other one is $\theta = 0$ – *and* (b) to make the "smooth" transition between them reasonably fast (how fast depends on the selection of the initial and target states u_0 and u_d). An important related open question is the wellposedness of the multiplicative controllability results with respect to the changes in control. This issue seems plausible, at least in the context of approximate controllability, based on the available existence and uniqueness results in the area of parabolic pde's with discontinuous coefficients.

- Another interesting open question is to try to approach the issue of bilinear approximate controllability for (4.1) as the "limit" of a sequence of suitable bilinear optimal control problems. In the framework of additive controls we refer to the pioneering paper [113] by J.L. Lions in this respect.

Below in section 4.5, we also discuss some other partial approximate controllability properties for system (4.1) with static both $\theta = \theta(x)$ and $\alpha = \alpha(x)$.

4.2   Proofs of Theorem 4.1 and Corollary 4.1

Proof of Theorem 4.1. The plan of the proof is rather simple:

(i) First, we will show that a "large" (in magnitude) negative constant control α can steer (4.1) arbitrarily quickly, say, at time T_1 to a state which lies within certain uniform distance, defined solely by θ, from the origin.

(ii) Then it is very easy to show that, within certain time-interval (T_1, T_*) the drift motion of (4.1) with $\alpha = 0$ will bring our system into a "desirably" small (for our purposes) neighborhood of the origin in $C(\bar{\Omega})$. We will show that T_* depends on θ and the radius of this neighborhood, but not on the initial datum u_0.

(iii) Next, by using the classical moment approach to the issue of controllability, we will explicitly describe a static control $v = v(x)$ that steers the linear system

$$\frac{\partial w}{\partial t} = \Delta w + v(x) \quad \text{in} \ \omega \times (T_*, T), \tag{4.8}$$
$$w = 0 \ \text{in} \ \partial \Omega \times (T_*, T), \quad w \mid_{t=T_*} = u(\cdot, T_*),$$

to the zero-state at any time $T > T_*$.

(iv) Finally, we will show that for any $u(\cdot, T_*)$, lying in the neighborhood described in (ii), the control

$$\alpha = \frac{v}{w - \theta} \quad \text{lies in} \ L^\infty(\Omega \times (T_*, T)), \tag{4.9}$$

and it will steer (4.1) from $u(\cdot, T_*)$, found in the above, to the zero-state at time T as well.

By making use of appropriate controllability and regularity results available for the linear parabolic equations, we intend to show that the above plan is indeed feasible.

Step 1. Fix any $u_0 \in L^2(\Omega)$. We will need the following estimate.

Lemma 4.1. *For any* $t > 0, \alpha \in L^\infty(Q_t)$:

$$\| u(\cdot, t) \|_{L^2(\Omega)}^2 \leq e^{\operatorname{ess\,sup}_{Q_t}(2\alpha + |\alpha|)t} \| u_0 \|_{L^2(\Omega)}^2 + \frac{e^{\operatorname{ess\,sup}_{Q_t}(2\alpha + |\alpha|)t} - 1}{\operatorname{ess\,sup}_{Q_t}(2\alpha + |\alpha|)}$$

$$\times \| \alpha \|_{L^\infty(Q_t)} \operatorname{ess\,sup}_{\tau \in (0,t)} \| \theta(\cdot, \tau) \|_{L^2(\Omega)}^2. \qquad (4.10)$$

Proof of Lemma 4.1. Multiplying of (4.1) by u and further integration by parts over Ω yields in a standard way:

$$\frac{d \| u(\cdot, t) \|_{L^2(\Omega)}^2}{dt} + 2 \int_\Omega \| \nabla u(x, t) \|_{R^n}^2 \, dx$$

$$= 2 \int_\Omega \alpha u^2 dx - 2 \int_\Omega \alpha \theta u \leq 2 \int_\Omega \alpha u^2 dx + \int_\Omega | \alpha | u^2 dx + \int_\Omega | \alpha | \theta^2 dx$$

$$\leq \operatorname{ess\,sup}_{Q_t}(2\alpha + | \alpha |) \int_\Omega u^2 dx + \| \alpha \|_{L^\infty(Q_t)} \int_\Omega \theta^2 dx. \qquad (4.11)$$

Hence,

$$\| u(\cdot, \tau) \|_{L^2(0,1)}^2 \leq r(\tau) \quad \forall \tau \in (0, t), \ \forall t > 0,$$

where

$$\frac{dr}{d\tau} = \operatorname{ess\,sup}_{Q_t}(2\alpha + | \alpha |) r + \| \alpha \|_{L^\infty(Q_t)} \int_\Omega \theta^2 dx, \tau \in (0, t),$$

$$r(0) = \| u(\cdot, 0) \|_{L^2(\Omega)}^2.$$

Since,

$$r(t) = e^{\operatorname{ess\,sup}_{Q_t}(2\alpha + |\alpha|)t} \| u_0 \|_{L^2(\Omega)}^2$$

$$+ \int_0^t e^{\operatorname{ess\,sup}_{Q_t}(2\alpha + |\alpha|)(t - \tau)} \| \alpha \|_{L^\infty(Q_t)} \left(\int_\Omega \theta^2 dx \right) d\tau,$$

one can easily obtain (4.10). This ends the proof of Lemma 4.1. $\qquad \square$

Let now α be a negative constant. Then, (4.10) becomes

$$\| u(\cdot,t) \|^2_{L^2(\Omega)} \le e^{\alpha t} \| u_0 \|^2_{L^2(\Omega)} + \frac{e^{\alpha t} - 1}{\alpha} \, | \alpha | \, \mathrm{ess\,sup}_{\tau \in (0,t)} \| \theta(\cdot,\tau) \|^2_{L^2(\Omega)} \, .$$

Hence, as $\alpha \to -\infty$, this yields that the state (4.1) can be steered arbitrarily quickly to a state, whose $L^2(\Omega)$-norm depends on θ only. Namely, for any $v > 0, T_1 > 0$ there is a negative constant control α, applying which on $(0, T_1)$ yields:

$$\| u(\cdot,T_1) \|_{L^2(\Omega)} \le \mathrm{ess\,sup}_{\tau \in (0,T_1)} \| \theta(\cdot,\tau) \|_{L^2(\Omega)} + v. \qquad (4.12)$$

For the case when $\theta = \theta(x)$ the following lemma refines (4.12) in terms of convergence in $L^2(\Omega)$.

Lemma 4.2. *Let* $T_1 > 0$ *be given,* $\theta \in L^2(\Omega)$ *be static, and* α *be a negative constant control on* $(0, T_1)$. *Then*

$$u(\cdot,T_1) \to \theta \quad \text{in } L^2(\Omega) \quad \text{as } \alpha \to -\infty. \qquad (4.13)$$

Proof of Lemma 4.2. Indeed, this follows from the Fourier representation of the solutions to (4.1) with constant control α:

$$u(x,t) = \sum_{k=1}^{\infty} e^{(\lambda_k + \alpha)t} \left(\int_{\Omega} u_0(r) \omega_k(r) dr \right) \omega_k(x)$$

$$- \sum_{k=1}^{\infty} \int_0^t e^{(\lambda_k + \alpha)(t-\tau)} \left(\alpha \int_{\Omega} \theta(r) \omega_k(r) dr \right) \omega_k(x)$$

$$= \sum_{k=1}^{\infty} e^{(\lambda_k + \alpha)t} \left(\int_{\Omega} u_0(r) \omega_k(r) dr \right) \omega_k(x)$$

$$- \sum_{k=1}^{\infty} \frac{e^{(\lambda_k + \alpha)t} - 1}{\lambda_k + \alpha} \left(\alpha \int_{\Omega} \theta(r) \omega_k(r) dr \right) \omega_k(x), \qquad (4.14)$$

where λ_k and $\omega_k(x)$, $k = 1,\ldots$ denote respectively the eigenvalues and orthonormalized in $L^2(\Omega)$ eigenfunctions of the spectral problem $\Delta \omega = \lambda \omega$, $\omega \in H_0^1(0,1)$. In particular, it is known that

$$0 \ge \lambda_1 \ge \lambda_2 \ge \ldots, \qquad (4.15)$$

and $\lambda_k \to -\infty$ as k increases. Setting $t = T_1$ and taking α to $-\infty$ yields (4.13).

Indeed, to show that note first that for any constant $\alpha < 0$ we have

$$u(x, T_1) - \theta(x) = \sum_{k=1}^{\infty} e^{(\lambda_k + \alpha)T_1} \left(\int_{\Omega} u_0(r) \omega_k(r) dr \right) \omega_k(x)$$

$$- \sum_{k=1}^{\infty} \frac{e^{(\lambda_k + \alpha)T_1} + \lambda_k/\alpha}{1 + \lambda_k/\alpha} \int_{\Omega} \theta(r) \omega_k(r) dr \omega_k(x).$$

Hence,

$$\| u(\cdot, T_1) - \theta \|_{L^2(\Omega)} \leq \left(\sum_{k=1}^{\infty} e^{2(\lambda_k + \alpha)T_1} \left(\int_{\Omega} u_0(r) \omega_k(r) dr \right)^2 \right)^{1/2}$$

$$+ \left(\sum_{k=1}^{N} \left(\frac{e^{(\lambda_k + \alpha)T_1} + \lambda_k/\alpha}{1 + \lambda_k/\alpha} \right)^2 \left(\int_{\Omega} \theta(r) \omega_k(r) dr \right)^2 \right)^{1/2}$$

$$+ \left(\sum_{k=N+1}^{\infty} 4 \left(\int_{\Omega} \theta(r) \omega_k(r) dr \right)^2 \right)^{1/2},$$

which holds for every positive integer N and every $\alpha \leq \lambda_{N+1}$.

Furthermore, for any $\xi > 0$ we can select $N = N(\xi)$ such that

$$\sum_{k=N+1}^{\infty} \left(\int_{\Omega} \theta(r) \omega_k(r) dr \right)^2 < \xi/36,$$

which is possible since $\theta \in L^2(\Omega)$. Then, since with $\alpha \to -\infty$,

$$e^{(\lambda_k + \alpha)T_1} \to 0 \quad \text{for all } k = 1, \ldots,$$

and

$$\frac{e^{(\lambda_k + \alpha)T_1} + \lambda_k/\alpha}{1 + \lambda_k/\alpha} \to 0 \quad \text{for all } k = 1, \ldots, N(\xi),$$

we can find an $\alpha^* < 0$ such that for all $\alpha < \min\{\alpha^*, N(\xi)\}$ we have:

$$\left(\sum_{k=1}^{\infty} e^{2(\lambda_k + \alpha)T_1} \left(\int_{\Omega} u_0(r)\omega_k(r)dr \right)^2 \right)^{1/2} \leq \frac{\xi}{3},$$

$$\left(\sum_{k=1}^{N} \left(\frac{e^{(\lambda_k + \alpha)T_1} + \lambda_k/\alpha}{1 + \lambda_k/\alpha} \right)^2 \left(\int_{\Omega} \theta(r)\omega_k(r)dr \right)^2 \right)^{1/2} \leq \frac{\xi}{3}.$$

Combining the above three estimates yields

$$\| u(\cdot, T_1) - \theta \|_{L^2(\Omega)} \leq \xi,$$

implying (4.13). This ends the proof of Lemma 4.2. $\qquad\square$

Step 2. As we showed in the above, for any $v > 0$, $T_1 > 0$ we can steer (4.1) to a state $u(\cdot, T_1)$ satisfying (4.12). Select, for certainty, some $T_1 > 0$ and

$$v = 1 \qquad\qquad (4.16)$$

for our further consideration.

Consider now any number $\delta > 0$ and apply next the zero-control α on an interval (T_1, T_2), where $T_2 = T(\delta) > T_1$ is such that

$$e^{\lambda_1(T_2 - T_1)} \left(\text{ess sup}_{\tau \in (0, T_1)} \| \theta(\cdot, \tau) \|_{L^2(\Omega)} + 1 \right) \leq \delta.$$

(Note that T_2 can be "large.")

Then, regardless of u_0, due to (4.14) and (4.12), (4.16),

$$\| u(\cdot, T_2) \|_{L^2(\Omega)} \leq \delta. \qquad\qquad (4.17)$$

Step 3. Take now any $T_3 > T_2(\delta)$. Then, again with $\alpha = 0$ on (T_2, T_3),

$$u(x, T_3) = \sum_{k=1}^{\infty} e^{\lambda_k(T_3 - T_2)} \left(\int_{\Omega} u(r, T_2)(r)\omega_k(r)dr \right) \omega_k(x),$$

which in turn implies that $u(\cdot, T_3) \in C(\bar{\Omega})$ and the following estimates hold:

$$\| u(\cdot, T_3) \|_{C(\bar{\Omega})} \leq C_1 \| u(\cdot, T_3) \|_{H^{1+[n/2]}(\Omega)}$$

$$\leq C_2 \left(\sum_{k=1}^{\infty} \lambda_k^{1+[n/2]} e^{2\lambda_k(T_3 - T_2)} \left(\int_{\Omega} u(r, T_2)(r)\omega_k(r)dr \right)^2 \right)^{1/2} \qquad (4.18)$$

for some positive constants C_1 and C_2. In other words, for some positive constant C_3 (depending though on $T_3 - T_2$), also using (4.17), we obtain from (4.18) that, regardless of u_0,

$$\| u(\cdot, T_3) \|_{C(\bar{\Omega})} \le C_1 \| u(\cdot, T_3) \|_{H^{1+[n/2]}(\Omega)} \le C_3 \| u(\cdot, T_2) \|_{L^2(\Omega)} \le C_3 \delta. \quad (4.19)$$

Step 4. Take any $T > T_3$ and consider system (4.8) with $T_* = T_3$ and

$$v(x) = -\sum_{k=1}^{\infty} \left(\int_{\Omega} u(r, T_3)(r) \omega_k(r) dr \right) \frac{e^{\lambda_k(T-T_3)} \lambda_k}{e^{\lambda_k(T-T_3)} - 1} \omega_k(x). \quad (4.20)$$

As in the above, one can see that $v \in H^{1+[n/2]}(\Omega)$. The corresponding solution admits a representation on $[T_3, T]$ like in (4.14):

$$w(x,t) = \sum_{k=1}^{\infty} \left(\int_{\Omega} u(r, T_3)(r) \omega_k(r) dr \right) \left[e^{(\lambda_k(t-T_3))} - e^{\lambda_k(T-T_3)} \frac{e^{\lambda_k(t-T_3)} - 1}{e^{\lambda_k(T-T_3)} - 1} \right] \omega_k(x).$$

Note that

$$w(x,T) = 0.$$

Accordingly, for $t \in [T_3, T]$ (and also using (4.18) and (4.19))

$$\| w(\cdot, t) \|_{C(\bar{\Omega})} \le C_1 \| w(\cdot, t) \|_{H^{1+[n/2]}(\Omega)}$$

$$\le C_2 \left\{ \sum_{k=1}^{\infty} \lambda^{1+[n/2]} \left(\int_{\Omega} u(r, T_3)) \omega_k(r) dr \right)^2 \right.$$

$$\left. \times \left[e^{\lambda_k(t-T_3)} - e^{\lambda_k(T-T_3)} \frac{e^{\lambda_k(t-T_3)} - 1}{e^{\lambda_k(T-T_3)} - 1} \right]^2 \right\}^{1/2}$$

$$\le C_4 \| u(\cdot, T_3) \|_{H^{[n/2]+1}(\Omega)}$$

$$\le C_5 \| u(\cdot, T_2) \|_{L^2(\Omega)} \le C_5 \delta \quad (4.21)$$

for some positive constants C_4 and C_5.

Step 5. Finally, recalling that all the above calculations are valid for any $\delta > 0$, select

$$\delta < \frac{v_0}{2C_5},$$

where v_0 is from (4.7). Then, in view of (4.2) and (4.21), we can select on (T_3, T) the bilinear control for (4.1) as in (4.9):

$$\alpha = \frac{v}{w - \theta}$$

as an element of $L^\infty(\Omega \times (T_3, T))$. With this control solutions to (4.1) and (4.8) become identical on (T_3, T), which ensures (4.3) and ends the proof of Theorem 4.1.   □

Proof of Corollary 4.1. This proof makes use of the above five steps with the following changes.

Step 1. Taking into account (4.6) in place of (4.11), we obtain the following inequality for solutions to (4.4) with

$$\alpha = \alpha_* s,$$

(α_* is from (4.6)), where s is a positive parameter to be selected later:

$$\frac{d \| y(\cdot, t) \|^2_{L^2(\Omega)}}{dt} + 2 \int_\Omega \| \nabla y(x, t) \|^2_{R^n} \, dx$$

$$= 2s \int_\Omega \alpha_* f(x, t, y) y \, dx - 2s \int_\Omega \alpha_* \theta y \, dx$$

$$\leq -2sp \, \| \alpha_* \|_{L^\infty(\Omega)} \int_\Omega y^2 dx + s \int_\Omega | \alpha_* | y^2 dx + s \int_\Omega | \alpha_* | \theta^2 dx$$

$$\leq -s(2p - 1) \, \| \alpha_* \|_{L^\infty(\Omega)} \int_\Omega y^2 dx + \| \alpha_* \|_{L^\infty(\Omega)} \, s \int_\Omega \theta^2 dx,$$

which eventually yields an estimate of type (4.10) for any $s > 0$:

$$\| y(\cdot, t) \|^2_{L^2(0,1)} \leq e^{-s(2p-1)\|\alpha_*\|_{L^\infty(\Omega)} t} \, \| u_0 \|^2_{L^2(\Omega)}$$

$$+ \frac{e^{-s(2p-1)\|\alpha_*\|_{L^\infty(\Omega)} t} - 1}{-s(2p-1) \, \| \alpha_* \|_{L^\infty(\Omega)}} \, \| \alpha_*$$

$$\times \|_{L^\infty(\Omega)} \, \operatorname{ess\,sup}_{\tau \in (0,t)} s \, \| \theta(\cdot, \tau) \|^2_{L^2(\Omega)} \, .$$

Hence, as $s \to \infty$, we obtain an estimate like in (4.12), (4.16) in the case of system (4.4)–(4.6) as well:

$$\| y(\cdot, T_1) \|_{L^2(\Omega)} \leq \frac{1}{(2p - 1)} \operatorname{ess\,sup}_{\tau \in (0, T_1)} \| \theta(\cdot, \tau) \|_{L^2(\Omega)} + 1. \qquad (4.22)$$

Remark 4.4. Condition (4.6) plays the crucial role in the derivation of (4.22). It ensures that the ("decelerating") effect of the nonlinear term on the reaction at hand is "non-degenerate" compared to the linear case (4.1), see Lemma 4.1 and (4.12), (4.16) in the above. In turn, condition (4.5) will be used in the next step.

Steps 2–4. We can further use Steps 2–4 in the above proof of Theorem 4.1 without any changes (except for using (4.22) in place of (4.12), (4.16)), because there we deal with $\alpha = 0$, eliminating nonlinearity.

Step 5. Since in (4.21)

$$\| w(\cdot,t) \|_{C(\bar{\Omega})} \leq C_5 \delta,$$

the inequality (4.5) yields:

$$\| f(\cdot,\cdot,w(\cdot,t)) \|_{L^\infty(\Omega)} \leq M \| w(\cdot,t) \|_{C(\bar{\Omega})} \leq MC_5 \delta.$$

Accordingly, we select now

$$\delta < \frac{v_0}{2MC_5}$$

and the bilinear control $\alpha \in L^\infty(\Omega \times (T_3,T))$ for (4.4) as follows:

$$\alpha(x,t) = \frac{v(x)}{f(x,t,w(x,t)) - \theta(x,t)}.$$

With this control solutions to (4.8) and (4.4) become identical on (T_3,T), which ensures the steering of the latter to the zero-state at time T and ends the proof of Corollary 4.1.   □

4.3   Proof of Theorem 4.2 and Corollary 4.2

Proof of Theorem 4.2. This proof follows the argument of Steps 3–5 of the proof of Theorem 4.1. Without loss of generality we can assume that ω is an open subinterval of $\Omega = (0,1)$.

Fix any $u_0 \in L^2(0,1)$. Take any $\xi > 0$. Then, applying the zero-control α on $(0,\xi)$ yields as in (4.19):

$$\| u(\cdot,\xi) \|_{C[0,1]} \leq C_3 \| u_0 \|_{L^2(0,1)} . \tag{4.23}$$

Take any $T > \xi$ and, in place of (4.8), consider the following 1-D linear system:

$$q_t = q_{xx} + v(x,t)\chi_\omega \quad \text{in} \ (0,1) \times (\xi,T), \tag{4.24}$$

$$q \mid_{x=0,1} = 0, \quad q \mid_{t=\xi} = u(\cdot,\xi) \in L^2(0,1),$$

where χ_ω is the characteristic function of ω and v is the locally-distributed control supported in ω.

Let us recall [45, 46, 50] that (4.24) is (globally) exactly null-controllable in $L^2(0,1)$. Namely, for any $u(\cdot,\xi) \in L^2(0,1)$ there exists a $v \in L^2((0,1) \times (\xi,T))$ such that

$$q(\cdot,T) = 0.$$

Moreover,

$$\| v \|_{L^2((0,1)\times(\xi,T))} \leq L \| u(\cdot,\xi) \|_{L^2(0,1)}, \tag{4.25}$$

where L is a positive non-decreasing function of $T - \xi$. Using the continuous embedding $H_0^1(0,1) \subset C[0,1]$ and the formula like (4.14) for solutions to (4.24), it is classical to show that for $t \in [\xi,T]$:

$$\| q(\cdot,t) \|_{C[0,1]} \leq C_1 \| q(\cdot,t) \|_{H_0^1(0,1)} \leq L_1 \left\{ \| u(\cdot,\xi) \|_{C[0,1]} + \| v \|_{L^2(Q_T)} \right\}$$

$$\leq L_1 \left\{ \| u(\cdot,\xi) \|_{C[0,1]} + L \| u(\cdot,\xi) \|_{L^2(0,1)} \right\}, \tag{4.26}$$

where L_1 is a positive function of $T - \xi$, C_1 is as in (4.18), and L is from (4.25). Combining (4.26) with (4.23) yields for $t \in [\xi,T]$:

$$\| q(\cdot,t) \|_{C[0,1]} \leq L_1(1+L)C_3 \| u_0 \|_{L^2(0,1)} . \tag{4.27}$$

We now select σ in Theorem 4.2 as follows:

$$\sigma < \frac{v_0}{2L_1(1+L)C_3}.$$

Then, analogously to Step 5 in the proof of Theorem 4.1, due to (4.7) and (4.27), the bilinear control

$$\alpha = \frac{v}{q - \theta} \chi_\omega \quad \text{lies in} \quad L^2((0,1) \times (\xi,T))$$

for any u_0 satisfying

$$\| u_0 \|_{L^2(0,1)} \leq \sigma.$$

It makes systems (4.1), (4.7) and (4.24) identical on (ξ,T), which ensures, in particular, that

$$u(\cdot,T) = 0.$$

This ends the proof of Theorem 4.2.   □

To prove Corollary 4.2, one just needs to add Step 2 of the proof of Theorem 4.1 to that of Theorem 4.2, which employs the drift motion with $\alpha = 0$ to steer the system at hand to a state whose $L^2(0,1)$-norm does not exceed σ.

4.4 Proof of Theorem 4.3

Step 1. Fix any $\bar{T} > T = T(\theta)$, found in the proof of Theorem 4.1 and select any $\varepsilon > 0$ and $u_0 \in L^2(\Omega)$. It is sufficient to consider any target state like

$$u_d = \sum_{k=1}^{K} d_k \omega_k, \quad K < \infty, \tag{4.28}$$

which we also fix in the further consideration.

We follow next the steps 1–4 in the argument of Theorem 4.1, except for the selection of the distributive static control v in (4.20), which we replace as with the following:

$$v(x) = -\sum_{k=1}^{\infty} \left(\int_{\Omega} u(r, T_3)(r) \omega_k(r) dr \right) \frac{e^{\lambda_k(T-T_3)} \lambda_k}{e^{\lambda_k(T-T_3)} - 1} \omega_k(x)$$

$$+ \sum_{k=1}^{K} \frac{\lambda_k}{e^{\lambda_k(T-T_3)} - 1} s d_k \omega_k(x), \tag{4.29}$$

where s is a positive constant parameter (to be selected precisely in Step 2 below).

The solution formula like (4.14) implies that

$$w(\cdot, T) = s u_d,$$

and

$$w(x,t) = \sum_{k=1}^{\infty} \left(\int_{\Omega} u(r, T_3)) \omega_k(r) dr \right) \left[e^{\lambda_k(t-T_3)} - e^{\lambda_k(T-T_3)} \frac{e^{\lambda_k(t-T_3)} - 1}{e^{\lambda_k(T-T_3)} - 1} \right] \omega_k(x)$$

$$+ \sum_{k=1}^{K} \frac{e^{\lambda_k(t-T_3)} - 1}{e^{\lambda_k(T-T_3)} - 1} s d_k \omega_k(x).$$

Step 2. Again, as in (4.21), for $t \in [T_3, T]$ we have:

$$\| w(\cdot, t) \|_{C(\bar{\Omega})} \leq C_1 \| w(\cdot, t) \|_{H^{1 + [n/2]}(\Omega)}$$

$$\leq C_2 \left\{ \sum_{k=1}^{\infty} \lambda^{1 + [n/2]} \left(\int_{\Omega} u(r, T_3)) \omega_k(r) dr \right)^2 \right.$$

$$\left. \times \left[e^{\lambda_k(t-T_3)} - e^{\lambda_k(T-T_3)} \frac{e^{\lambda_k(t-T_3)} - 1}{e^{\lambda_k(T-T_3)} - 1} \right]^2 \right\}^{1/2}$$

$$+ C_2 \left\{ \sum_{k=1}^{\infty} \lambda^{1 + [n/2]} \left[\frac{e^{\lambda_k(t-T_3)} - 1}{e^{\lambda_k(T-T_3)} - 1} s d_k \right]^2 \right\}^{1/2}$$

$$\leq C_4 \| u(\cdot, T_3) \|_{H^{[n/2]+1}(\Omega)} + C_6 s \| u_d \|_{H^{1+[n/2]}(\Omega)}$$

$$\leq C_5 \delta + C_6 s \| u_d \|_{H^{1+[n/2]}(\Omega)} \qquad (4.30)$$

for some positive constant C_6 and $C_{1,2,4,5}$ like in (4.21).

Now, in place of the choice of δ as in Step 5 in the proof of Theorem 4.1, we select now any $\delta > 0$ and $s > 0$ such that

$$\delta \leq \frac{v_0}{4C_5}, \quad s \leq \frac{v_0}{4C_6 \| u_d \|_{H^{[n/2]+1}(\Omega)}}. \qquad (4.31)$$

(Recall that such choice of δ defines T_3 and eventually $T = T(\theta)$ in the proof of Theorem 4.1.) Then we can select on (T_3, T) the bilinear control

$$\alpha = \frac{v}{w - \theta}$$

lying in $L^{\infty}(\Omega \times (T_3, T))$ due to (4.30) and (4.2). With this control, again, solutions to (4.1) and (4.8) become identical on (T_3, T), which ensures that

$$u(\cdot, T) = su_d. \qquad (4.32)$$

Step 3. Next, with s selected in (4.31), we consider a time interval $(T, T + \mu)$, $\mu \in (0, \bar{T} - T)$ and the constant bilinear control

$$\alpha = a, \quad \text{where} \quad e^{a\mu} = \frac{1}{s} \quad \text{or} \quad a = -\frac{\ln s}{\mu}.$$

With this control, in view of (4.32), we have:

$$u(x, T + \mu) = \sum_{k=1}^{K} e^{\lambda_k \mu} e^{a\mu} s d_k \omega_k(x) = \sum_{k=1}^{K} e^{\lambda_k \mu} d_k \omega_k(x).$$

Hence, we can find a sufficiently small $\mu = \mu_*$, for which

$$\| u(\cdot, T + \mu_*) - u_d \|_{L^2(\Omega)} \leq \varepsilon.$$

It remains only to set $\hat{T} = T + \mu_*$. This ends the proof of Theorem 4.3. $\qquad \square$

4.5   Further Discussion of Approximate Controllability

The general case when $\theta \neq 0$ seems of more difficult nature, if not doubtful, to achieve the global approximate controllability for a system like (4.1). To somewhat illustrate the difficulty here, consider a "more explicit" situation when

$$\theta \in H^2(\Omega) \bigcap H_0^1(\Omega),$$

in which case the substitution $z = u - \theta$ transforms (4.1) into

$$\frac{\partial z}{\partial t} = \Delta z + \alpha z + \Delta \theta \quad \text{in } Q_T, \tag{4.33}$$

$$z = 0 \quad \text{in } \Sigma_T, \quad z \mid_{t=0} = z_0 \in L^2(\Omega).$$

Though, at the first glance, the source term does not have now the factor α, its contribution to the solution still depends highly nonlinearly on α (defining the corresponding solution mapping). In other words, one cannot structurally separate the contribution of the source term from the effect of bilinear control (in contrast to the classical controllability framework with additive controls).

From this standpoint, it seems more reasonable to focus first on the issue of the partial approximate controllability, i.e., to try to identify at least some set of states that can (approximately) be reached by using available bilinear controls. Along these lines, in addition to Lemma 4.2 in the above, let us make the following observations, invoking some classical results dealing with the steady-state solutions to (4.1).

Consider the static case when

$$\theta = \theta(x) \quad \text{and} \quad \alpha = \alpha(x).$$

Then, the corresponding solution to (4.33) admits the following representation:

$$z(x,t) = \sum_{k=1}^{\infty} e^{\beta_k t} \left(\int_{\Omega} z_0(r) w_k(r) dr \right) w_k(x)$$

$$+ \sum_{k=1}^{\infty} \frac{e^{\beta_k t} - 1}{\beta_k} \left(\int_{\Omega} \Delta \theta(r) w_k(r) dr \right) w_k(x),$$

where β_k's and w_k's are the solutions to the spectral problem $\Delta w - \alpha(x)w = \beta w$. Hence, if all β_k's are negative, which is guaranteed when

$$\text{ess sup}_{\Omega} \alpha(x) \leq c < 0, \tag{4.34}$$

then

$$z(\cdot,t) \to \phi = -\sum_{k=1}^{\infty} \frac{1}{\beta_k} \int_{\Omega} \Delta \theta(r) w_k(r) dr w_k \in H^2(\Omega) \bigcap H_0^1(\Omega)$$

as $t \to \infty$. One can see that ϕ satisfies the Dirichlet problem

$$\Delta\phi + \alpha\phi = -\Delta\theta \quad \text{in } \Omega, \tag{4.35}$$

$$\phi_{\partial\Omega} = 0,$$

which admits a unique solution when (4.34) holds. Hence, we have the following simple property.

Proposition 4.1. *Let* $\theta \in H^2(\Omega) \cap H_0^1(\Omega)$ *and* $\alpha = \alpha(x)$ *be static in (4.1). Then any of the elements of the set*

$$\theta + \{\phi \mid \phi \text{ satisfies (4.35) for some } \alpha \text{ like in (4.34)}\}$$

can arbitrarily close be approximated in $L^2(\Omega)$ *by solutions to (4.1) with properly selected bilinear controls* $\alpha(x)$, *regardless of the choice of* u_0.

Note that (4.34) and (4.35) imply that the set of states described in Proposition 4.1 is bounded in $H_0^1(\Omega)$.

In the case when $\theta \notin H^2(\Omega) \cap H_0^1(\Omega)$, instead of Proposition 4.1, we accordingly have

Proposition 4.2. *Let* $\theta = \theta(x)$ *and* $\alpha = \alpha(x)$ *be static in (4.1). Then any of the elements of the set*

$$\left\{ p \mid \frac{\Delta p}{p - \theta} \in L^\infty(\Omega), \ \frac{\Delta p(x)}{p(x) - \theta(x)} \geq h > 0 \text{ a.e. in } \Omega, h \in R \right\} \subset H^2(\Omega) \cap H_0^1(\Omega)$$

can arbitrarily close be approximated in $L^2(\Omega)$ *by solutions to (4.1) with properly selected bilinear controls* $\alpha(x)$, *regardless of the choice of* u_0.

Indeed, consider any p as in the above. To prove Proposition 4.2 one just needs to select

$$\alpha = -\frac{\Delta p}{p - \theta},$$

which is then strictly negative and thus all the β_k's are negative as well. For this α the corresponding p is the point of the only equilibrium for (4.1) as $t \to \infty$. More precisely, since

$$\Delta p + \alpha(p - \theta) = 0 \quad \text{in } Q_T,$$

$$p = 0 \quad \text{in } \Sigma_T, \quad p \mid_{t=0} = p,$$

we have the following mixed problem for $(u - p)$:

$$\frac{\partial(u - p)}{\partial t} = \Delta(u - p) + \alpha(u - p) \quad \text{in} \quad Q_T,$$

$$u - p = 0 \quad \text{in} \quad \Sigma_T, \quad u \mid_{t=0} = u_0 - p,$$

whose solution converges in $L^2(\Omega)$ to the zero-state as $t \to \infty$.

To qualitatively illustrate Proposition 4.2, let us assume that $\Omega = (0, 1)$ and consider $\theta(x) \equiv 2$. Then, regardless of u_0, according to Proposition 4.2, the states of system (4.1) can (with properly selected static α) approximate in $L^2(\Omega)$ any non-negative function from $C^2[0, 1]$, vanishing at $x = 0$ and $x = 1$, whose values are strictly less than 2 and whose second derivative is negative in $[0, 1]$ (e.g., $p = -4(x - 0.5)^2 + 1$).

However, these states do not, of course, fully describe all possible "approximately reachable" states for our system. For example, all other states through which the system will drift with, say, $\alpha = 0$, after any of the above-described p's is (approximately) achieved, are also achievable. Also, solutions to (4.1) can grow unboundedly when, e.g., "large" positive constant controls α are applied.

Chapter 5
Classical Controllability for the Semilinear Parabolic Equations with Superlinear Terms

Abstract In this chapter we establish the global approximate controllability of the semilinear heat equation with superlinear term, governed in a bounded domain by a pair of controls: (I) the traditional internal either locally distributed or lumped control *and* (II) the lumped control entering the system as a time-dependent coefficient. The motivation for the latter is due to the well known lack of global controllability properties for this class of pde's when they are steered solely by the former controls. Our approach involves an asymptotic technique allowing us to *separate and combine* the impacts generated by the above-mentioned two types of controls. In particular, the addition of multiplicative control allows us to reduce the use of the additive one to the local controllability technique only.

5.1 Introduction

5.1.1 Problem Formulation and Motivation

We consider now the following Dirichlet boundary problem, governed in a bounded domain $\Omega \subset R^n$ by the lumped multiplicative control $k = k(t)$ and the *additive* locally distributed control $v(x,t)\chi_\omega(x)$, supported in the given subdomain $\omega \subset \Omega$:

$$\frac{\partial u}{\partial t} = \Delta u + k(t)u - f(x,t,u,\nabla u) + v(x,t)\chi_\omega(x) \text{ in } Q_T = \Omega \times (0,T), \qquad (5.1)$$

$$u = 0 \text{ in } \Sigma_T = \partial\Omega \times (0,T), \quad u \mid_{t=0} = u_0 \in L^2(\Omega), \quad k \in L^\infty(0,T), \ v \in L^2(Q_T).$$

In the one space dimension we will also consider the case when both controls are lumped, that is, they are the functions of time only: $k = k(t)$ and $v = v(t)$.

The problem of our interest is the approximate controllability of system (5.1) in the system's phase-space $L^2(\Omega)$. Namely, given the initial state u_0, we want to know *whether the range of the solution mapping*

$$L^\infty(0,T) \times L^2(Q_T) \ni (k,v) \ \to \ u(\cdot,T) \in L^2(\Omega) \qquad (5.2)$$

A.Y. Khapalov, *Controllability of Partial Differential Equations Governed by Multiplicative Controls*, Lecture Notes in Mathematics 1995, DOI 10.1007/978-3-642-12413-6_5, © Springer-Verlag Berlin Heidelberg 2010

is dense in $L^2(\Omega)$. (In fact, due to the possible nonuniqueness of solutions to (5.1) the situation here is more complex.)

It is well known [36, 44, 50, 138] that a rather general semilinear parabolic equation, governed in a bounded domain by the classical either boundary or additive locally distributed controls *only* (i.e., no "changeable" bilinear control $k(\cdot)$ in (5.1)) is globally approximately controllable in $L^2(\Omega)$, provided that the nonlinearity is globally Lipschitz. The methods of these works make use of the fixed point argument and the fact that such semilinear equations can be viewed as "linear equations" with the coefficients uniformly bounded in some sense. Alternative approach employs the global inverse function theorem – we refer in this respect to the work [102] on the semilinear wave and plate equations.

However, the situation is principally different if nonlinear terms admit polynomial superlinear growth at infinity.

Given $T > 0$, we further assume that $f(x,t,u,p)$ is Lebesgue's measurable in x, t, u, p, and continuous in u, p for almost all $(x,t) \in Q_T$, and is such that

$$| f(x,t,u,p) | \leq \beta \, | u \, |^{r_1} + \beta \, \| \, p \, \|_{R^n}^{r_2} \text{ a.e. in } Q_T \text{ for } u \in R, \ p \in R^n, \quad (5.3)$$

$$\int_\Omega f(x,t,\phi,\nabla\phi)\,\phi dx \geq (v-1) \int_\Omega \| \, \nabla\phi \, \|_{R^n}^2 \, dx$$

$$-\rho \int_\Omega (1+\phi^2)dx \ \forall\phi \in H_0^1(\Omega), \quad (5.4)$$

where $\beta, v, \rho > 0$, $T\rho \leq \beta$, and

$$r_1 \in (1, 1+\frac{4}{n}), \quad r_2 \in (1, 1+\frac{2}{n+2}). \quad (5.5)$$

The reader can notice that conditions (5.3)–(5.5) on the nonlinear term f coincide with conditions (2.2)–(2.4) in Chapter 2 for $n = 1$ and are more restrictive than the respective conditions in Chapter 3. More precisely, in this section, as in Chapter 2, we require f to be superlinear near the origin as well (see also the discussion in the end of Chapter 2). Once again we refer, e.g., to [97] (p. 466), where it was shown that system (5.1), (5.3)–(5.5) admits at least one generalized solution in $C([0,T];L^2(\Omega))\cap H_0^{1,0}(Q_T)\cap L^{2+4/n}(Q_T)$, while its uniqueness is not guaranteed.

It turns out that in the superlinear case like (5.3)–(5.5) the impact from the *sole* additive control $v(t)\chi_\omega(x)$ does not propagate "effectively" from its support to the rest of the space domain: regardless of how large the control applied on ω is, the corresponding solutions remain uniformly bounded on any closed subset of $\Omega\backslash\bar{\omega}$. (In other words, given the initial state u_0, there exist target states, namely, "sufficiently large" on $\Omega\backslash\omega$, which are strictly separated from the range of the corresponding solution mapping.) This is true in any of the (phase-) spaces $L^p(\Omega)$, $1 \leq p < \infty$ at any positive time, e.g., for the functions $f = f(x,t,u)$ such that

$f(x,t,u)u > c_1 \mid u \mid^{2+r} -c_2$ for some constants $c_1, c_2, r > 0$ ([50], and the references therein). On the other hand, for certain refinements of conditions (5.3)–(5.5) a number of positive "superlinear" controllability results were obtained in [61,64–66], see Remark 5.1 below for details.

In this section our goal is to show that the above-outlined principal difficulty with propagation of control impact can be overcome by using an additional bilinear lumped control $k = k(t)$, entering the equation (5.1) as a coefficient and thus affecting the qualitative behavior of system (5.1), (5.3)–(5.5) in the entire space domain.

Remark 5.1. • In spite of the lack of global controllability of (5.1), (5.3)–(5.5) discussed in the above, it was shown in [64] that this equation is actually globally approximately controllable at any time $T > 0$ *solely* by means of the additive locally distributed controls in the spaces that are *weaker* than any of $L^p(\Omega)$, $1 \le p < \infty$. Moreover, under the additional assumption that the superlinear term is locally Lipschitz (which ensures the uniqueness of solutions in $C([0,T];L^2(\Omega))$) the global finite dimensional exact controllability of (5.1), (5.3)–(5.5) (i.e., not necessarily to the equilibrium) at any positive time $T > 0$ was also established in [64].

• The global approximate controllability of (5.1), (5.3)–(5.5) with $k = 0$ was shown in [61] for the *static* controls $v = v(x)$ supported in the *entire* Ω.

• For the one dimensional version (5.7), (5.8)–(5.10) (see below) of system (5.1) it was shown in [65] that, if in (5.8) $\beta = \beta(t) \to 0$ faster than any $e^{-v/t}$, $v > 0$ as $t \to 0$, then (5.7), (5.8)–(5.10) with $k = 0$ is globally approximately controllable in $L^2(\Omega)$ at any time only by means of the lumped control $v = v(t)$, provided that the endpoints of the interval (a,b) are the irrational numbers. This result was recently extended to the case of several dimensions and locally distributed controls in [66].

• The method of [61, 64–66] is based on the idea to "suppress" the effect of nonlinearity by applying the actual control action only during asymptotically short period of time. Similar idea is used in this chapter and in [86] in the context of bilinear ode's.

• Though in Chapter 5 we discuss the global approximate controllability of the equation (5.1), we would also like to mention here some related works on the very close global exact null-controllability property (i.e., the exact steering to the origin) by means of the additive locally distributed controls only. In [43] the latter property was shown in $L^2(\Omega)$ (or appropriate Sobolev space) with the reaction term $f = f(x,t,u)$ only, assuming that f can grow superlinearly at the *logarithmic rate* like $\lim_{|p| \to \infty} f(p)/(p \log \mid p \mid) = 0$. Assuming the dissipativity condition, this result was improved in [9] to the rate $\lim_{|p| \to \infty} f(p)/(p(\log \mid p \mid)^{3/2}) = 0$. Also in [9] some interesting non-global exact null-controllability results were given.

5.1.2 Main Results: Combining Additive and Multiplicative Controls

The multidimensional case with additive locally distributed controls.

Theorem 5.1. *Let conditions (5.3)–(5.5) hold. Then the range of the solution mapping (5.2) is dense in* $L^2(\Omega)$.

Note now that, since the boundary problem (5.1), (5.3)–(5.5) admits multiple solutions, this result is qualitatively different from the classical understanding of the approximate controllability as steering (associated with applications in the first place), which is as follows: (5.1) is said to be globally approximately controllable in $L^2(\Omega)$ at time T if for any $u_0, u_T \in L^2(\Omega), \xi > 0$ there is a control pair (k, v) such that

$$\| u(\cdot, T) - u_T \|_{L^2(Q_T)} \le \xi.$$

Clearly, this classical definition is ill-posed in our case of possible multiple solutions. Therefore, we will also use its adjustment which requires one to find a control pair which steers all the possible realizations of a solution to (5.1) in a uniform fashion. This type of controllability was investigated in [61, 64–66] and the previous Chapter 4 of this monograph.

Definition 5.1. We will say that the system (5.1), (5.3)–(5.5), admitting multiple solutions, is globally approximately controllable in $L^2(\Omega)$ at time T if for every $\xi > 0$ and $u_0, u_T \in L^2(\Omega)$ there is a control pair $(k, v) \in L^\infty(Q_T) \times L^2(Q_T)$ such that for all (i.e., possibly multiple) solutions of (5.1), (5.3)–(5.5); corresponding to it

$$\| u(\cdot, T) - u_T \|_{L^2(\Omega)} \le \xi. \tag{5.6}$$

Theorem 5.2. *System (5.1), (5.3)–(5.5) is globally approximately controllable in* $L^2(\Omega)$ *at any time* $T > 0$ *in the sense of Definition 5.1.*

The 1-*D* case with all lumped controls.

Consider now the one dimensional version of problem (5.1), (5.3)–(5.5) with all lumped controls:

$$u_t = u_{xx} + k(t)u - f(x, t, u, u_x) + v(t)\chi_{(a,b)}(x) \text{ in } Q_T = (0, 1) \times (0, T), \tag{5.7}$$

$$u\big|_{x=0,1} = 0, \quad u\big|_{t=0} = u_0 \in L^2(0, 1), \ k \in L^\infty(0, T), \ v \in L^2(0, T).$$

Here both $k = k(t)$ and $v = v(t)$ are the functions of time only.

By distinguishing the 1-*D* case we pursue two goals. Firstly, the positive result for the case of lumped additive controls implies the same for the locally distributed ones (since the former controls are a degenerate subclass of the latter ones). Accordingly, our proof of Theorems 5.1/5.2 is given below as the immediate consequence of the 1-*D*-"lumped" case. Secondly, lumped controls are of special interest being closer to the engineering applications. Focusing on them, we can give somewhat more "explicit" feeling of our method and of the general conditions (5.3)–(5.5), which for the equation (5.7) are as follows:

$$| f(x,t,u,p) | \leq \beta \, | u |^{r_1} + \beta \, | p |^{r_2} \quad \text{a.e. in } Q_T \text{ for } u,p \in R, \quad (5.8)$$

$$\int_0^1 f(x,t,\phi,\phi_x) \, \phi dx \geq (v-1) \int_0^1 \phi_x^2 dx - \rho \int_0^1 \phi^2 dx \quad \forall \phi \in H_0^1(0,1), \quad (5.9)$$

where $\beta, v, \rho > 0$, $\rho T \leq \beta$, and

$$r_1 \in (1,5), \quad r_2 \in (1,5/3). \quad (5.10)$$

Theorem 5.3. *If $a < b$ are any irrational numbers from $(0,1)$, then system (5.7)–(5.10) is globally approximately controllable in $L^2(\Omega)$ at any time $T > 0$ in the sense of Definition 5.1.*

Clearly the assumption that the endpoints of the interval (a,b) are the irrational numbers makes the result of Theorem 5.1 unstable with respect to the choice of control support (a,b). We stress however that it is well known that this assumption is intrinsic for lumped controls even in the linear case. In this respect one may prefer its immediate "stable" corollary – Theorem 5.2.

The rest of Chapter 5 is organized as follows. In the next two its sections we prove and recall some auxiliary results for the linear version of system (5.7). Then we give the proof of Theorem 5.3, followed by the proofs of Theorems 5.1 and 5.2.

5.2 Preliminary Estimates

Consider the boundary problem (5.7), (5.8)–(5.10) assuming that bilinear control k is constant, i.e., $k(t) \equiv \alpha$ on $(0,T)$:

$$w_t = w_{xx} + \alpha w - f(x,t,w,w_x) + v(t)\chi_{(a,b)}(x) \quad \text{in } Q_T, \quad (5.11)$$

$$w \mid_{x=0,1} = 0, \quad w \mid_{t=0} = w_0 \in L^2(0,1), \, v \in L^2(0,T).$$

We have the following a priori estimate.

Lemma 5.1. *Given $T > 0, \alpha \geq 0$ and a positive number μ, any solution to (5.11), (5.8)–(5.10) (if there are multiple solutions) satisfies the following two estimates:*

$$\| w \|_{\mathscr{B}(0,T)}, \| w \|_{L^6(Q_T)} \leq C(\mu) e^{(\alpha+\rho+\mu/2)T} \left(\| w_0 \|_{L^2(0,1)} + \| v \|_{L^2(0,T)} \right),$$
(5.12)

where $C(\mu)$ does not depend on α.

Here and below we routinely use symbols c and C to denote (different) *generic* positive constants or positive-valued functions.

Proof. It is similar to that of Lemma 3.1 in Chapter 3. Indeed, recall ([97]) that $f(\cdot, \cdot, w, w_x) \in L^{6/5}(Q_T)$ and that the following energy equality holds for (5.11) treated as a linear equation with the source term $f(x, t, w, w_x) + v(t)\chi_{(a,b)}(x)$, e.g., [97] (p. 142):

$$\frac{1}{2} \| w \|_{L^2(0,1)}^2 |_0^t + \int_0^t \int_0^1 (w_x^2 - \alpha w^2 + f(x, s, w, w_x)w$$
$$- v(t)\chi_{(a,b)}(x)w)dxds = 0 \quad \forall t \in [0,T].$$
(5.13)

Here and everywhere below, if there exist several solutions to (5.7), we always deal separately with a selected one, while noticing that all the estimates hold uniformly.

Combining (5.13) and (5.9) yields:

$$\| w(\cdot, t) \|_{L^2(0,1)}^2 + 2\nu \int_0^t \int_0^1 w_x^2(x, s) \, dxds$$

$$\leq \| w_0 \|_{L^2(0,1)}^2 + 2(\alpha + \rho) \int_0^t \int_0^1 w^2 dxds + 2 \int_0^t \int_0^1 v(t)\chi_{(a,b)}(x)w dxds$$

$$\leq \left(\| w_0 \|_{L^2(0,1)}^2 + \frac{b-a}{\mu} \| v \|_{L^2(0,T)}^2 \right) + 2(\alpha + \rho + \mu/2) \int_0^t \| w(\cdot, \tau) \|_{L^2(0,1)}^2 \, d\tau$$

$$\leq \left(\| w_0 \|_{L^2(0,1)}^2 + \frac{b-a}{\mu} \| v \|_{L^2(0,T)}^2 \right) + 2(\alpha + \rho + \mu/2)$$

$$\times \int_0^t \left(\| w(\cdot, \tau) \|_{L^2(0,1)}^2 + 2\nu \int_0^\tau \int_0^1 w_x^2(x, s) \, dxds \right) d\tau \quad \forall t \in [0,T].$$
(5.14)

In the above we have used Young's inequality

$$2wv\chi_{(a,b)} \leq \frac{1}{\mu}(v\chi_{(a,b)})^2 + \mu w^2,$$

which holds for any positive μ. Applying Gronwall-Bellman inequality to (5.14) yields the first estimate in (5.12) with respect to the $\mathcal{B}(0,T)$-norm. The second estimate (with properly arranged generic constant) follows by the continuity of the embedding $\mathcal{B}(0,T)$ into $L^6(Q_T)$ (e.g., [97], pp. 467, 75). This ends the proof of Lemma 5.11.   □

Remark 5.2. Note that if $\alpha < 0$ in (5.11), then, as (5.14) implies, (5.12) holds with no α in it.

We now intend to evaluate the difference between the solution w to (5.11) and that to its truncated version

$$y_t = y_{xx} + \alpha y + \chi_{(a,b)}v(t) \quad \text{in } Q_T, \ v \in L^2(0,T), \tag{5.15}$$

$$y\,|_{x=0,1} = 0, \quad y\,|_{t=0} = y_0 \in L^2(0,1),$$

assuming that $w_0 = y_0$.
  Denote $z = w - y$, then

$$z_t = z_{xx} + \alpha z - f(x,t,w,w_x) \quad \text{in } Q_T,$$

$$z\,|_{x=0,1} = 0, \quad z\,|_{t=0} = 0.$$

Similar to (5.13) and (5.14) we have,

$$\| z(\cdot,t) \|^2_{L^2(0,1)} + 2 \int_0^t \int_0^1 z_x^2(x,s)\,dxds$$

$$\leq 2\alpha \int_0^t \int_0^1 z^2 dxds + 2 \int_0^t \int_0^1 zf(x,s,w,w_x)dxds$$

$$\leq 2\alpha \int_0^t \int_0^1 z^2 dxds + 2\,\| z \|_{L^6(Q_t)}\| f(\cdot,\cdot,w,w_x) \|_{L^{6/5}(Q_t)}$$

$$\leq 2\alpha \int_0^t \int_0^1 z^2 dxds + 2c\,\| z \|_{\mathcal{B}(0,t)}\| f(\cdot,\cdot,w,w_x) \|_{L^{6/5}(Q_T)}$$

$$\leq 2\alpha \int_0^t \| z \|^2_{\mathcal{B}(0,s)}\,ds + \delta\,\| z \|^2_{\mathcal{B}(0,t)}$$

$$+ \frac{c^2}{\delta}\,\| f(\cdot,\cdot,w,w_x) \|^2_{L^{6/5}(Q_T)} \quad \forall t \in [0,T], \tag{5.16}$$

where we have used Hölder's and Young's inequalities and the continuity of the embedding $\mathscr{B}(0,T)$ into $L^6(Q_T)$, due to which,

$$\| z \|_{L^6(Q_T)} \leq c \| z \|_{\mathscr{B}} (0,T). \tag{5.17}$$

From (5.16), we have

$$\max_{\tau \in (0,t)} \| z(\cdot,\tau) \|^2_{L^2(0,1)} \leq 2\alpha \int_0^t \| z \|^2_{\mathscr{B}(0,s)} \, ds + \delta \| z \|^2_{\mathscr{B}(0,t)}$$

$$+ \frac{c^2}{\delta} \| f(\cdot,\cdot,w,w_x) \|^2_{L^{6/5}(Q_T)} \quad \forall t \in [0,T].$$

Hence, again from (5.16),

$$\| z \|^2_{\mathscr{B}(0,t)} \leq 4\alpha \int_0^t \| z \|^2_{\mathscr{B}(0,s)} \, ds + 2\delta \| z \|^2_{\mathscr{B}(0,t)}$$

$$+ \frac{2c^2}{\delta} \| f(\cdot,\cdot,w,w_x) \|^2_{L^{6/5}(Q_T)} \quad \forall t \in [0,T]$$

and

$$(1-2\delta) \| z \|^2_{\mathscr{B}(0,t)} \leq 4\alpha \int_0^t \| z \|^2_{\mathscr{B}(0,s)} \, ds + \frac{2c^2}{\delta} \| f(\cdot,\cdot,w,w_x) \|^2_{L^{6/5}(Q_T)}. \tag{5.18}$$

Making use of Gronwall-Bellman inequality, we derive from (5.18) that

$$\| z \|_{\mathscr{B}(0,T)} \leq e^{\frac{2\alpha}{1-\delta}T} \frac{\sqrt{2}c}{\sqrt{\delta}} \| f(\cdot,\cdot,w,w_x) \|_{L^{6/5}(Q_T)}, \tag{5.19}$$

provided that

$$0 < \delta < \frac{1}{2}. \tag{5.20}$$

Now, using (5.8) and Hölder's inequality (as in [97], p. 469; [61], p. 863), we obtain:

$$\| f(\cdot,\cdot,w,w_x) \|_{L^{6/5}(Q_T)} \leq \beta T^{\frac{5}{6}(1-\frac{r_1}{5})} \| w \|^{r_1}_{L^6(Q_T)} + \beta T^{\frac{5}{6}(1-\frac{3r_2}{5})} \| w_x \|^{r_2}_{L^2(Q_T)}. \tag{5.21}$$

Combining (5.21), (5.19)–(5.20) and (5.17) yields

Lemma 5.2. *Given $T > 0, \alpha \geq 0, \delta \in (0,1/2)$, and $w_0 = y_0$, we have the following two estimates for the difference $z = w - y$ between any corresponding solution w*

to (5.11), (5.8)–(5.10) (if there are multiple ones) and the unique corresponding solution to (5.15):

$$\| z \|_{\mathscr{B}(0,T)}, \| z \|_{L^6(Q_T)} \leq Ce^{\frac{2\alpha T}{1-\delta}} \frac{1}{\sqrt{\delta}} (T^{\frac{5}{6}(1-\frac{r_1}{5})} \| w \|_{L^6(Q_T)}^{r_1}$$

$$+ T^{\frac{5}{6}(1-\frac{3r_2}{5})} \| w_x \|_{L^2(Q_T)}^{r_2}) \quad \forall \delta \in (0, 1/2), \qquad (5.22)$$

where C does not depend on α.

5.3 Controllability Properties of the Truncated Linear System (5.15)

Here we would like to remind the reader some controllability properties of the linear system (5.15).

Denote by $\lambda_k = -(\pi k)^2 + \alpha$, $\omega_k(x) = \sqrt{2} \sin \pi k x$, $k = 1, \ldots$ the eigenvalues and orthonormalized in $L^2(0,1)$ eigenfunctions of the spectral problem: $\omega_{xx} + \alpha \omega = \lambda \omega$, $\omega \in H_0^1(0,1)$.

It is well known that the general solution to (5.15) admits the following representation:

$$y(x,t) = \sum_{k=1}^{\infty} e^{\lambda_k t} \left(\int_0^1 y_0(r) \omega_k(r) dr \right) \omega_k(x)$$

$$+ \sum_{k=1}^{\infty} \int_0^t e^{\lambda_k(t-\tau)} \left(\int_0^1 v(\tau) \chi_{(a,b)}(r) \omega_k(r) dr \right) d\tau \, \omega_k(x), \qquad (5.23)$$

where the series converge in $L^2(0,1)$ uniformly over $t \geq 0$.

Let $\{q_{T,k}\}_{k=1}^{\infty}$ be a biorthogonal sequence to $\{e^{\lambda_k \tau}\}_{k=1}^{\infty}$ in the closed subspace

$$\mathrm{cl} \left(\mathrm{span}\{e^{\lambda_k \tau} \mid k = 1, \ldots\} \right)$$

of $L^2(0,T)$ [38, 115]:

$$\int_0^T e^{\lambda_k \tau} q_{T,l}(\tau) d\tau = \begin{cases} 1, & \text{if } k = l, \\ 0, & \text{if } k \neq l, \end{cases}$$

where

$$\| q_{T,k} \|_{L^2(0,T)} - \frac{1}{d_k(\alpha,T)}, \tag{5.24}$$

$$d_k(\alpha,T) = \inf\{\| e^{\lambda_k t} + \sum_{i=1, i \neq k}^{I} b_i e^{\lambda_i t} \|_{L^2(0,T)} |\ b_i \in R, I = 1,2,\ldots\}.$$

Assume that $a \pm b$ are the irrational numbers. We need this to ensure that $\int_a^b \sin \pi k x\, dx \neq 0$ for all $k - 1, \ldots$.

Denote

$$v_{T,k}(\tau) = q_{T,k}(T - \tau)\,(\sqrt{2} \int_a^b \sin \pi k x\, dx)^{-1}, \quad \tau \in (0,T), \tag{5.25}$$

so that

$$\int_0^T \int_0^1 e^{\lambda_k(T-\tau)} v_{T,l}(\tau) \chi_{(a,b)}(r)\, \omega_k(r)\, dr\, d\tau = \begin{cases} 1, & \text{if } k = l, \\ 0, & \text{if } k \neq l. \end{cases} \tag{5.26}$$

From (5.23) and (5.26) it follows that, given the positive integer L and the real numbers $a_1, \ldots, a_L$, if one applies control

$$\hat{v}_T(t) = \sum_{k=1}^{L} a_k v_{T,k}(t), \quad t \in (0,T) \tag{5.27}$$

in (5.15), then

$$y(x,T) = \sum_{k=1}^{L} a_i \omega_k(x) + \sum_{k=1}^{\infty} e^{(-(\pi k t)^2 + \alpha)T} \left(\int_0^1 y_0(r)\omega_k(r)dr \right) \omega_k(x), \tag{5.28}$$

where, by (5.24) and (5.25),

$$\| \hat{v}_T \|_{L^2(0,T)} \leq \gamma(\alpha,T) = \sum_{k=1}^{L} \frac{1}{d_k(\alpha,T)}\ |\ a_k(\sqrt{2} \int_a^b \sin \pi k x\, dx)^{-1}\ |. \tag{5.29}$$

Remark 5.3. It follows from (5.24) that $\gamma(\alpha,T)$ in (5.29) is nonincreasing in $T > 0$.

Also, from (5.28),

$$\| y(\cdot,T) - \sum_{k=1}^{L} a_i \omega_k(\cdot) \|_{L^2(0,1)} \leq e^{\alpha T}\ \| y_0 \|_{L^2(0,1)}. \tag{5.30}$$

Now we are ready to prove Theorem 5.3.

5.4 Proof of Theorem 5.3: The 1-D Case with All Lumped Controls

The scheme of the proof is as follows.

(1) Given the initial and target state u_0 and u_T, we steer the system at hand "close" to the zero-state (equilibrium) employing the constant negative bilinear controls only.

(2) Using (a sort of) locally controllability technique with only additive controls active, we steer the system "close" to a state su_T for some small parameter $s > 0$.

(3) Again, employing only constant positive bilinear controls, we "stretch" the latter state to the desirable length u_T.

Step 1. Approximate null-controllability. Take any $T^* > 0$. Then it follows from the proof of Lemma 5.1 that if one applies control pair $k(t) \equiv \alpha < 0, v(t) = 0$ on $(0, T^*)$, the corresponding solution(s) to (5.7), (5.8)–(5.10) can be made arbitrarily small in $L^2(0,1)$ by selecting appropriately small negative α.

Indeed, it follows from (5.14) that

$$2(-\alpha - \rho - \frac{\mu}{2}) \int_0^T \int_0^1 w^2 dx ds \leq \| w_0 \|^2_{L^2(0,1)} + \frac{b-a}{\mu} \| v \|^2_{L^2(0,T)} .$$

As $\alpha \to -\infty$, this implies that we can make $\| u(\cdot, t_*) \|_{L^2(0,1)}$ as small as we wish for some $t_* \in (0, T)$ (in general, t_* can be different for different multiple solutions). This "smallness" is preserved on $[t_*, T]$ by Remark 5.2, applied with $v = 0$ and the same α on (t_*, T). In other words, we have the global approximate controllability to the origin in the sense of Definition 5.1, just by using constant bilinear controls.

Step 2. From Step 1 it follows that, without loss of generality, we may further assume that the initial state u_0 in (5.1) is arbitrarily small in $L^2(0,1)$. (Otherwise, we need to apply the argument of Step 1 on some "small" time-interval $(0, T^*), T^* < T$.)

To prove Theorem 5.3, it is sufficient to show that any function like

$$u_T(x) = \sum_{k=1}^{L} a_i \omega_k(x)$$

can be approached by $u(\cdot, T)$ arbitrarily close in the sense of (5.6).

Fix any positive integer L and the real numbers $a_1, \ldots, _L$.

Given $T > 0$, select a parameter $s \in (0,1)$ and also μ and δ in Lemmas 5.1 and 5.2. By Step 1, without loss of generality we may assume that

$$\| u_0 \|_{L^2(0,1)} \leq s^2.$$

Consider any $\varepsilon \in [0, T/2]$ and apply on the interval $(0, T - \varepsilon)$ the control pair (see (5.25)–(5.27) for notations)

$$k(t) = \alpha = 0, \quad v_{s,T-\varepsilon}(t) = s\hat{v}_{T-\varepsilon} = s\sum_{k=1}^{L} a_k v_{k,T-\varepsilon}(t), \quad t \in (0, T - \varepsilon).$$

Then, in notations of section 5.2 with $\alpha = 0$, $u = w = y + z$, and, see (5.28),

$$u(\cdot, T - \varepsilon) = s u_T + (y(\cdot, T - \varepsilon) - s u_T) + z(\cdot, T - \varepsilon), \tag{5.31}$$

where in view of (5.30) and Lemmas 5.1 and 5.2, applied with $\alpha = 0$ on $(0, T - \varepsilon)$,

$$\| y(\cdot, T - \varepsilon) - s u_T \|_{L^2(0,1)} \leq \| u_0 \|_{L^2(0,1)} \leq s^2, \tag{5.32}$$

$$\| z(\cdot, T - \varepsilon) \|_{L^2(0,1)} \leq C(T) \left(\| u_0 \|_{L^2(0,1)} \right.$$

$$\left. + s \| \hat{v}_{T-\varepsilon} \|_{L^2(0,T-\varepsilon)} \right)^{\min\{r_1, r_2\}}, \tag{5.33}$$

as $s \to 0$, where $C(T)$ does not depend on ε.

Since $r_1, r_2 > 1$, (5.31)–(5.33) yields that

$$u(\cdot, T - \varepsilon) = s u_T + p(\cdot, T - \varepsilon), \tag{5.34}$$

where, as it follows from Remark 5.3, uniformly over $\varepsilon \in [0, T/2]$,

$$\| p(\cdot, T - \varepsilon) \|_{L^2(0,1)} = o(s). \tag{5.35}$$

Step 3. On the interval $(T - \varepsilon, T)$ we apply controls

$$k(t) = \alpha > 0, \quad v(t) = 0, \quad t \in (T - \varepsilon, T).$$

Then, again in notations of section 5.2, applied now on the interval $(T - \varepsilon, T)$,

$$u(\cdot, T) = y(\cdot, T) + z(\cdot, T). \tag{5.36}$$

Here, according to (5.23) and (5.34), applied on $(T - \varepsilon, T)$,

$$y(x, T) = \sum_{k=1}^{\infty} e^{(-(\pi k)^2 + \alpha)\varepsilon} \left(\int_0^1 u(r, T - \varepsilon) \omega_k(r) dr \right) \omega_k(x)$$

$$= e^{\alpha\varepsilon} \sum_{k=1}^{\infty} e^{-(\pi k)^2 \varepsilon} \left(\int_0^1 u(r, T - \varepsilon) \omega_k(r) dr \right) \omega_k(x)$$

$$= e^{\alpha\varepsilon} \left(s u_T + p(\cdot, T - \varepsilon) + h(\varepsilon)(s u_T + p(\cdot, T - \varepsilon)) \right), \tag{5.37}$$

where $h(\varepsilon) \to 0$ as $\varepsilon \to 0$ (by continuity of solutions to (5.11) in time). In other words, in view of (5.35),

$$y(\cdot, T) = se^{\alpha \varepsilon} u_T + se^{\alpha \varepsilon} g(\cdot, s, \varepsilon), \tag{5.38}$$

where

$$\| g(\cdot, s, \varepsilon) \|_{L^2(0,1)} \to 0 \text{ as } s, \varepsilon \to 0. \tag{5.39}$$

On the other hand, by Lemma 5.2, applied on $(T - \varepsilon, T)$ with $v = 0$:

$$\| z(\cdot, T) \|_{L^2(0,1)} \leq Ce^{\frac{2\alpha \varepsilon}{1-\delta}} (\varepsilon^{\frac{5}{6}(1-\frac{r_1}{3})} \| w \|^{r_1}_{L^6(Q_{T-\varepsilon,T})}$$

$$+ \varepsilon^{\frac{5}{6}(1-\frac{3r_2}{5})} \| w_x \|^{r_2}_{L^2(Q_{T-\varepsilon,T})}), \tag{5.40}$$

where $Q_{T-\varepsilon,T} = (0, 1) \times (T - \varepsilon, T)$. In turn by Lemma 5.1, applied on $(T - \varepsilon, T)$ with $v = 0$:

$$\| w \|_{L^6(Q_{T-\varepsilon,T})}, \| w_x \|_{L^2(Q_{T-\varepsilon,T})}$$

$$\leq Ce^{(\alpha+\rho+\mu/2)\varepsilon} \| u(\cdot, T - \varepsilon) \|_{L^2(0,1)}, \tag{5.41}$$

for some constant $C > 0$.

Step 4. Summarizing the above estimates, we select parameters α, ε, and s so that

(A) $s \to 0+$;

(B) $e^{\alpha \varepsilon} = \dfrac{1}{s}$;

(C) $\varepsilon \to 0$ in such a way that also

$$e^{\frac{2\alpha \varepsilon}{1-\delta}} \varepsilon^{\min\{\frac{5}{6}(1-\frac{r_1}{3}); \frac{5}{6}(1-\frac{3r_2}{5})\}} = \varepsilon^{\min\{\frac{5}{6}(1-\frac{r_1}{3}); \frac{5}{6}(1-\frac{3r_2}{5})\}} s^{-\frac{2}{1-\delta}} \to 0.$$

Under these conditions we have, firstly, that, in view of (A), (B) and (5.34), (5.35), the right-hand side of (5.41) is bounded above by a constant and, secondly, that, by (C) and (5.40),

$$\| z(\cdot, T) \|_{L^2(0,1)} \to 0.$$

Then from (5.36), (5.38) and (5.39) this yields that

$$\| u(\cdot, T) - u_T \|_{L^2(0,1)} \to 0, \tag{5.42}$$

which completes the proof of Theorem 5.3.   □

5.5   Proof of Theorems 5.1/5.2: Locally Distributed Additive Controls

This proof, in fact, is identical to that of Theorem 5.3, with the following minor differences.

- In the proof of Theorem 5.3, in Step 2, we can select control $\hat{v}_{T/2}$ first and then, as $\varepsilon \to 0$, apply it *only* on the interval $(T - \varepsilon, T - \varepsilon - 0.5)$, i.e., the same additive control (but shifted in time) for all $\varepsilon \in [0, T/2]$. In this way Remark 5.3 is not necessary to use in (5.35). Analogously, in the proof of Theorem 5.1/5.2, in place of $\hat{v}_t$ in the above, we can select any function $v = \hat{v}(x,t), t \in (0, T/2), x \in \omega$. Then the argument of Theorem 5.3 will lead us to the convergence as in (5.42) to $u_T = y(\cdot, T/2)$, which is the state of the truncated multidimensional linear version of (5.15) with $\alpha = 0$ generated by the selected $\hat{v}(x,t)$. It remains to recall that the latter is approximately controllable in $L^2(\Omega)$ at time $T/2$ (or any other positive time, due to the dual unique continuation property from an open set $\omega \times (0, T/2)$), i.e., the set of such $y(\cdot, T/2)$ is dense in $L^2(\Omega)$.
- In several space dimensions Lemmas 5.1 and 5.2 are principally no different from the one dimensional case.

5.6   Concluding Remarks

- It seems quite possible that the results of Chapter 5 can be extended at no extra cost to boundary controls in place of the additive ones.
- In the proof of Theorem 5.3 we followed the Fourier series approach, which is due to the delicate nature of the lumped additive controls involving the Riesz's basis properties of the sequence of exponentials (see section 5.3). However, as we showed it in the (sketch of the) proof of Theorems 5.1/5.2, this approach can be avoided in part when we are dealing with the "stable" locally distributed controls. From this viewpoint it seems very plausible that these theorems can be extended to the case of more general parabolic equations with variable coefficients.

Part II
Multiplicative Controllability
of Hyperbolic Equations

Chapter 6
Controllability Properties of a Vibrating String with Variable Axial Load and Damping Gain

Abstract We show, in a constructive way, that the set of equilibrium states like $(y_d, 0)$ of a vibrating string that can approximately be reached in $H_0^1(0,1) \times L^2(0,1)$ by varying its axial load and the gain of damping is dense in the subspace $H_0^1(0,1) \times \{0\}$ of this space.

6.1 Introduction

6.1.1 Problem Setting

In this opening section of Chapter 6 we consider the following initial and boundary-value problem for the one dimensional wave equation modeling oscillations of a vibrating string with clapped ends:

$$y_{tt} = y_{xx} + v(x,t)y - \gamma(t)y_t, \quad x \in (0,1), \ t > 0, \qquad (6.1)$$
$$y(0,t) = y(1,t) = 0, \quad y(x,0) = y_0(x), \ y_t(x,0) = y_1(x).$$

Here the multiplicative controls are the coefficients $v(x,t)$ and $\gamma(t)$. The former we interpret as the axial load applied to the string at point x and time t, and the latter is the gain of the viscous (motion-activated) damping acting upon the string at time t. Our goal is to investigate the controllability properties of model (6.1). Namely, *given the (non-zero) initial state (y_0, y_1), we would like to know what states $(y(\cdot,t), y_t(\cdot,t))$ can be achieved by system (6.1) at times $t > 0$ by applying various aforementioned multiplicative controls v and γ.*

In terms of applications, a problem like this arises, e.g., in the context of so-called "smart materials", whose properties can be altered by applying various external factors such as temperature, electrical current or magnetic field.

Remark 6.1. One can easily notice that the zero state $(y_0, y_1) = (0,0)$ is the fixed point for the solution mappings of system (6.1), regardless of the choice of controls v and γ. Hence, it cannot be steered anywhere from this state. Respectively, everywhere below we will consider only non-zero initial states (y_0, y_1).

A.Y. Khapalov, *Controllability of Partial Differential Equations Governed by Multiplicative Controls*, Lecture Notes in Mathematics 1995, DOI 10.1007/978-3-642-12413-6_6, © Springer-Verlag Berlin Heidelberg 2010

To our knowledge, the problem of bilinear controllability was addressed for the first time by J.M. Ball, J.E. Mardsen, and M. Slemrod in [8] for the system like (6.1) but with no damping, namely, as follows:

$$u_{tt} = u_{xx} + v(t)u \quad x \in (0,1), \ t > 0 \tag{6.2}$$
$$u(0,t) = u(1,t) = 0, \quad u(x,0) = u_0(x), \ u_t(x,0) = u_1(x).$$

The authors of [8] managed to prove the global approximate controllability of (6.2) in $H_0^1(0,1) \times L^2(0,1)$ by means of time-dependent controls $v \in L^r([0,\infty);R)$, $r \geq 1$ only, assuming that all the modes in the initial datum are active. More precisely, if we denote by $\lambda_k = -(\pi k)^2$, $\omega_k = \sqrt{2}\sin \pi kx$, $k = 1,\ldots$ the eigenvalues and the orthonormal in $L^2(0,1)$ eigenfunctions associated with the spectral problem $\omega_{xx} = \lambda \omega$ in $H_0^1(0,1)$, then, assuming that

$$\left(\int_0^1 u_0 \sin \pi kxdx \right)^2 + \left(\int_0^1 u_1 \sin \pi kxdx \right)^2 \neq 0 \ \forall k = 1,\ldots,$$

it was shown in [8] that the set of all states which system (6.2) can achieve at times $T > 1$, namely, the set

$$\bigcup_{T \geq 1, \ v \in L^2([0,\infty);R)} (u(\cdot,T), u_t(\cdot,T)),$$

is dense in $H_0^1(0,1) \times L^2(0,1)$, see [8], p. 594. The methods of [8] make use of the inverse function theorem and involve dealing with the associated Riesz basis of exponential time-dependent functions. They are not constructive.

Another potential inconvenience of this result is the fact that it is achieved for controls v lying in the space $L^2([0,\infty);R)$. In other words, they can be unbounded or just "very large" (if, say, we approximate L^2-controls by "smooth" functions) time-dependent functions of rather complex nature. Having in mind that control v models an axial load, such controls can be difficult to implement in practice. We discuss the results of [8] in Chapter 9 below.

6.1.2 Main Results

In this chapter our goal is to investigate the issue of controllability properties of system (6.1), *while trying to use "explicit and constructive" bilinear controls* v and γ. We employ quite different (compared to [8]) methods to investigate the steering properties of system (6.1) to, what we call, "*target states of equilibrium type*" (or just "equilibrium states") like

$$(y_d,0) \in H_0^1(0,1) \times \{0\} \subset H_0^1(0,1) \times L^2(0,1), \tag{6.3}$$

which are of importance in many applications. To justify this name, let us note (we discuss this below in detail) that, in fact, the set of "actual" equilibrium states, namely, those satisfying (6.3) and an equation like

$$y_{dxx} + \bar{v}(x)y_d = 0, \quad x \in (0,1) \tag{6.4}$$

for some $\bar{v} \in L^\infty(0,1)$ is dense in $H_0^1(0,1)$.

Firstly, we will show that almost any non-negative (or, alternatively, non-positive) target equilibrium state can approximately be achieved by applying suitable time-independent bilinear control $v = v(x)$ and a constant control $\gamma(t) = \gamma$. (In terms of applications, this means that *one does not have to change the axial load and the gain of damping during the duration of steering.*)

Theorem 6.1. *Consider any* <u>*nonnegative non-zero*</u> *initial state of equilibrium type*

$$(y_0, y_1 = 0), \quad y_0 \in H_0^1(0,1), \quad y_0 \neq 0, \quad y_0 \geq 0 \text{ in } (0,1)$$

and any <u>*nonnegative*</u> *target equilibrium state*

$$(y_d, 0), \quad y_d \in H_0^1(0,1), \quad y_d \geq 0 \text{ a.e. in } (0,1).$$

Then for every $\varepsilon > 0$ *there are a time* $T = T(\varepsilon, y_0, y_d)$, *a static (x-dependent only) control* $v \in C^1[0,1]$ *and a constant control* $\gamma > 0$ *such that*

$$\| (y_d, 0) - (y(\cdot, T), y_t(\cdot, T)) \|_{H^2(0,1) \times L^2(0,1)} \leq \varepsilon. \tag{6.5}$$

The next result asserts that the set of equilibrium states of system (6.1) with nonzero initial datum that can be achieved by piecewise constant-in-time bilinear controls v and γ is dense in $H_0^1(0,1) \times \{0\}$.

Theorem 6.2. *Consider any initial state* $(y_0, y_1) \in H_0^1(0,1) \times L^2(0,1)$, $(y_0, y_1) \neq (0,0)$ *and any target equilibrium state* $(y_d, 0)$, $y_d \in H_0^1(0,1)$. *Then for every* $\varepsilon > 0$ *there are a time* $T = T(\varepsilon, y_0, y_1, y_d)$ *and piecewise constant-in-time controls* (v, γ) *of the form*

$$v(x,t) = \bar{v}(x) + h(t), \quad \gamma = \gamma(t), \tag{6.6}$$

where $\bar{v} \in C^1[0,1]$ *and* $h(t)$ *and* $\gamma(t)$ *are piecewise constant functions with finitely many points of discontinuity* (γ, *in fact, may have only one such point*), *such that*

$$\| (y_d, 0) - (y(\cdot, T), y_t(\cdot, T)) \|_{H_0^1(0,1) \times L^2(0,1)} \leq \varepsilon. \tag{6.7}$$

Remark 6.2. Note that, in contrast to the above-cited result of [8] that requires all the modes in the initial datum be active, Theorem 6.2 simply requires the given initial state to be a non-zero state.

In the proofs of Theorems 6.1 and 6.2 we explicitly describe how bilinear controls, mentioned in their respective statements, can be constructed.

6.2   Proof of Theorem 6.1: Controllability to Nonnegative Equilibrium Target States

Consider any $v \in L^\infty(0,1)$ and denote by λ_k and ω_k, $k = 1, \ldots$ respectively the eigenvalues and the orthonormal in $L^2(0,1)$ eigenfunctions associated with the spectral problem

$$\omega_{xx} + v(x)\omega = \lambda\omega, \quad \omega \in H_0^1(0,1). \tag{6.8}$$

It is well-known that

$$\| v \|_{L^\infty(0,1)} \geq \lambda_1 > \lambda_2 > \ldots \quad \text{and} \quad \lambda_k \to -\infty \text{ as } k \to \infty. \tag{6.9}$$

Our plan of the proof of Theorem 6.1 is as follows:

- We intend to show that for "almost any" given non-negative target state $y_d \in H_0^1(0,1)$ we can select a $\bar{v}(x)$ such that (6.4) holds and $y_d / \| y_d \|_{L^2(0,1)}$ is the first orthonormal eigenfunction, associated with the largest eigenvalue in the corresponding representation (6.8)–(6.9) with $\bar{v}$ in place of v, i.e.,

$$\omega_1(x) = \frac{y_d(x)}{\| y_d \|_{L^2(0,1)}}. \tag{6.10}$$

- Then, having in mind that the general solution to (6.1) is described by the series (6.12)–(6.18) below, we will show that the actual control pair (v, γ) can be selected of the form

$$v(x) = \bar{v}(x) + a, \quad \gamma(t) = \gamma > 0, \tag{6.11}$$

where constant a is selected to ensure that the first term in (6.12)–(6.18) will converge to the desirable target state $(y_d, 0)$ as t increases, while a positive constant γ is selected to guarantee that at the same time the remainder of the series (6.12)–(6.18) will become "negligibly small."

Step 1. Solution formula for (6.1). It can be shown by the classical methods that system (6.1) has a unique solution in $C([0,T]; H_0^1(0,1)) \cap C^1([0,T]; L^2(0,1))$ on every time-interval $[0,T]$, $T > 0$.

Consider any $v \in L^\infty(0,1)$, $\gamma > 0$ in (6.1). Then, in the notations of (6.8)–(6.9), we have the following formula for solutions to the boundary problem (6.1):

$$y(x,t) = \sum_{k=1}^{\infty} c_k(t)\omega_k(x), \quad x \in (0,1), \ t > 0, \tag{6.12}$$

where

$$\frac{d^2 c_k}{dt^2} = \lambda_k c_k - \gamma \frac{dc_k}{dt}, \quad t > 0,$$

$$c_k(0) = \int_0^1 y_0 \omega_k dx, \quad \dot{c}_k(0) = \int_0^1 y_1 \omega_k dx. \tag{6.13}$$

Solving the system of ordinary differential equations (6.13) yields:

- In the case when $\gamma^2 + 4\lambda_k = 0$:

$$c_k(t) = e^{-\frac{\gamma}{2}t} c_k(0) + t\left(\dot{c}_k(0) + \frac{c_k(0)\gamma}{2}\right) e^{-\frac{\gamma}{2}t}; \tag{6.14}$$

- For $\gamma^2 + 4\lambda_k > 0$:

$$c_k(t) = e^{(-\frac{\gamma}{2} + \sqrt{\gamma^2/4 + \lambda_k})t} \frac{\dot{c}_k(0) - c_k(0)(-\frac{\gamma}{2} - \sqrt{\gamma^2/4 + \lambda_k})}{\sqrt{\gamma^2 + 4\lambda_k}}$$

$$+ e^{(-\frac{\gamma}{2} - \sqrt{\gamma^2/4 + \lambda_k})t} \frac{-\dot{c}_k(0) + c_k(0)(-\frac{\gamma}{2} + \sqrt{\gamma^2/4 + \lambda_k})}{\sqrt{\gamma^2 + 4\lambda_k}}, \tag{6.15}$$

$$\dot{c}_k(t) = e^{(-\frac{\gamma}{2} + \sqrt{\gamma^2/4 + \lambda_k})t} \left(-\frac{\gamma}{2} + \sqrt{\gamma^2/4 + \lambda_k}\right) \frac{\dot{c}_k(0) - c_k(0)(-\frac{\gamma}{2} - \sqrt{\gamma^2/4 + \lambda_k})}{\sqrt{\gamma^2 + 4\lambda_k}}$$

$$+ e^{(-\frac{\gamma}{2} - \sqrt{\gamma^2/4 + \lambda_k})t} \left(-\frac{\gamma}{2} - \sqrt{\gamma^2/4 + \lambda_k}\right)$$

$$\times \frac{-\dot{c}_k(0) + c_k(0)(-\frac{\gamma}{2} + \sqrt{\gamma^2/4 + \lambda_k})}{\sqrt{\gamma^2 + 4\lambda_k}}; \tag{6.16}$$

- In the case when $\gamma^2 + 4\lambda_k < 0$:

$$c_k(t) = c_k(0) e^{-\frac{\gamma}{2}t} \cos\left(t\sqrt{-\gamma^2/4 - \lambda_k}\right)$$

$$+ e^{-\frac{\gamma}{2}t} \sin\left(t\sqrt{-\gamma^2/4 - \lambda_k}\right) \frac{\dot{c}_k(0) + \gamma c_k(0)/2}{\sqrt{-\gamma^2/4 - \lambda_k}}, \tag{6.17}$$

$$\dot{c}_k(t) = c_k(0) \frac{-\gamma}{2} e^{-\frac{\gamma}{2}t} \cos\left(t\sqrt{-\gamma^2/4 - \lambda_k}\right)$$

$$- c_k(0) e^{-\frac{\gamma}{2}t} \sqrt{-\gamma^2/4 - \lambda_k} \sin\left(t\sqrt{-\gamma^2/4 - \lambda_k}\right)$$

$$+ \frac{-\gamma}{2} e^{-\frac{\gamma}{2}t} \sin\left(t\sqrt{-\gamma^2/4 - \lambda_k}\right) \frac{\dot{c}_k(0) + \gamma c_k(0)/2}{\sqrt{-\gamma^2/4 - \lambda_k}}$$

$$+ e^{-\frac{\gamma}{2}t} (\dot{c}_k(0) + \gamma c_k(0)/2) \cos\left(t\sqrt{-\gamma^2/4 - \lambda_k}\right). \tag{6.18}$$

Recall now that the norm

$$\| \phi \|_v = \left(\int_0^1 (\phi_x^2 + (-v(x)+v)\phi^2)dx \right)^{1/2} = \left(\sum_{k=1}^{\infty} (-\lambda_k + v) \left(\int_0^1 \phi \omega_k dx \right)^2 \right)^{1/2},$$

(6.19)

where v is any number exceeding $\| v \|_{L^{\infty}(0,1)}$, is equivalent to the standard norm in $H_0^1(0,1)$. Hence, the series (6.12)–(6.18) converges in the space $C([0,];H_0^1(0,1)) \cap C^1([0,T];L^2(0,1))$ for every $T > 0$.

Step 2. Selection of "approximate" target state. To prove Theorem 6.1, it is sufficient to consider any set of non-negative target states y_d which is dense in the set of all non-negative elements of $H_0^1(0,1)$. To this end, we will consider only (a) nonzero non-negative continuously differentiable functions $y_d = y_d(x), x \in [0,1]$ that (b) vanish at $x = 0,1$ and (c) whose second derivatives are piecewise continuous with finitely many discontinuities of the first kind (hence, $y_d \in H^2(0,1) \cap H_0^1(0,1) \subset C^1[0,1]$) and are such that (d)

$$y_d(x) > 0 \text{ in } (0,1) \quad \text{and} \quad \frac{y_{dxx}}{y_d} \in L^{\infty}(0,1). \qquad (6.20)$$

To ensure the last condition in (6.20), it is sufficient to select y_d such that, in addition to the preceding conditions, it is linear near the endpoints $x = 0,1$. This would guarantee that $y_{dxx} = 0$ near $x = 0,1$, while elsewhere, due to the first condition in (6.20), the denominator in (6.20) is strictly separated from 0.

Let us give an example how one can approximate any non-negative element $g \in H_0^1(0,1)$ in this space by a sequence of functions described in the above.

(A) First of all, recall that the set of all infinitely many times differentiable functions of finite support in $(0,1)$ is dense in $H_0^1(0,1)$. Hence, one can expect that a non-negative g can be approximated in this space by a sequence of also non-negative infinitely many times differentiable functions of finite support. Let us show this explicitly.

Consider any non-negative function $g \in H_0^1(0,1)$ and for any $r \in (0,1/4)$ introduce the function

$$g_r(x) = \begin{cases} g(x) & \text{for } x \in (r, 1-r), \\ 0 & \text{for } x \in (-\infty, r) \cup (1-r, \infty). \end{cases}$$

Making use of the non-negative kernel function

$$\theta(t) = \begin{cases} \kappa e^{\frac{1}{t^2-1}} & \text{for } t^2 < 1, \\ 0 & \text{for } t^2 \geq 1, \end{cases}$$

where

$$\kappa = \left(\int\limits_{-1}^{1} e^{\frac{1}{t^2-1}} ds \right)^{-1},$$

also introduce the set of "averaged" functions

$$g_{rh}(x) = \frac{1}{h} \int\limits_{|x-s|<h} \theta(\frac{|x-s|}{h}) g_r(s) ds$$

for any $h \in (0,r)$. Note that every function $g_{rh}(x)$ is non-negative and vanishes in $(0, r-h) \cup (1-r+h, 1)$.

We claim now that $g(x)$ is a limit point in $H_0^1(0,1)$ of the above-described set of averaged functions $g_{rh}(x)$, $h \in (0,r), r \in (0,1/4)$.

Indeed, (for our given $g(x)$) let us evaluate the distance between g and g_{rh} in $H^1(0,1)$ (recall that $H_0^1(0,1)$ is a proper subspace of it):

$$\| g - g_{rh} \|_{H^1(0,1)}^2 \leq \| g - g_{rh} \|_{H^1(r,1-r)}^2 + 2 \| g \|_{H^1(0,r)}^2 + 2 \| g \|_{H^1(1-r,1)}^2$$

$$+ 2 \| g_{rh} \|_{H^1(0,r)}^2 + 2 \| g_{rh} \|_{H^1(1-r,1)}^2 . \tag{6.21}$$

Note now that, for our (that is, fixed) g,

$$2 \| g \|_{H^1(0,r)}^2 + 2 \| g \|_{H^1(1-r,1)}^2 \to 0 \text{ as } r \to 0.$$

Hence, for any $\varepsilon > 0$ we can find a sufficiently small $r_1 \in (0,1/4)$ such that

$$2 \| g \|_{H^1(0,r)}^2 + 2 \| g \|_{H^1(1-r,1)}^2 < \varepsilon/3 \quad \forall r \in (0, r_1].$$

To evaluate the fourth and fifth terms in (6.21), recall (as it follows from the definition of θ, see, e.g., [136]) that

$$g_{rhx}(x) = \frac{1}{h} \int\limits_{|x-s|<h} \theta_x(\frac{|x-s|}{h}) g_r(s) ds$$

and

$$\frac{| \theta_x(\frac{|x-s|}{h}) |}{h} = \frac{\kappa}{h} e^{\frac{1}{(\frac{x-s}{h})^2-1}} 2 \frac{|x-s|}{h} \left(\frac{1}{(\frac{x-s}{h})^2 - 1} \right)^2 \leq C \frac{1}{h}$$

for some constant $C > 0$ when $|x - s| < h$. Hence,

$$2 \| g_{rh} \|^2_{H^1(0,r)} + 2 \| g_{rhx} \|^2_{H^1(1-r,1)} < \hat{C} \left(\| g_{rhx} \|^2_{L^2(0,r)} + \| g_{rh} \|^2_{L^2(1-r,1)} \right)$$

$$\leq \hat{C} \| g \|^2_{C[0,1]} \left[\int_0^r \left(\int_{x-h}^{x+h} \frac{ds}{h} \right)^2 + \int_{1-r}^1 \left(\int_{x-h}^{x+h} \frac{ds}{h} \right)^2 \right]$$

$$\leq 8\hat{C} \| g \|^2_{C[0,1]} \, r < \varepsilon/3 \quad \forall r \in (0, r_2]$$

for some $r_2 \in (0, r_1]$ and constant $\hat{C} > 0$.

It is also known that the first term on the right in (6.21) tends to zero as $h \to 0$ for any fixed r (e.g., [136]). Hence, we can find an $h = h(r_2) \in (0, r_2)$ such that

$$\| g - g_{r_2h} \|^2_{H^1(r_2, 1-r_2)} < \varepsilon/3 \quad \forall h \in (0, h(r_2)].$$

Combining all the above, yields that for every $\varepsilon > 0$ we can find a pair $(r_2, h(r_2))$ such that

$$\| g - g_{r_2h(r_2)} \|^2_{H^1(0,1)} < \varepsilon,$$

that is, any non-negative element g of $H_0^1(0, 1)$ can be approximated by non-negative infinitely many times differentiable functions like g_{rh} of finite support in $(0, 1)$ in this space.

(B) Furthermore, each of above-described functions g_{rh} can in turn be approximated in $H^1(0, 1)$ by a sequence of functions

$$g_{rhm}(x) = g_{rh}(x) + 1/m$$

as $m \to \infty$. Note that functions $g_{rhm}(x)$ do not vanish at the endpoints $x = 0, 1$ and are not elements of $H_0^1(0, 1)$ anymore.

(C) It remains to show that each of the functions g_{rhm} can be approximated in $H^1(0, 1)$ by a sequence of functions satisfying the conditions (a)–(d) described in the beginning of Step 2.

Consider any function $g_{rhm}(x)$, which, as we recall, is constant in $(0, r_*)$ and in $(1 - r_*, 1)$, $r_* = r - h$.

Firstly, we approximate this function on the just-mentioned intervals by continuous piecewise linear functions whose graphs connect respectively the point $(0, 0)$ to $(r_*/2, 1/m)$ to $(r_*, 1/m)$ for $x \in [0, r_*]$ and the point $(1, 0)$ to $((2 - r_*)/2, 1/m)$ to $(1 - r_*, 1/m)$ for $x \in [1 - r_*, 1]$.

Secondly, the obtained broken lines can then be "smoothened" at the corners by using pieces of circles of radii, say, $1/(4m + k)$ as $k \to \infty$ with centers located on the bisectors of the angles generated by the corresponding adjacent straight lines of the graphs (so that these lines are tangent to the aforementioned circles). These "smoothened" lines are the graphs of the functions satisfying all the conditions (a)–(d), described in the above in the beginning of Step 2, and approximate in $H^1(0,1)$ the function g_{rhm} as $k \to \infty$.

Step 3. Selection of γ and the form of $v(x)$. Fix any initial datum $(y_0, 0)$, $y_0 \in H_0^1(0,1)$, $y_0 \neq 0$, $y_0(x) \geq 0$ in $(0,1)$. To prove Theorem 6.1, it is sufficient to consider any target equilibrium state $(y_d, 0)$ with y_d as in (6.20).
Select

$$\bar{v}(x) = -\frac{y_{dxx}(x)}{y_d(x)}, \quad x \in (0,1). \tag{6.22}$$

Note that (6.4) holds for this y_d with our selection of $\bar{v}$ (recall also that $\bar{v}$ is not identically zero in $L^\infty(0,1)$ in view of (6.20)).

Denote the eigenvalues and the orthonormal eigenfunctions associated with $v(x) = \bar{v}(x)$ respectively by $\lambda_1 > \lambda_2 > \ldots$ and by $\omega_1, \omega_2, \ldots$ (that is, here we use the same notations as in the generic case described in Step 1).

Then the eigenvalues in (6.8)–(6.9) corresponding to $v(x) = \bar{v}(x) + a$ as in (6.11), where a is any constant, will be as follows:

$$\lambda_1 + a, \ \lambda_2 + a, \ldots,$$

while the eigenfunctions will remain the same as in the case $v = \bar{v}$.

Remark 6.3. • It was shown in Chapter 2 of this monograph that the function $y_d / \|y_d\|_{L^2(0,1)}$ associated with any function y_d satisfying the conditions (a)–(d), given in the beginning of step 2 in the above, is the first eigenfunction of the spectral problem (6.8)–(6.9) with $\bar{v}$ as in (6.22).
• Note that (6.22) implies that

$$\lambda_1 = 0.$$

In turn, in the notations of (6.13), condition (6.20) yields that

$$c_1(0) = \int_0^1 y_0 \omega_1 dx > 0,$$

since y_0 is non-negative and is not identically zero.

Consider now any control pair of the form $(\bar{v} + a, \gamma)$ as in (6.11), where a and γ are some constants, whose values we intend to select below.

First we select γ in $(0, \sqrt{-\lambda_2})$, for example,

$$\gamma = \frac{\sqrt{-2\lambda_2}}{2}. \tag{6.23}$$

This (selected) value of γ will remain fixed till the end of the proof of Theorem 6.1.

Assume now that

$$a \in (-\gamma^2/8, \gamma^2/8). \tag{6.24}$$

The precise value(s) of a will be selected a little bit later.

For these γ and a we have

$$\frac{\gamma^2}{4} + \lambda_1 + a = \frac{\gamma^2}{4} + a > 0 \quad \text{and} \quad \frac{\gamma^2}{4} + \lambda_k + a < 0 \text{ for } k = 2, \dots \tag{6.25}$$

Select any

$$v = \| \bar{v} \|_{L^\infty(0,1)} + \gamma^2, \tag{6.26}$$

which ensures that the norm $\| (\cdot) \|_{\bar{v}}$ in (6.19) is equivalent to the standard norm in $H_0^1(0,1)$ for any a satisfying (6.24).

Step 4. Evaluation of $y(\cdot, t) - y_d$ and $y_t(\cdot, t)$. The solution formulas (6.12)–(6.18), (6.11) yield for our initial datum $(y_0, 0)$:

$$\|y(\cdot,t) - y_d\|_v^2 \le 2(-a+v) \left(e^{(-\frac{\gamma}{2} + \sqrt{\gamma^2/4+a})t} \frac{c_1(0)(\frac{\gamma}{2} + \sqrt{\gamma^2/4+a})}{\sqrt{\gamma^2+4a}} - \int_0^1 y_d \omega_1 dx \right)^2$$

$$+ 2(-a+v) \left(e^{(-\frac{\gamma}{2} - \sqrt{\gamma^2/4+a})t} \frac{c_1(0)(-\frac{\gamma}{2} + \sqrt{\gamma^2/4+a})}{\sqrt{\gamma^2+4a}} \right)^2$$

$$+ \sum_{k=2}^\infty 2(-\lambda_k - a + v) \left(c_k(0) e^{-\frac{\gamma}{2}t} \cos\left(t\sqrt{-\gamma^2/4 - \lambda_k - a} \right) \right)^2$$

$$+ \sum_{k=2}^\infty 2(-\lambda_k - a + v)$$

$$\times \left(e^{-\frac{\gamma}{2}t} \sin\left(t\sqrt{-\gamma^2/4 - \lambda_k - a} \right) \frac{\gamma c_k(0)/2}{\sqrt{-\gamma^2/4 - \lambda_k - a}} \right)^2. \tag{6.27}$$

Making use of (6.23)–(6.26), we derive for $k = 2, \dots$:

$$0 < -\lambda_k - a + v \le -\lambda_k + \| \bar{v} \|_{L^\infty(0,1)} + \frac{9\gamma^2}{8},$$

$$0 < -\frac{3\gamma^2}{8} - \lambda_2 < -\frac{3\gamma^2}{8} - \lambda_k < -\gamma^2/4 - \lambda_k - a \le -\gamma^2/8 - \lambda_k,$$

and also

$$0 < -a + v \le \| \bar{v} \|_{L^\infty(0,1)} + \frac{9\gamma^2}{8}, \quad 0 < \gamma^2/8 < \gamma^2/4 + a \le 3\gamma^2/8.$$

In turn, making use of these estimates, we further obtain from (6.27):

$$\| y(\cdot,t) - y_d \|_v^2 \le 2(\| \bar{v} \|_{L^\infty(0,1)} + \frac{9\gamma^2}{8})$$

$$\times \left(e^{(-\frac{\gamma}{2} + \sqrt{\gamma^2/4+a})t} \frac{c_1(0)(\frac{\gamma}{2} + \sqrt{\gamma^2/4+a})}{\sqrt{\gamma^2+4a}} - \int_0^1 y_d \omega_1 dx \right)^2$$

$$+ C_1(\gamma) e^{-\gamma t} \| y_0 \|_{H_0^1(0,1)}^2 \tag{6.28}$$

for some constant $C_1(\gamma)$ independent of a satisfying (6.24) (recall that γ is already fixed as in (6.23)).

Analogously, using the formulas (6.12)–(6.18), we can obtain a similar estimate for the velocity component of the solution:

$$\| y_t(\cdot,t) \|_{L^2(0,1)}^2 \le 2\left(e^{(-\frac{\gamma}{2} + \sqrt{\gamma^2/4+a})t} \left(-\frac{\gamma}{2} + \sqrt{\gamma^2/4+a} \right) \frac{c_1(0)(\frac{\gamma}{2} + \sqrt{\gamma^2/4+a})}{\sqrt{\gamma^2+4a}} \right)^2$$

$$+ C_2(\gamma) e^{-\gamma t} \| y_0 \|_{H_0^1(0,1)}^2 \tag{6.29}$$

for some constant $C_2(\gamma)$ independent of a satisfying (6.24).

Step 5. Selection of a. Denote

$$\beta_1(a) = -\frac{\gamma}{2} + \sqrt{\gamma^2/4+a}$$

and

$$\beta_2(a) = \frac{\frac{\gamma}{2} + \sqrt{\gamma^2/4+a}}{\sqrt{\gamma^2+4a}}.$$

Then

$$\lim_{a \to 0} \beta_1(a) = 0, \quad \lim_{a \to 0} \beta_2(a) = 1. \tag{6.30}$$

In the new notations we have:

$$\| y(\cdot,t) - y_d \|_v^2 + \| y_t(\cdot,t) \|_{L^2(0,1)}^2 \le 2(\| \bar{v} \|_{L^\infty(0,1)} + \frac{9\gamma^2}{8})$$

$$\times \left(e^{\beta_1(a)t} c_1(0) \beta_2(a) - \int_0^1 y_d \omega_1 dx \right)^2$$

$$+ 2 \left(e^{\beta_1(a)t} \beta_1(a) c_1(0) \beta_2(a) \right)^2$$

$$+ (C_1(\gamma) + C_2(\gamma)) e^{-\gamma t} \| y_0 \|_{H_0^1(0,1)}^2 . \tag{6.31}$$

Since, regardless of a satisfying (6.23), the last two terms on the right of (6.31) tend to zero as t increases, to prove Theorem 6.1 (that is, to establish (6.5)), it remains to find a within the constraints (6.24) which at the same time would make the first two terms on the right of (6.31) also "small."

Recall (see Remark 6.3) that we have:

$$c_1(0) = \int_0^1 y_d \omega_1 dx = \| y_d \|_{L^2(0,1)} > 0.$$

Furthermore, since Theorem 6.1 deals with the issue of approximate controllability, without loss of generality, we may assume that we have two cases:

either

$$c_1(0) > \int_0^1 y_d \omega_1 dx > 0 \tag{6.32}$$

or

$$0 < c_1(0) < \int_0^1 y_d \omega_1 dx. \tag{6.33}$$

Assume, for example, that (6.32) holds, in which case we will consider now only negative a's satisfying (6.24), for which

$$\beta_1(a) < 0 \quad \text{and} \quad \beta_2(a) > 1.$$

Hence

$$c_1(0)\beta_2(a) > c_1(0).$$

Set

$$t_*(a) = \frac{1}{\beta_1(a)} \left(\ln \frac{\int_0^1 y_d \omega_1 dx}{c_1(0)\beta_2(a)} \right).$$

Then

$$\lim_{a \to 0-} t_*(a) = \infty, \tag{6.34}$$

$$e^{\beta_1(a)t_*(a)} c_1(0) \beta_2(a) - \int_0^1 y_d \omega_1 dx = 0, \tag{6.35}$$

that is, the first term on the right in (6.31) vanishes for $t = t_*(a)$.

Note that (6.30), (6.34) and (6.35) mean that the second term on the right of (6.31) tends to zero when $t = t_*(a)$ and $a \to 0-$.

Hence,

$$\lim_{a \to 0-} \left(\| y(\cdot, t_*(a)) - y_d \|_v^2 + \| y_t(\cdot, t_*(a)) \|_{L^2(0,1)}^2 \right) = 0,$$

which implies (6.5) (in view of (6.19)).

The case (6.33) can be dealt analogously. This ends the proof of Theorem 6.1.

□

6.3 Proof of Theorem 6.2: Controllability to Any Equilibrium Target States

One can notice in the proof of Theorem 6.1 that the condition that the initial state y_0 and the target state y_d are non-negative is only used to ensure that y_d (or its approximation) can be made co-linear to the first eigenfunction associated with some bilinear control $\bar{v}(x)$ as in (6.4). In other words, the result of Theorem 6.1 remains true for the given (non-zero) target state y_d if it happens that we have the following:

(A) There can be found a $\bar{v}(x)$ such that $y_d / \| y_d \|_{L^2(0,1)}$ is an eigenfunction, say, ω_{k_*}, for the spectral problem (6.8)–(6.9) with $v = \bar{v}$.

(B) The initial datum is of equilibrium type like $(y_0, 0)$.

(C) In the solution expansion (6.12)–(6.18) the k_*-th term is the first present, i.e., all the coefficients $c_k(0), \dot{c}_k(0), k = 1, \ldots k_* - 1$ are equal to zero in (6.13).

(D) If conditions (A)–(C) hold, then the argument of Theorem 6.1 applies in this case with no changes (except for minor straightforward adjustments). Recall now that the controls v and γ found in the proof of Theorem 6.1 and the duration of their engagement depend only on the (magnitudes of the) given initial and target states. Therefore, we can as well replace condition (C) with the assumption that all the above-mentioned coefficients $c_k(0), \dot{c}_k(0), k = 1, \ldots k_* - 1$ are "negligibly small," namely, are such that the effect of the aforementioned bilinear controls v and γ would not result in the "substantial" increase of that part of solution to (6.1) which is contributed by the first $k_* - 1$ terms in its expansion (6.12).

Respectively, our plan of the proof of Theorem 6.2 is as follows:

- We intend to show that for "almost any" given target state $y_d \in H_0^1(0,1)$ we can select a $\bar{v}(x)$ such that (6.4) holds and, hence, $y_d / \| y_d \|_{L^2(0,1)}$ is one of the orthonormal eigenfunction in the representation (6.8)–(6.9) (i.e., not necessarily the first one as it was in the proof of Theorem 6.1). In other words, in (6.8)–(6.9)

$$\omega_{k_*}(x) = \frac{y_d(x)}{\| y_d \|_{L^2(0,1)}} \tag{6.36}$$

for some integer k_*.

- We will show that the actual bilinear controls v and γ can be selected as in (6.6), where

$$h(t) = \begin{cases} h_1^* = -\lambda_1 - 1 & \text{for } t \in (t_0 = 0, t_1^*), \\ h_1 = -\lambda_1 & \text{for } t \in (t_1^*, t_1), \\ \vdots \\ h_{k_*-1}^* = -\lambda_{k_*-1} - 1 & \text{for } t \in (t_{k_*-2}, t_{k_*-1}^*), \\ h_{k_*-1} = -\lambda_{k_*-1} & \text{for } t \in (t_{k_*-1}^*, t_{k_*-1}), \\ h_{k_*}^* = -\lambda_{k_*} - 1 & \text{for } t \in (t_{k_*-1}, t_{k_*}), \\ h_{k_*} = a & \text{for } t \in (t_{k_*}, T), \end{cases} \tag{6.37}$$

$$\gamma(t) = \begin{cases} 0 & \text{for } t \in (0, t_{k_*}), \\ \gamma & \text{for } t \in (t_{k_*}, T), \end{cases} \tag{6.38}$$

where $0 = t_0 \le t_1^* \le t_1 \le \ldots \le t_{k_*}$ are some moments selected so that on each of the time-intervals $(t_0, t_1), \ldots, (t_{k_*-1}, t_{k_*})$ applying controls described by (6.6), (6.37)–(6.38) would make the corresponding coefficients $c_k(t), \dot{c}_k(t), k = 1, \ldots k_* - 1$ in the solution formula (6.42) (see below) for (6.1)/(6.2) "negligibly small" at time t_{k_*} (this ensures condition (D)), while making $\dot{c}_{k_*}(t_{k_*}) = 0$ (this ensures condition (B)). On the last "time-leg" (t_{k_*}, T) we will apply suitable controls $v = \bar{v} + a$ and γ, as found in the proof of Theorem 6.1, to "adjust" the magnitude of the k_*-th term in the solution representation in order to approximate our target state y_d at time $t = T$.

The form (6.37) represents the "worst case scenario," namely, when all the modes $k = 1, \ldots, k_* - 1$ in the initial datum are active (for the selected v), that is, $c_k^2(0) + \dot{c}_k^2(0) \neq 0, k = 1, \ldots, k_* - 1$ in (6.43). If this is not the case, then the corresponding pair of controls h_k^*, h_k in (6.37) should be omitted.

Step 1. Selection of ("approximate") target state. To prove Theorem 6.2, it is sufficient to consider any set of target states y_d which is dense in $H_0^1(0, 1)$. To this end, we will consider only (a) nonzero continuously differentiable functions $y_d = y_d(x), x \in [0, 1]$ that (b) vanish at $x = 0, 1$ and (c) whose second derivatives are piecewise continuous with finitely many discontinuities of the first kind (hence, $y_d \in H^2(0, 1) \cap H_0^1(0, 1) \subset C^1[0, 1]$) such that (d)

$$\begin{cases} \frac{y_{dxx}}{y_d}, & \text{where } y_d(x) \neq 0, \\ 0, & \text{where } y_d(x) = 0, \end{cases} \in L^\infty(0, 1). \tag{6.39}$$

To ensure (6.39) it is sufficient to select y_d which, in addition to the above, has finitely many zeros and is linear near the points x in [0,1], where it vanishes. This would guarantee that $y_{dxx} = 0$ near the above mentioned "zero" points, while elsewhere, the denominator in (6.39) is strictly separated from 0. This can be achieved, for example, as follows.

Consider any non-negative element $g \in H_0^1(0,1)$. Let us show how it can be approximated in this space by a sequence of functions described in the above.

First of all, recall that the sequence $\{\sin \pi kx, \ k = 1, \ldots\}$ provides a basis for $H_0^1(0,1)$. Hence, the set of finite sums

$$\left\{ \phi(x) = \sum_{k=1}^{K} a_k \sin \pi kx \ \Big| \ a_k \in R, \ k = 1, \ldots, K, \ K = 1, \ldots \right\}$$

is dense in this space. Each of such finite sums can in turn be approximated by a sequence of functions satisfying the conditions (a)–(d) in the above, by making use of the technique described in Step 2 in the proof of Theorem 6.1.

Indeed, consider any finite sum ϕ as in the above. Then it can have only finitely many zero points, say, $x_0 = 0, x_1, \ldots, x_n = 1$. On each of the intervals $[x_{i-1}, x_i], i = 1, \ldots, n$ we can apply the just-mentioned technique from the proof of Theorem 6.1 to approximate ϕ in $H_0^1(x_{i-1}, x_i)$ by a sequence of functions that satisfying the conditions (a)–(d) from the above on $[x_{i-1}, x_i]$ only, whose graphs are pieces of straight lines near the endpoints x_{i-1} and x_i. If one selects these lines to be the same on the right and on the left of any of the "interior" points $x_i, i = 1, \ldots, n-1$, then the resulting sequence of functions will approximate ϕ in $H_0^1(0,1)$ and will satisfy all the required conditions (a)–(d).

Step 2. Solution formula for (6.2). Note that on the time-interval $(0, t_{k_*})$ system (6.1) governed by controls (6.6), (6.37)–(6.38) becomes (6.2). So, *below we have in mind that $u(x,t) = y(x,t)$ on this interval.*

Consider any nonzero initial datum $(y_0, y_1) \in H_0^1(0,1) \times L^2(0,1)$ and select any target state y_d satisfying the conditions (a)–(d) given in Step 1.

Without loss of generality, we may assume that

$$\left(\int_0^1 y_d y_0 dx \right)^2 + \left(\int_0^1 y_d y_1 dx \right)^2 \neq 0.$$

(Indeed, since the set of all y_d, described in Step 1, is dense in $H_0^1(0,1)$, its subset for which the above inequality holds, is dense in this space as well.)

Let

$$\bar{v} = \begin{cases} -\frac{y_{dxx}}{y_d}, & \text{where } y_d(x) \neq 0, \\ 0, & \text{where } y_d(x) = 0, \end{cases} \in L^\infty(0,1). \tag{6.40}$$

(compare with (6.39)).

We denote, as before, the eigenvalues of the spectral problem (6.8)–(6.9) with $v = \bar{v}$ as in (6.40) by $\lambda_k, \ k = 1, \ldots$ with

$$\lambda_{k_*} = 0.$$

Respectively, the eigenvalues of this spectral problem associated with

$$v = \bar{v} + a$$

will be

$$\lambda_k + a, \ k = 1, \dots .$$

Recall, however that for any value of a the corresponding eigenfunctions remain the same in both cases and we denote them again by $\omega_k, k = 1, \dots$.

As (6.12)–(6.18) implies, we have the following solution formula for system (6.2) with

$$v(x) = \bar{v}(x) + a = -\frac{y_{dxx}}{y_d} + a, \tag{6.41}$$

where a is any constant:

$$u(x,t) = \sum_{k=1}^{\infty} c_k(t)\omega_k(x), \ x \in (0,1), \ t > 0, \tag{6.42}$$

where

$$\frac{d^2 c_k}{dt^2} = (\lambda_k + a)c_k, \quad t > 0,$$

$$c_k(0) = \int_0^1 y_0 \omega_k dx, \quad \dot{c}_k(0) = \int_0^1 y_1 \omega_k dx, \tag{6.43}$$

where in the case when $\lambda_k + a = 0$:

$$c_k(t) = t\dot{c}_k(0) + c_k(0), \quad \dot{c}_k(t) = \dot{c}_k(0), \tag{6.44}$$

and when $\lambda_k + a > 0$:

$$c_k(t) = \frac{1}{2}e^{t\sqrt{\lambda_k+a}}\left(c_k(0) + \frac{\dot{c}_k(0)}{\sqrt{\lambda_k+a}}\right)$$

$$+ \frac{1}{2}e^{-t\sqrt{\lambda_k+a}}\left(c_k(0) - \frac{\dot{c}_k(0)}{\sqrt{\lambda_k+a}}\right), \tag{6.45}$$

$$\dot{c}_k(t) = e^{t\sqrt{\lambda_k+a}}\frac{\dot{c}_k(0) + c_k(0)\sqrt{\lambda_k+a}}{2}$$

$$- e^{-t\sqrt{\lambda_k+a}}\frac{-\dot{c}_k(0) + c_k(0)\sqrt{\lambda_k+a}}{2}, \tag{6.46}$$

and in the case when $\lambda_k + a < 0$:

$$c_k(t) = c_k(0)\cos\left(t\sqrt{-\lambda_k-a}\right) + \sin\left(t\sqrt{-\lambda_k-a}\right)\frac{\dot{c}_k(0)}{\sqrt{-\lambda_k-a}}, \tag{6.47}$$

$$\dot{c}_k(t) = -c_k(0)\sqrt{-\lambda_k-a}\sin\left(t\sqrt{-\lambda_k-a}\right) + \dot{c}_k(0)\cos\left(t\sqrt{-\lambda_k-a}\right). \tag{6.48}$$

Note that our selection of approximate target state y_d in the beginning of this step ensures that in (6.43)

$$c_{k_*}^2(0) + \dot{c}_{k_*}^2(0) \neq 0.$$

Step 3. Selection of bilinear controls. Without loss of generality, we can assume that $k_* > 1$ in (6.36) for our selected $\bar{v}$ as in (6.39)–(6.40).

We intend now to select control $h(t)$ of the form (6.37) in such a way that on $(0, t_{k_*})$ applying control (6.6), (6.39), (6.37)–(6.38) would make the corresponding terms containing $c_k(t_{k_*}), \dot{c}_k(t_{k_*})$, $k < k_*$, in the solution formula (6.42) "small."

Step 3.1. Control on $(0, t_{k_*-1})$. We begin with the selection of h_1^* as in (6.37), namely,

$$h_1^* = -\lambda_1 - 1.$$

Then, as (6.47)–(6.48) imply (with $a = h_1^*$), the point $(c_1(t), \dot{c}_1(t))$ moves with period 2π in the associated two dimensional space around the origin along the circle

$$c_1^2(t) + \dot{c}_1^2(t) \; = \; c_1^2(0) + \dot{c}_1^2(0) = r_1^2.$$

Hence, *within one such rotation around the origin*, we can select some moment t_1^* such that

$$c_1(t_1^*) = \mu_{10}, \; \dot{c}_1(t_1^*)) = \mu_{11}, \quad \mu_{10}^2 + \mu_{11}^2 = r_1^2,$$

where $\mu_{10} > 0$ and $\mu_{11} < 0$. Note that $|\mu_{11}|$ can be made arbitrarily small and that $t_1^* \in [0, 2\pi]$.

Since $\lambda_k - \lambda_1 - 1 < 0$ for $k \geq 2$, due to (6.47)–(6.48), we have:

$$|\lambda_k - \lambda_1 - 1| \, c_k^2(t) + \dot{c}_k^2(t) \; = |\lambda_k - \lambda_1 - 1| \, c_k^2(0) + \dot{c}_k^2(0), \; k = 2, \ldots, \; t \in [0, t_1^*].$$
(6.49)

Then, at time t_1^* we "switch" (see (6.37)) to the control

$$h_1 = -\lambda_1.$$

Making use of the solution formula (6.44), applied with $a = h_1$ and (μ_{10}, μ_{11}) in place of $(c_1(0), \dot{c}_1(0))$, at time

$$t_1 = t_1^* - \frac{\mu_{10}}{\mu_{11}}$$
(6.50)

we will have

$$c_1(t_1) = 0, \; \dot{c}_1(t_1)) = \mu_{11}.$$
(6.51)

Again, since $\lambda_k - \lambda_1 - 1 < \lambda_k - \lambda_1 < 0$ for $k \geq 2$, due to (6.47)–(6.48) and (6.49),

$$|\lambda_k - \lambda_1| \, c_k^2(t) + \dot{c}_k^2(t) = |\lambda_k - \lambda_1| \, c_k^2(t_1^*) + \dot{c}_k^2(t_1^*) \leq |\lambda_k - \lambda_1 - 1| \, c_k^2(t_1^*) + \dot{c}_k^2(t_1^*)$$
$$= |\lambda_k - \lambda_1 - 1| \, c_k^2(0) + \dot{c}_k^2(0), \; k = 2, \ldots, \; t \in [t_1^*, t_1].$$
(6.52)

On the other hand, it follows from (6.52) that

$$c_k^2(t_1) \leq \frac{1}{|\lambda_k - \lambda_1|} \left(|\lambda_k - \lambda_1 - 1| \, c_k^2(0) + \dot{c}_k^2(0) \right), \quad k = 2, \ldots$$

Hence,

$$
\begin{aligned}
|\lambda_k - \lambda_1 - 1| \, c_k^2(t_1) + \dot{c}_k^2(t_1) &= (-\lambda_k + \lambda_1 + 1) c_k^2(t_1) + \dot{c}_k^2(t_1) \\
&= |\lambda_k - \lambda_1| \, c_k^2(t_1) + \dot{c}_k^2(t_1) + c_k^2(t_1) \\
&\leq \left(1 + \frac{1}{|\lambda_k - \lambda_1|} \right) \\
&\quad \times \left(|\lambda_k - \lambda_1 - 1| \, c_k^2(0) + \dot{c}_k^2(0) \right), \, k = 2, \ldots \quad (6.53)
\end{aligned}
$$

We can repeat this procedure on (t_1, t_2) with h as in (6.37) to steer the point $(c_2(t), \dot{c}_2(t))$ from its position $(c_2(t_1), \dot{c}_2(t_1))$ to a position like

$$c_2(t_2) = 0, \quad \dot{c}_2(t_2)) = \mu_{21}$$

(compare with (6.50)–(6.51)).

Remark 6.4. • Let us stress here that the estimates (6.53) imply that the magnitudes of values of $c_2(t_1)$ and $\dot{c}_2(t_1)$ are bounded above by a number which does not depend on the value of μ_{11} in (6.51) (recall that $\mu_{10} = \sqrt{r_1^2 - \mu_{11}^2}$). Hence, given the choice of the "target values"

$$\mu_{20} = c_2(t_2^*), \quad \mu_{21} = \dot{c}_2(t_2^*),$$

where, analogously to the above (see the control structure (6.37)),

$$c_2^2(t_2^*) + \dot{c}_2^2(t_2^*) = c_2^2(t_1) + \dot{c}_2^2(t_1),$$

the durations of time-intervals (t_1, t_2^*) and (t_2^*, t_2) (where the instants t_2^* and t_2 are calculated analogously to t_1^* and t_1 as in the above, see (6.50), in particular) are bounded above respectively by the numbers 2π and by $|\mu_{20}/\mu_{21}|$ which do not depend on the duration of $(0, t_1)$ and the choice of μ_{10}, μ_{11}.
• Respectively, in view of (6.50)–(6.51), given the choice of μ_{20} and μ_{21}, at the instant $t = t_2$ the values of $(c_1(t_2), \dot{c}_1(t_2))$ can be made arbitrarily small *merely* by suitable selection of ("small") μ_{11} on the previous interval, regardless of our controls actions on (t_1, t_2) as they are described in (6.6), (6.37)–(6.38) (and in this step).
• The above two remarks can be extended to other time-intervals, making use of the estimate (6.54), derived below (further discussion of this type of recursive evaluation is given in the next Chapter 7).

Analogously to the derivation of (6.53), we will have on (t_1, t_2) for $k = 3, \ldots$:

$$| \lambda_k - \lambda_2 - 1 | c_k^2(t_2) + \dot{c}_k^2(t_2) \leq \left(1 + \frac{1}{| \lambda_k - \lambda_2 |} \right) \left(| \lambda_k - \lambda_2 - 1 | c_k^2(t_1) + \dot{c}_k^2(t_1) \right).$$

Note now that, in view of (6.9), $| \lambda_k - \lambda_2 - 1 | < | \lambda_k - \lambda_1 - 1 |$, which, combined with (6.53), yields:

$$\begin{aligned}
| \lambda_k - \lambda_2 - 1 | c_k^2(t_2) + \dot{c}_k^2(t_2) &\leq \left(1 + \frac{1}{| \lambda_k - \lambda_2 |} \right) \left(| \lambda_k - \lambda_2 - 1 | c_k^2(t_1) + \dot{c}_k^2(t_1) \right) \\
&\leq \left(1 + \frac{1}{| \lambda_k - \lambda_2 |} \right) \left(| \lambda_k - \lambda_1 - 1 | c_k^2(t_1) + \dot{c}_k^2(t_1) \right) \\
&= \left(1 + \frac{1}{| \lambda_k - \lambda_1 |} \right) \left(1 + \frac{1}{| \lambda_k - \lambda_2 |} \right) \\
&\quad \times \left(| \lambda_k - \lambda_1 - 1 | c_k^2(0) + \dot{c}_k^2(0) \right), \quad k = 3, \ldots
\end{aligned}$$

Repeating this procedure for all time-intervals (t_{k-1}, t_k), $k = 1, \ldots, k_* - 1$, we will have for $k \geq k_*$:

$$\begin{aligned}
| \lambda_k - \lambda_{k_* - 1} - 1 | c_k^2(t_{k_* - 1}) + \dot{c}_k^2(t_{k_* - 1}) &\leq \left(1 + \frac{1}{| \lambda_k - \lambda_1 |} \right) \cdots \left(1 + \frac{1}{| \lambda_k - \lambda_{k_* - 1} |} \right) \\
&\quad \times \left(| \lambda_k - \lambda_1 - 1 | c_k^2(0) \right. \\
&\quad \left. + \dot{c}_k^2(0) \right), \quad k = k_*, \ldots
\end{aligned} \qquad (6.54)$$

Step 3.2. Control on $(t_{k_* - 1}, t_{k_*})$. On this time-interval (with t_{k_*} to be selected a little bit later) we engage control

$$h_{k_*}^* = -\lambda_{k_*} - 1$$

(see (6.37)–(6.38)), which forces the point $(c_{k_*}(t), \dot{c}_{k_*}(t))$ to move with period 2π in the associated two dimensional space around the origin along the circle:

$$c_{k_*}^2(t) + \dot{c}_{k_*}^2(t) = c_1^2(t_{k_* - 1}) + \dot{c}_{k_*}^2(t_{k_* - 1}). \qquad (6.55)$$

Without loss of generality we can further assume that

$$\int_0^1 y_d \omega_{k_*} dx > 0.$$

Then, we can select the moment t_{k_*} as the first moment when

$$c_{k_*}(t_{k_*}) > 0, \quad \dot{c}_{k_*}(t_{k_*}) = 0.$$

Now, we are in the position to apply the controls found in Theorem 6.1 to finish the steering on some interval (t_{k_*}, T) arbitrarily close to the desirable target state $(y_d, 0)$, provided that we can show that condition (D) mentioned in the beginning of this section is satisfied, namely, we need to show that the magnitude of the part of the solution to (6.2), (6.6), (6.37)–(6.38)) contributed by the first $k_* - 1$ terms can be made arbitrarily small at time $t = t_{k_*}$ in the norm of $H_0^1(0,1) \times L^2(0,1)$, independently on the magnitude of the remainder of the series in the expansion (6.42).

Step 3.3. Verification of condition (D). Since on the time-interval (t_{k_*-1}, t_{k_*}) the points $(c_k(t), \dot{c}_k(t)), k \geq k_*$ move in the corresponding two dimensional space around the origin along the respective ellipses, we have for $k \geq k_*$:

$$
\begin{aligned}
| \lambda_k - \lambda_{k_*} - 1 | \, c_k^2(t_{k_*}) + \dot{c}_k^2(t_{k_*}) &= | \lambda_k - \lambda_{k_*} - 1 | \, c_k^2(t_{k_*-1}) + \dot{c}_k^2(t_{k_*-1}) \\
&\leq | \lambda_k - \lambda_{k_*-1} - 1 | \, c_k^2(t_{k_*-1}) + \dot{c}_k^2(t_{k_*-1}) \\
&\leq \left(1 + \frac{1}{| \lambda_k - \lambda_1 |} \right) \cdots \left(1 + \frac{1}{| \lambda_k - \lambda_{k_*-1} |} \right) \\
&\quad \times \left(| \lambda_k - \lambda_1 - 1 | \, c_k^2(0) + \dot{c}_k^2(0) \right), \quad (6.56)
\end{aligned}
$$

where we also used (6.54).

Thus we established the following.

Proposition 6.1. *The estimate (6.56) does not depend on the choice of μ_{k0}, μ_{k1}, $k = 1, \ldots, k_* - 1$ (see also Remark 6.4). Respectively, the value of the part of solution to (6.2), (6.6), (6.37)–(6.38) at time $t = t_{k_*}$ contributed by the first $k_* - 1$ terms in (6.42), namely, the value*

$$
M_{k_*} = \sum_{k=1}^{k_*-1} (| \lambda_k | \, c_k^2(t_{k_*}) + \dot{c}_k^2(t_{k_*})
$$

can be made as small as we wish in the norm of $H_0^1(0,1) \times L^2(0,1)$ (recall (6.19)), independently of the value in the right-hand side of estimate (6.56) representing in turn an upper bound for the contribution of the remainder of the solution expansion (6.42) at $t = t_{k_*}$.

Remark 6.5. The value M_{k_*} can be made arbitrarily small by selecting respectively small values for $\mu_{k1}, k = 1, \ldots, k_* - 1$ in the above argument *backwards in time*. More precisely, as Remark 6.4 implies, first the value of μ_{k_*-11} should be selected to make the corresponding $(k_* - 1)$-th term of M_{k_*} "small." This, given the control structure in (6.37), will define the control action, namely, the duration of the time-interval (t_{k_*-2}, t_{k_*-1}), which, by Remark 6.4, is bounded by a number independent of our control auctions on the preceding time-intervals. Then, based on this information, one can select $\mu_{(k_*-2)1}$ so that the "known future" control auctions on the time-interval (t_{k_*-2}, t_{k_*}) will not make the values of $c_{k_*-2}, \dot{c}_{k_*-2}$ "too large" at time t_{k_*}. This can be achieved due to the formulas (6.42)–(6.48). After that one may further select the values for $\mu_{k_*-31}, \mu_{k_*-41}, \ldots, \mu_{11}$ to make M_{k_*} as small as we wish.

Step 3.4. Evaluation of $c_{k_*}(t_1^*)$. It remains to show that the magnitude of $c_{k_*}(t_{k_*})$ (recall that $\dot{c}_{k_*}(t_{k_*}) = 0$) is separated from zero and bounded above by numbers which are independent of $\mu_{k0}, \mu_{k1}, k = 1, \ldots, k_* - 1$.

First of all, (6.56) provides the required estimate from above, which is independent of $\mu_{k0}, \mu_{k1}, k = 1, \ldots, k_* - 1$.

It remains to obtain a similar estimate from below. To this end, let us recall that due to the equality in (6.49),

$$
\begin{aligned}
\max\{1, |\lambda_{k_*} - \lambda_1 - 1|\}\left(c_{k_*}^2(t_1^*) + \dot{c}_{k_*}^2(t_1^*)\right) &\geq |\lambda_{k_*} - \lambda_1 - 1| \, c_{k_*}^2(t_1^*) + \dot{c}_{k_*}^2(t_1^*) \\
&= |\lambda_{k_*} - \lambda_1 - 1| \, c_{k_*}^2(0) + \dot{c}_{k_*}^2(0) \\
&= (|\lambda_{k_*} - \lambda_1| + 1) c_{k_*}^2(0) + \dot{c}_{k_*}^2(0) \\
&\geq \left(c_{k_*}^2(0) + \dot{c}_{k_*}^2(0)\right).
\end{aligned}
$$

Hence,

$$
\left(c_{k_*}^2(t_1^*) + \dot{c}_{k_*}^2(t_1^*)\right) \geq \frac{1}{\max\{1, |\lambda_{k_*} - \lambda_1 - 1|\}}\left(c_{k_*}^2(0) + \dot{c}_{k_*}^2(0)\right), \qquad (6.57)
$$

where the right-hand side is independent of μ_{10}, μ_{11}.

Analogously, making use of the first *equality* in (6.52) and inequality

$$
\frac{|\lambda_{k_*} - \lambda_1|}{|\lambda_{k_*} - \lambda_1 - 1|} < 1,
$$

we obtain:

$$
\begin{aligned}
\max\{1, |\lambda_{k_*} - \lambda_1|\}\left(c_{k_*}^2(t_1) + \dot{c}_{k_*}^2(t_1)\right) &\geq |\lambda_{k_*} - \lambda_1| \, c_{k_*}^2(t_1) + \dot{c}_{k_*}^2(t_1) \\
&= |\lambda_{k_*} - \lambda_1| \, c_{k_*}^2(t_1^*) + \dot{c}_{k_*}^2(t_1^*) \\
&\geq \frac{|\lambda_{k_*} - \lambda_1|}{|\lambda_{k_*} - \lambda_1 - 1|} \\
&\quad \times \left(|\lambda_{k_*} - \lambda_1 - 1| \, c_{k_*}^2(t_1^*) + \dot{c}_{k_*}^2(t_1^*)\right) \\
&= \frac{|\lambda_{k_*} - \lambda_1|}{|\lambda_{k_*} - \lambda_1 - 1|} \\
&\quad \times \left((|\lambda_{k_*} - \lambda_1| + 1) c_{k_*}^2(t_1^*) + \dot{c}_{k_*}^2(t_1^*)\right) \\
&\geq \frac{|\lambda_{k_*} - \lambda_1|}{|\lambda_{k_*} - \lambda_1 - 1|}\left(c_{k_*}^2(t_1^*) + \dot{c}_{k_*}^2(t_1^*)\right).
\end{aligned}
$$

Hence, also using (6.57),

$$\left(c_{k_*}^2(t_1)+\dot{c}_{k_*}^2(t_1)\right) \geq \frac{1}{\max\{1,|\lambda_{k_*}-\lambda_1|\}}\frac{1}{\max\{1,|\lambda_{k_*}-\lambda_1-1|\}}$$
$$\times \frac{|\lambda_{k_*}-\lambda_1|}{|\lambda_{k_*}-\lambda_1-1|}\left(c_{k_*}^2(0)+\dot{c}_{k_*}^2(0)\right),$$

where the right-hand side is again independent of μ_{10},μ_{11}.

Repeating this procedure on each of the intervals $(t_1,t_2),\dots,(t_{k_*-2},t_{k_*-1})$, we obtain the following estimate:

$$c_{k_*}^2(t_{k_*-1})+\dot{c}_{k_*}^2(t_{k_*-1}) \geq m\left(c_{k_*}^2(0)+\dot{c}_{k_*}^2(0)\right), \tag{6.58}$$

where $m>0$ is some positive constant independent of $\mu_{k0},\mu_{k1},k=1,\dots,k_*-1$.

Making use of (6.55), (6.58) can be extended to the desirable estimate from below:

$$c_k^2(t_{k_*})+\dot{c}_k^2(t_{k_*})=c_k^2(t_{k_*}) \geq m\left(c_{k_*}^2(0)+\dot{c}_{k_*}^2(0)\right). \tag{6.59}$$

Proposition 6.1 and estimates (6.56) and (6.59) verify condition (D). This completes the proof of Theorem 6.2.   □

Chapter 7
Controllability Properties of a Vibrating String with Variable Axial Load Only

Abstract We show that the set of equilibrium-like states $(y_d, 0)$ of a vibrating string which can approximately be reached in the energy space $H_0^1(0,1) \times L^2(0,1)$ from *almost* any non-zero initial datum, namely, $(y_0, y_1) \in (H^2(0,1) \cap H_0^1(0,1)) \times H^1(0,1)$, $(y_0, y_1) \neq (0,0)$ by varying its *axial load only* is dense in the subspace $H_0^1(0,1) \times \{0\}$ of this space. This result is based on a constructive argument and makes use of piecewise constant-in-time control functions (loads) only, which enter the model equation as coefficients.

7.1 Introduction

In this chapter we consider the following reduced version of model (6.1) from Chapter 6:

$$y_{tt} = y_{xx} + v(x,t)y, \quad x \in (0,1), \ t > 0, \tag{7.1}$$
$$y(0,t) = y(1,t) = 0, \quad y(x,0) = y_0(x), \ y_t(x,0) = y_1(x).$$

The difference between (7.1) and model (6.1) in Chapter 6 is that the damping term is omitted now. In other words, we have a principal reduction in the means to control the system at hand. This time, given the non-zero initial state (y_0, y_1), we would like to know what states $(y(\cdot,t), y_t(\cdot,t))$ can be achieved by system (6.1) at times $t > 0$ by varying the load v *only*. The "trade-off" here is an extra restriction on the regularity of the initial datum. The main result of this chapter is as follows.

Theorem 7.1. *Consider any initial state* $(y_0, y_1) \in (H^2(0,1) \cap H_0^1(0,1)) \times H^1(0,1)$, $(y_0, y_1) \neq (0,0)$ *and any target equilibrium-like state* $(y_d, 0)$, $y_d \in H_0^1(0,1)$. *Then for every* $\varepsilon > 0$ *there are a time* $T = T(\varepsilon, y_0, y_1, y_d)$ *and a piecewise constant-in-time control* v *of the form*

$$v(x,t) = \bar{v}(x) + h(t), \tag{7.2}$$

A.Y. Khapalov, *Controllability of Partial Differential Equations Governed by Multiplicative Controls*, Lecture Notes in Mathematics 1995,
DOI 10.1007/978-3-642-12413-6_7, © Springer-Verlag Berlin Heidelberg 2010

where $\bar{v} \in L^\infty(0,1)$ and $h(t)$ is a piecewise constant function with finitely many points of discontinuity such that

$$\| (y_d,0) - (y(\cdot,T), y_t(\cdot,T)) \|_{H_0^1(0,1) \times L^2(0,1)} \leq \varepsilon. \tag{7.3}$$

We emphasize that the duration of control time T, required to achieve (7.3), depends on the given initial and target states in a nonlinear fashion (which, in turn, is due to the nonlinear nature of bilinear controls) and in some cases it can be "very long." In the proof of Theorem 7.1 we explicitly describe how suitable bilinear controls can be constructed and evaluate the left-hand side in (7.3) explicitly as well (see (7.2), (7.16), (7.19)–(7.22), and (7.51) below).

Remark 7.1. • Again, in contrast to the above-cited result of [8] requiring that all the modes in the initial datum be active, Theorem 7.1 requires only the given initial datum be a non-zero one.
- The model (7.1) is similar to model (6.2) from Chapter 6. However, in the former v is the function of both x and t and in the latter v is the function of t only. As we mentioned it in the introduction of Chapter 6, the global approximate controllability of (6.2) was established in [8] in a nonconstructive way for rather complex lumped controls $v = v(t)$. In turn, the controllability results of this section are constructive and the actual controls $v = v(x,t)$ applied in the proofs (even though they formally depend on two variables) are of rather simple, "more applicable" form.
- As the reader noticed, in Theorem 7.1 we require the initial datum in (7.1) be of a higher regularity compared to the target state. This "discrepancy" arises within the framework of our method (see the argument leading to the estimate (7.51) along the lines (7.49)–(7.50) below).

7.2 Preliminary Results: Solution Formulas

$\bar{v} \in L^\infty(0,1)$ and, as in Chapter 6, denote by λ_k and ω_k, $k = 1,\ldots$ respectively the eigenvalues and the orthonormal in $L^2(0,1)$ eigenfunctions associated with the spectral problem

$$\omega_{xx} + \bar{v}(x)\omega = \lambda\omega, \quad \omega \in H_0^1(0,1). \tag{7.4}$$

It is well-known that

$$\| \bar{v} \|_{L^\infty(0,1)} \geq \lambda_1 > \lambda_2 > \ldots \quad \text{and} \quad \lambda_k \to -\infty \text{ as } k \to \infty. \tag{7.5}$$

Respectively, the eigenvalues of the spectral problem like (7.4) with

$$v = \bar{v} + a, \tag{7.6}$$

in place of $\bar{v}$, where a is any number, will be as follows: $\lambda_k + a$, $k = 1, \ldots$. However, for any value of a the corresponding eigenfunctions remain the same in both of the above cases.

Consider any initial datum $(y_0, y_1) \in (H^2(0,1) \cap H_0^1(0,1)) \times H^1(0,1)$. Once again, we have the following solution formula for system (7.1), (7.6):

$$y(x,t) = \sum_{k=1}^{\infty} c_k(t) \omega_k(x), \quad x \in (0,1), \quad t > 0, \tag{7.7}$$

where

$$\frac{d^2 c_k}{dt^2} = (\lambda_k + a) c_k, \quad t > 0,$$

$$c_k(0) = \int_0^1 y_0 \omega_k dx, \quad \dot{c}_k(0) = \int_0^1 y_1 \omega_k dx. \tag{7.8}$$

In the case when $\lambda_k + a = 0$ we have:

$$c_k(t) = t \dot{c}_k(0) + c_k(0), \quad \dot{c}_k(t) = \dot{c}_k(0); \tag{7.9}$$

in turn, for $\lambda_k + a > 0$:

$$c_k(t) = \frac{1}{2} e^{t\sqrt{\lambda_k + a}} \left(c_k(0) + \frac{\dot{c}_k(0)}{\sqrt{\lambda_k + a}} \right)$$
$$+ \frac{1}{2} e^{-t\sqrt{\lambda_k + a}} \left(c_k(0) - \frac{\dot{c}_k(0)}{\sqrt{\lambda_k + a}} \right), \tag{7.10}$$

$$\dot{c}_k(t) = e^{t\sqrt{\lambda_k + a}} \frac{\dot{c}_k(0) + c_k(0)\sqrt{\lambda_k + a}}{2}$$
$$- e^{-t\sqrt{\lambda_k + a}} \frac{-\dot{c}_k(0) + c_k(0)\sqrt{\lambda_k + a}}{2}; \tag{7.11}$$

and in the case when $\lambda_k + a < 0$:

$$c_k(t) = c_k(0) \cos\left(t\sqrt{-\lambda_k - a}\right)$$
$$+ \sin\left(t\sqrt{-\lambda_k - a}\right) \frac{\dot{c}_k(0)}{\sqrt{-\lambda_k - a}}, \tag{7.12}$$

$$\dot{c}_k(t) = -c_k(0)\sqrt{-\lambda_k - a} \sin\left(t\sqrt{-\lambda_k - a}\right)$$
$$+ \dot{c}_k(0) \cos\left(t\sqrt{-\lambda_k - a}\right). \tag{7.13}$$

Since the norm

$$\| \phi \|_{\lambda_0} = \left(\int_0^1 (\phi_x^2 + (-v(x) + \lambda_0)\phi^2)dx \right)^{1/2}$$

$$= \left(\sum_{k=1}^{\infty} (-\lambda_k + \lambda_0) \left(\int_0^1 \phi \omega_k dx \right)^2 \right)^{1/2}, \tag{7.14}$$

where λ_0 is any number exceeding λ_1, is equivalent to the standard norm in $H_0^1(0,1)$, the series (7.7)–(7.13) converges in $C([0,T];H_0^1(0,1)) \cap C^1([0,T]; L^2(0,1))$ for every $T > 0$.

Remark 7.2. In fact, since our initial datum lies in $(H^2(0,1) \cap H_0^1(0,1)) \times H^1(0,1)$, we have even better regularity for solutions, which we employ below.

7.3 Proof of Theorem 7.1

Step 1. Selection of "approximate target state." To prove Theorem 7.1, similar to the proof of Theorem 6.1 in Chapter 6, it is sufficient to consider any set of target states y_d which is dense in $H_0^1(0,1)$. To this end, we will further consider only *non-zero* y_d's that satisfy the following conditions: (a) $y_d \in C^1[0,1]$; (b) $y_d(0) = y_d(1) = 0$; (c) the second derivatives of y_d are piecewise continuous with finitely many discontinuities of the first kind (hence, $y_d \in H^2(0,1) \cap H_0^1(0,1)$) such that

(d)

$$\begin{cases} \frac{y_{dxx}}{y_d}, & \text{where } y_d(x) \neq 0, \\ 0, & \text{where } y_d(x) = 0, \end{cases} \in L^\infty(0,1). \tag{7.15}$$

Step 2. Further plan of the proof. Consider any non-zero initial datum $(y_0, y_1) \in (H^2(0,1) \cap H_0^1(0,1)) \times H^1(0,1)$ and select any target state y_d satisfying the conditions (a)–(d) given in Step 1.

Without loss of generality, we may assume that

$$\left(\int_0^1 y_d y_0 dx \right)^2 + \left(\int_0^1 y_d y_1 dx \right)^2 \neq 0.$$

Indeed, since the set of all y_d, described in Step 1, is dense in $H_0^1(0,1)$, its subset for which the above inequality holds, is dense in this space as well.

Select (see (7.15))

$$\bar{v} = \begin{cases} -\frac{y_{dxx}}{y_d}, & \text{where } y_d(x) \neq 0, \\ 0, & \text{where } y_d(x) = 0, \end{cases} \in L^\infty(0,1). \qquad (7.16)$$

Our further plan of the proof of Theorem 7.1 is as follows:

- Note first that for our selected target state $y_d \in H^2(0,1) \cap H_0^1(0,1)$ the equation (6.4) holds with $\bar{v}(x)$ as in (7.16). Hence, $y_d / \| y_d \|_{L^2(0,1)}$ is one of the eigenfunction, associated with the representation (7.4)–(7.5), i.e., for some positive integer k_* in (7.4)–(7.5) we have:

$$\lambda_{k_*} = 0 \text{ and } \omega_{k_*}(x) = \frac{y_d(x)}{\| y_d \|_{L^2(0,1)}}. \qquad (7.17)$$

Note that our selection of approximate target state y_d in the beginning of this step ensures that $c_{k_*}^2(0) + \dot{c}_{k_*}^2(0) \neq 0$ in (7.8).

- Select any number λ_0 larger than λ_1. Then, since $(y_0, y_1) \in (H^2(0,1) \cap H_0^1(0,1)) \times H^1(0,1)$,

$$\gamma^2(N) = \sum_{k=N+1}^{\infty} \left((\lambda_0 - \lambda_k)^2 c_k^2(0) + (\lambda_0 - \lambda_k) \dot{c}_k^2(0) \right) \to 0 \text{ as } N \to \infty. \qquad (7.18)$$

- Next, given a positive integer $N > k_*$ (without loss of generality, in view of (7.18), we further assume that $N > k_*$), in the following steps 3–10 we will construct the actual control $v(x,t)$ of the form (7.2), (7.16) with a piecewise constant function $h(t)$ of the following structure:

$$h(t) = \begin{cases} h_k^* = -\lambda_{k-1}, & t \in (t_{k-1}, t_k^*) \\ h_k = -\lambda_k, & t \in (t_k^*, t_k) \end{cases} \text{ for } k = 1, \ldots, k_* - 1, \qquad (7.19)$$

while for $k = k_*$, associated with our target state y_d,

$$h(t) = h_{k_*}^* = -\lambda_{k_*-1}, \quad t \in (t_{k_*-1}, t_{k_*}), \qquad (7.20)$$

then, again,

$$h(t) = \begin{cases} h_k^* = -\lambda_{k-1}, & t \in (t_{k-1}, t_k^*) \\ h_k = -\lambda_k, & t \in (t_k^*, t_k) \end{cases} \text{ for } k = k_* + 1, \ldots, N, \qquad (7.21)$$

and, finally,

$$h(t) = \begin{cases} h_{N+1}^* = -\lambda_{k_*-1}, & t \in (t_N, t_{N+1}), \\ h_{N+1} = -\lambda_{k_*}, & t \in (t_{N+1}, T). \end{cases} \qquad (7.22)$$

In the above the moments $0 = t_0 \leq t_1^* \leq t_1 \leq \ldots \leq T$ are to be selected. The form (7.19)–(7.22) represents the "worst case scenario," namely, when all modes in the initial datum are active (relative to the selected $\bar{v}$ in (7.16)), that is, $c_k^2(0) + \dot{c}_k^2(0) \neq 0$ in (7.8). If this is not the case, then the corresponding pair of controls (h_k^*, h_k) should be omitted.

Our *control strategy*, implemented in steps 3–10 below, is the following:

(I) Given N, we intend to control each of the first N harmonics in the expansion (7.7)–(7.8) *individually and subsequently in time* as given in (7.2), (7.16), (7.19)–(7.22), while simultaneously evaluating the effect of each such ("piecewise constant") action on the remaining infinitely many harmonics. Our goal is, on the one hand, making use of (7.19) and (7.21), to make all the coefficients $c_k(t), \dot{c}_k(t), k = 1, \ldots, N, k \neq k_*$ in the solution formula "negligibly small", while, on the other hand, making use of (7.20) and (7.22), $c_{k_*}(t)$ and $\dot{c}_{k_*}(t)$ are to be adjusted to approximate the target state $(y_d, 0)$.

We intend to act here in such a way that, given the pre-assigned precision of steering for the first N harmonics (due to (7.3)), each of the aforementioned subsequent individual control action will preserve the precision of steering already achieved for "prior harmonics" till the final moment T.

(II) The key observation allowing us to achieve the above goal is the fact that formulas (7.9)–(7.13) describe the following *three qualitative motions* of the points $(c_k(t), \dot{c}_k(t))$ in the associated two dimensional space when constant controls (like in (7.19)–(7.22)) are used: (i) the *ellipsoidal motions* around the origin, due to (7.12)–(7.13); (ii) the *horizontal motions* towards and away from the origin, due to (7.9); and (iii) the motions due to (7.10)–(7.11).

(III) To make any of the first $N - 1$ pairs of coefficients $(c_k(t), \dot{c}_k(t))$ $(k \neq k_*)$ as small as we wish, we will employ the ellipsoidal motion (i) on each of the corresponding time-intervals (t_{k-1}, t_k^*), and then the horizontal motion (ii) on each of the subsequent intervals (t_k^*, t_k) (see (7.19) and (7.21)).

(IV) A suitable combination of the aforementioned motions is used on (t_N, T) to "adjust" the magnitude of the k_*-th term in (7.7) to approximate the target state $(y_d, 0)$ (see (7.20) and (7.22)).

(V) After that, making use of (7.18), we will show that, within our strategy, a suitably large N will ensure that all the coefficients $c_k(t), \dot{c}_k(t), k = 1, \ldots, k \neq k_*$ are, in fact, "small" at time T.

We begin now our construction of control h as it was outlined in (7.19)–(7.22). Without loss of generality, we can assume that $k_* > 1$ in (7.17) for our selected $\bar{v}$ in (7.16).

Step 3. Control on $(0, t_{k_*-1})$.

We begin with the selection of h_1^* as in (7.19)–(7.22), namely, $h_1^* = -\lambda_0$. Then, as (7.12)–(7.13) imply (with $a = h_1^*$), the point $(c_1(t), \dot{c}_1(t))$ moves with period $2\pi/(\lambda_0 - \lambda_1)^{1/2}$ in the associated two dimensional space around the origin along the ellipse

$$| \lambda_1 - \lambda_0 | c_1^2(t) + \dot{c}_1^2(t) = | \lambda_1 - \lambda_0 | c_1^2(0) + \dot{c}_1^2(0) = r_{10}^2.$$

Hence, *within one rotation around the origin*, we can select some moment t_1^* such that

$$c_1(t_1^*) = \mu_{10} > 0, \ \dot{c}_1(t_1^*) = \mu_{11} < 0,$$
$$| \lambda_1 - \lambda_0 | \mu_{10}^2 + \mu_{11}^2 = r_{10}^2, \ t_1^* \in [0, \frac{2\pi}{\sqrt{\lambda_0 - \lambda_1}}], \tag{7.23}$$

where, note, $| \mu_{11} |$ *can be made arbitrarily small*.

Then, at time t_1^* we "switch" according to (7.19)–(7.22) to the control $h_1 = -\lambda_1$. Making use of the solution formula (7.9), applied with $a = h_1$ and (μ_{10}, μ_{11}) in place of $(c_1(0), \dot{c}_1(0))$, at time

$$t_1 = t_1^* - \frac{\mu_{10}}{\mu_{11}} \tag{7.24}$$

we will have

$$c_1(t_1) = 0, \ \dot{c}_1(t_1) = \mu_{11}. \tag{7.25}$$

Simultaneously, since $\lambda_k - \lambda_0 < 0$ for $k \geq 2$, due to (7.12)–(7.13), we have:

$$| \lambda_k - \lambda_0 | c_k^2(t) + \dot{c}_k^2(t) = | \lambda_k - \lambda_0 | c_k^2(0) + \dot{c}_k^2(0)$$
$$= r_{k0}^2, k = 2, \ldots, \ t \in [0, t_1^*]. \tag{7.26}$$

Again, since $\lambda_k - \lambda_0 < \lambda_k - \lambda_1 < 0$ for $k \geq 2$, due to (7.12)–(7.13) and (7.26),

$$| \lambda_k - \lambda_1 | c_k^2(t_1) + \dot{c}_k^2(t_1) = | \lambda_k - \lambda_1 | c_k^2(t_1^*) + \dot{c}_k^2(t_1^*)$$
$$\leq | \lambda_k - \lambda_0 | c_k^2(t_1^*) + \dot{c}_k^2(t_1^*)$$
$$= | \lambda_k - \lambda_0 | c_k^2(0) + \dot{c}_k^2(0)$$
$$= r_{k0}^2, \quad k = 2, \ldots. \tag{7.27}$$

We can repeat this procedure on (t_1, t_2) with h as in (7.19)–(7.22) to steer the point $(c_2(t), \dot{c}_2(t))$ from its position $(c_2(t_1), \dot{c}_2(t_1))$ to a position like

$$c_2(t_2) = 0, \ \dot{c}_2(t_2)) = \mu_{21} < 0$$

at some instant t_2 calculated along the lines similar to (7.23)–(7.25).

Repeating the above procedure on $(t_2, t_3), \ldots, (t_{k_*-1}, t_{k_*})$ will result in steering the corresponding coefficients $(c_k(t), \dot{c}_k(t))$ to positions like

$$c_k(t_k) = 0, \ \dot{c}_k(t_k)) = \mu_{k1} < 0, \ k = 1, \ldots, k_* - 1, \tag{7.28}$$

where

$$c_k(t_k^*) = \mu_{k0}, \ \dot{c}_k(t_k^*)) = \mu_{k1}, \ | \lambda_k - \lambda_{k-1} | \mu_{k0}^2 + \mu_{k1}^2 \leq r_{k0}^2, \ k = 1, \ldots, k_* - 1, \tag{7.29}$$

and μ_{k1}'s can be selected arbitrarily small.

Evaluation of $(c_k(t_{k_*-1}), \dot{c}_k(t_{k_*-1})), k \geq k_*$. Similar to (7.26) and using (7.27), we have:

$$| \lambda_k - \lambda_1 | \, c_k^2(t_2^*) + \dot{c}_k^2(t_2^*) \; = | \lambda_k - \lambda_1 | \, c_k^2(t_1) + \dot{c}_k^2(t_1) \; \leq \; r_{k0}^2, \quad k-3, \dots \quad (7.30)$$

Then, similar to (7.27) and using (7.30), we derive:

$$\begin{aligned}
| \lambda_k - \lambda_2 | \, c_k^2(t_2) + \dot{c}_k^2(t_2) &= | \lambda_k - \lambda_2 | \, c_k^2(t_2^*) + \dot{c}_k^2(t_2^*) \\
&\leq | \lambda_k - \lambda_1 | \, c_k^2(t_2^*) + \dot{c}_k^2(t_2^*) \\
&\leq r_{k0}^2, \quad k = 3, \dots.
\end{aligned}$$

Repeating this procedure up to the moment t_{k_*-1}, we obtain:

$$| \lambda_k - \lambda_{k_*-1} | \, c_k^2(t_{k_*-1}) + \dot{c}_k^2(t_{k_*-1}) \; \leq \; r_{k0}^2, \quad k = k_*, \dots. \quad (7.31)$$

Evaluation of $(c_k(t_{k_*-1}), \dot{c}_k(t_{k_*-1})), k < k_*$. Let us stress here that the estimates (7.27) imply that the magnitudes of values of $c_2(t_1)$ and $\dot{c}_2(t_1)$ are bounded above by a number which does not depend on the value of μ_{11} in (7.23)–(7.25) (recall that $\mu_{10} = \sqrt{(r_{10}^2 - \mu_{11}^2)/(\lambda_0 - \lambda_1)}$).

Hence, given the choice of the "target values" $\mu_{20} = c_2(t_2^*)$, $\mu_{21} = \dot{c}_2(t_2^*)$ in (7.28)–(7.29), the durations of time-intervals (t_1, t_2^*) and $(t_2^*, t_2 = t_2^* - \mu_{20}/\mu_{21})$ are bounded above respectively by $2\pi/\sqrt{| \lambda_2 - \lambda_1 |}$ and by

$$\sqrt{\frac{r_{20}^2 - \mu_{21}^2}{\lambda_1 - \lambda_2} \frac{1}{\mu_{21}}} \quad \text{(or by } 0 \text{ if } r_{20} = 0\text{)},$$

which do *not* depend on the duration of $(0, t_1)$ and the choice of μ_{10}, μ_{11}.

In turn, on the time-interval (t_1, t_2) the evolution of $(c_1(t), \dot{c}_1(t))$ is defined by (7.10)–(7.11). Therefore, the above argument implies that within our control strategy, given the choice of μ_{20} and μ_{21}, the values of $(c_1(t_2), \dot{c}_1(t_2))$ can be made arbitrarily small *merely* by suitable selection of ("small") μ_{11} on the previous interval.

The above argument can be extended to the subsequent time-intervals, making use of the estimates (7.30)–(7.31). Namely, given the initial datum, we can show that the choice of $\mu_{k1}, k < k_*$ (note μ_{k0} is determined by μ_{k1}) defines an upper bound for the duration of the time-interval (t_{k-1}, t_k), which is independent of our control actions as in (7.19)–(7.22) on (t_0, t_{k-1}). Hence, we may conclude that we have the following property.

Proposition 7.1. *Within our control strategy, we can select* $\mu_{k0}, \mu_{k1}, k = 1, \dots,$ $k_* - 1$ *so that at time* $t = t_{k_*-1}$ *the value of*

$$M_{k_*} = \sum_{k=1}^{k_*-1} \left(c_k^2(t_{k_*-1}) + \dot{c}_k^2(t_{k_*-1}) \right)$$

is as small as we wish.

More precisely, to prove Proposition 7.1, one should apply a "backward" induction to select suitable values for $\mu_{k0}, \mu_{k1}, k = 1, \ldots, k_* - 1$ along our above-described "forward" constructions (7.19)–(7.22), *valid for any* μ_{k0}'s *and* μ_{k1}'s, as follows. First the value of $\mu_{k_*-1,1}$ should be selected to make the corresponding $(k_* - 1)$-th term of M_{k_*} "as small as we wish." This will define an upper bound for the duration of the time-interval (t_{k_*-2}, t_{k_*-1}), which, as we showed it above, is independent of our above-described control actions on (t_0, t_{k_*-2}). Then, based on this information, we can select $\mu_{k_*-2,1}$ (which, in turn, will define an upper bound for the length of (t_{k_*-3}, t_{k_*-2}) independently of the choice of $\mu_{k1}, k = 1, \ldots, k_* - 3$) so "small" that the "known future" control actions along (7.19)–(7.22) on the time-interval (t_{k_*-2}, t_{k_*-1}) will not make the values of $c_{k_*-2}(t), \dot{c}_{k_*-2}(t)$ "too large" at time t_{k_*-1}. Then, based on the information about the duration of (t_{k_*-3}, t_{k_*-1}), we can select $\mu_{k_*-3,1}$ so that the "known future" control actions on the time-interval (t_{k_*-3}, t_{k_*-1}) will keep the values of $c_{k_*-3}(t), \dot{c}_{k_*-3}(t)$ "sufficiently small" at time t_{k_*-1}. After that one has to select the remaining values for $\mu_{k_*-4,1}, \ldots, \mu_{11}$ in a similar fashion to make M_{k_*} as small as we wish.

Step 4. Control on (t_{k_*-1}, t_{k_*}). On this next time-interval (with t_{k_*} to be selected a little bit later) we engage control $h_{k_*}^* = -\lambda_{k_*-1}$, which forces the point $(c_{k_*}(t), \dot{c}_{k_*}(t))$ to move with period $2\pi/(\lambda_{k_*-1} - \lambda_{k_*})^{1/2}$ in the associated two dimensional space around the origin along the ellipse:

$$| \lambda_{k_*} - \lambda_{k_*-1} | \, c_{k_*}^2(t) + \dot{c}_{k_*}^2(t) = | \lambda_{k_*} - \lambda_{k_*-1} | \, c_1^2(t_{k_*-1}) + \dot{c}_{k_*}^2(t_{k_*-1}). \qquad (7.32)$$

Then, we select the moment t_{k_*} as the first moment when

$$c_{k_*}(t_{k_*}) > 0, \quad \dot{c}_{k_*}(t_{k_*}) = 0, \quad t_{k_*} \in [0, \frac{2\pi}{\sqrt{\lambda_{k_*-1} - \lambda_{k_*}}}]. \qquad (7.33)$$

Step 5. Evaluation of $c_k(t_{k_*}), \dot{c}_k(t_{k_*}), k \geq k_*$.

Evaluation of $c_{k_*}(t_{k_*})$. We will now show that the value of $c_{k_*}(t_{k_*}) > 0$ (recall that $\dot{c}_{k_*}(t_{k_*}) = 0$) is bounded above and separated from zero by numbers which are independent of $\mu_{k0}, \mu_{k1}, k = 1, \ldots, k_* - 1$.

Indeed, first of all, (7.31)–(7.33) provide such an estimate from above:

$$0 < c_{k_*}(t_{k_*}) \leq \sqrt{\frac{r_{k_*0}^2}{\lambda_{k_*-1} - \lambda_{k_*}}}, \qquad (7.34)$$

where r_{k0}^2 is defined in (7.26).

Let us now recall that, due to (7.26),

$$\max\{1, | \lambda_{k_*} - \lambda_0 |\} \, (c_{k_*}^2(t_1^*) + \dot{c}_{k_*}^2(t_1^*)) \geq | \lambda_{k_*} - \lambda_0 | \, c_{k_*}^2(t_1^*) + \dot{c}_{k_*}^2(t_1^*)$$
$$= | \lambda_{k_*} - \lambda_0 | \, c_{k_*}^2(0) + \dot{c}_{k_*}^2(0)$$
$$\geq \min\{1, | \lambda_{k_*} - \lambda_0 |\} \, (c_{k_*}^2(0) + \dot{c}_{k_*}^2(0)).$$

Hence,

$$\left(c_{k_*}^2(t_1^*) + \dot{c}_{k_*}^2(t_1^*)\right) \geq \frac{\min\{1, |\lambda_{k_*} - \lambda_0|\}}{\max\{1, |\lambda_{k_*} - \lambda_0|\}} \left(c_{k_*}^2(0) + \dot{c}_{k_*}^2(0)\right),$$

where note that the right-hand side is independent of μ_{10}, μ_{11}.

Analogously, making use of the first *equality* in (7.27), we obtain:

$$\max\{1, |\lambda_{k_*} - \lambda_1|\}\left(c_{k_*}^2(t_1) + \dot{c}_{k_*}^2(t_1)\right) \geq |\lambda_{k_*} - \lambda_1| \, c_{k_*}^2(t_1) + \dot{c}_{k_*}^2(t_1)$$
$$= |\lambda_{k_*} - \lambda_1| \, c_{k_*}^2(t_1^*) + \dot{c}_{k_*}^2(t_1^*)$$
$$\geq \min\{1, |\lambda_{k_*} - \lambda_1|\}\left(c_{k_*}^2(t_1^*) + \dot{c}_{k_*}^2(t_1^*)\right).$$

Hence,

$$\left(c_{k_*}^2(t_1) + \dot{c}_{k_*}^2(t_1)\right) \geq \frac{\min\{1, |\lambda_{k_*} - \lambda_0|\}}{\max\{1, |\lambda_{k_*} - \lambda_0|\}} \frac{\min\{1, |\lambda_{k_*} - \lambda_1|\}}{\max\{1, |\lambda_{k_*} - \lambda_1|\}} \left(c_{k_*}^2(0) + \dot{c}_{k_*}^2(0)\right),$$

where the right-hand side is again independent of μ_{10}, μ_{11}.

Repeating this procedure on each of the intervals $(t_1, t_2), \ldots, (t_{k_*-2}, t_{k_*-1})$, we obtain the following estimate:

$$c_{k_*}^2(t_{k_*-1}) + \dot{c}_{k_*}^2(t_{k_*-1}) \geq m\left(c_{k_*}^2(0) + \dot{c}_{k_*}^2(0)\right), \tag{7.35}$$

where $m > 0$ is some constant independent of $\mu_{k0}, \mu_{k1}, k = 1, \ldots, k_* - 1$.

Making use of (7.32)–(7.33), the estimate (7.35) can be extended to the desirable estimate for $c_k^2(t_{k_*})$ from below:

$$c_k^2(t_{k_*}) \geq \frac{m \min\{1, |\lambda_{k_*} - \lambda_{k_*-1}|\}}{|\lambda_{k_*} - \lambda_{k_*-1}|} \left(c_{k_*}^2(0) + \dot{c}_{k_*}^2(0)\right). \tag{7.36}$$

Evaluation of $c_k(t_{k_*}), \dot{c}_k(t_{k_*}), k > k_*$. We have, making use of (7.31) and (7.19)–(7.22) (compare to (7.32)),

$$|\lambda_k - \lambda_{k-1}| \, c_k^2(t_{k_*}) + \dot{c}_k^2(t_{k_*}) = |\lambda_k - \lambda_{k-1}| \, c_k^2(t_{k_*-1}) + \dot{c}_k^2(t_{k_*-1})$$
$$\leq r_{k0}^2, \quad k = k_* + 1, \ldots . \tag{7.37}$$

Step 6. Control on (t_{k_*}, t_N). It is not hard to see that, similar to (7.30)–(7.31), (7.37) can further be extended to the estimates:

$$|\lambda_k - \lambda_N| \, c_k^2(t_N) + \dot{c}_k^2(t_N) \leq r_{k0}^2, \quad k = N + 1, \ldots . \tag{7.38}$$

In turn, (7.28)–(7.29) can be extended to all $k \leq N, k \neq k_*$:

$$c_k(t_k) = 0, \quad \dot{c}_k(t_k)) = \dot{c}_k(t_k^*) = \mu_{k1}, \quad \dot{c}_k(t_k^*)) = \mu_{k0}, \quad k = 1, \ldots, N, k \neq k_*, \tag{7.39}$$

as well as the conclusion of Proposition 7.1, which yields the following:

Proposition 7.2. *For any given positive integers* k_*, N, $k_* < N$ *in (7.19)–(7.22), the value of*

$$\sum_{k=1, k\neq k_*}^{N} \left(c_k^2(t_N) + \dot{c}_k^2(t_N) \right)$$

can be made as small as we wish by suitable selection of $\mu_{k0}, \mu_{k1}, k = 1, \ldots, N, k \neq k_*$ *in (7.39).*

Step 7. Evaluation of $c_{k_*}(t_N), \dot{c}_{k_*}(t_N)$. As it follows from Steps 4 and 5, $c_{k_*}(t_{k_*}) > 0$, $\dot{c}_{k_*}(t_{k_*}) = 0$, and the value of $c_{k_*}(t_{k_*})$ is bounded above and separated from zero by numbers which do not depend on the parameters in (7.28)–(7.29) and the duration of $(0, t_{k_*})$. Furthermore, on the time-interval (t_{k_*}, t_N) the evolution of $(c_{k_*}(t), \dot{c}_{k_*}(t))$ is described by formulas (7.10)–(7.11), where the choice of parameter a is as in (7.19)–(7.22). The duration of this time-interval depends on the selection of parameters μ_{k0} and $\mu_{k1}, k = k_* + 1, \ldots, N$ as described in Step 6.

Note now that on any (generic) interval (τ_0, τ_1), if $c_{k_*}(\tau_0), \dot{c}_{k_*}(t_{\tau_1}) \geq 0$, and $\lambda_{k_*} + a > 0$, the formulas (7.10)–(7.11) yield:

$$c_k(\tau_1) = \frac{c_k(\tau_0)}{2} \left(e^{(\tau_1 - \tau_0)\sqrt{\lambda_k + a}} + e^{-(\tau_1 - \tau_0)\sqrt{\lambda_k + a}} \right)$$
$$+ \frac{\dot{c}_k(\tau_0)}{2\sqrt{\lambda_k + a}} \left(e^{(\tau_1 - \tau_0)\sqrt{\lambda_k + a}} - e^{-(\tau_1 - \tau_0)\sqrt{\lambda_k + a}} \right) \geq \frac{c_k(\tau_0)}{2},$$

$$\dot{c}_k(\tau_1) = \frac{c_k(\tau_0)\sqrt{\lambda_k + a}}{2} \left(e^{(\tau_1 - \tau_0)\sqrt{\lambda_k + a}} - e^{-(\tau_1 - \tau_0)\sqrt{\lambda_k + a}} \right)$$
$$+ \frac{\dot{c}_k(\tau_0)}{2} \left(e^{(\tau_1 - \tau_0)\sqrt{\lambda_k + a}} + e^{-(\tau_1 - \tau_0)\sqrt{\lambda_k + a}} \right) \geq \frac{\dot{c}_k(\tau_0)}{2}.$$

Applying the above estimates on each of the intervals (t_k, t_{k+1}^*) and (t_{k+1}^*, t_{k+1}), $k = k_*, \ldots, N-1$ with the choice of a as in (7.19)–(7.22), we obtain, making use of (7.33) and (7.36), that

$$c_k^2(t_N) + \dot{c}_k^2(t_N) \geq \left(\frac{c_k(t_{k_*})}{2^{2(N-k_*)}} \right)^2 + \left(\frac{\dot{c}_k(t_{k_*})}{2^{2(N-k_*)}} \right)^2 = \frac{c_k^2(t_{k_*})}{2^{4(N-k_*)}}$$

$$\geq \frac{m \min\{1, |\lambda_{k_*} - \lambda_{k_*-1}|\}}{2^{4(N-k_*)} |\lambda_{k_*} - \lambda_{k_*-1}|} \left(c_{k_*}^2(0) + \dot{c}_{k_*}^2(0) \right) > 0. \quad (7.40)$$

In turn, the same formulas (7.10)–(7.11), (7.33) and (7.34), also imply that

$$c_k^2(t_N) + \dot{c}_k^2(t_N) \leq L c_k^2(t_{k_*}) \leq L \frac{r_{k_*0}^2}{|\lambda_{k_*} - \lambda_{k_*-1}|}, \quad (7.41)$$

where L depends on $\mu_{k0}, \mu_{k1}, k = k_* + 1, \ldots, N$, but *not* on $\mu_{k0}, \mu_{k1}, k = 1, \ldots, k_* - 1$.

Remark 7.3. The estimates (7.40)–(7.41) can be refined but our main concern here is only to show their independence from the choice of the parameters μ_{k0}, μ_{k1}, $k = 1, \ldots, k_* - 1$ in (7.28)–(7.29).

Step 8. Control on (t_N, T). Here we apply the strategy similar to that of Step 3 to steer $(c_{k_*}(t_N), \dot{c}_{k_*}(t_N))$ close enough to the desirable target state $(y_d, 0)$. We begin with the selection of h^*_{N+1} as in (7.19)–(7.22), namely, $h^*_{N+1} = -\lambda_{k_*-1}$.

Then, as (7.12)–(7.13) imply (with $a = h^*_{N+1}$), the point $(c_{k_*}(t), \dot{c}_{k_*}(t))$ moves with period $2\pi/\sqrt{\lambda_{k_*-1} - \lambda_{k_*}}$ in the associated two dimensional space around the origin along the ellipse

$$| \lambda_{k_*} - \lambda_{k_*-1} | \, c^2_{k_*}(t) + \dot{c}^2_{k_*}(t) = | \lambda_{k_*} - \lambda_{k_*-1} | \, c^2_{k_*}(t_N) + \dot{c}^2_{k_*}(t_N).$$

In particular, at some instant, say, $t_* \in [t_N, t_N + 2\pi/\sqrt{\lambda_{k_*-1} - \lambda_{k_*}}]$, we will have

$$(c_{k_*}(t_*), \dot{c}_{k_*}(t_*)) = \left(\sqrt{c^2_{k_*}(t_N) + \frac{\dot{c}^2_{k_*}(t_N)}{| \lambda_{k_*} - \lambda_{k_*-1} |}}, 0 \right).$$

Denote

$$A = \sqrt{c^2_{k_*}(t_N) + \frac{\dot{c}^2_{k_*}(t_N)}{| \lambda_{k_*-1} - \lambda_{k_*} |}}.$$

Recall that, by (7.17),

$$y_d = \alpha \omega_{k_*}, \quad \alpha = \| y_d \|_{L^2 0,1} > 0. \tag{7.42}$$

We may assume without loss of generality that we have two options: either $\| y_d \|_{L^2 0,1} < A$ or $\| y_d \|_{L^2 0,1} > A$.

Step 9. Option #1: $\| y_d \|_{L^2 0,1} < A$. If we have this option, select a moment

$$t^*_{N+1} \in \left[t_N, t_N + \frac{2\pi}{\sqrt{\lambda_{k_*-1} - \lambda_{k_*}}} \right]$$

such that

$$c_{k_*}(t^*_{N+1}) = v_{k_*0} \in (\| y_d \|_{L^2 0,1}, A), \quad \dot{c}_{k_*}(t^*_{N+1})) = v_{k_*1} < 0, \tag{7.43}$$
$$| \lambda_{k_*} - \lambda_{k_*-1} | \, v^2_{k_*0} + v^2_{k_*1} = A^2 \, | \lambda_{k_*} - \lambda_{k_*-1} |.$$

Note that $| v_{k_*1} |$ can be made arbitrarily small.

Then, at time t^*_{N+1} we "switch" according to (7.19)–(7.22) to the control $h_{N+1} = -\lambda_{k_*}$. Making use of the solution formula (7.9), applied with $a = h_{N+1}$ and (v_{k_*0}, v_{k_*1}) in place of $(c_{k_*}(0), \dot{c}_{k_*}(0))$, at time

$$T = t^*_{N+1} + \frac{\| y_d \|_{L^2(0,1)} - v_{k_*0}}{v_{k_*1}} \tag{7.44}$$

we will have

$$c_{k_*}(T) = \| y_d \|_{L^2(0,1)}, \quad \dot{c}_{k_*}(T) = v_{k_*1}. \tag{7.45}$$

Evaluation of $c_k(T), \dot{c}_k(T), k = 1, \ldots, N, k \neq k_*$.

Case $k = k_* + 1, \ldots, N$: Since $\lambda_k - \lambda_{k_*-1} < \lambda_k - \lambda_{k_*} < 0$ for $k > k_*$, we have the extension of estimates (7.31) and (7.38) as follows.

Similar to (7.37) and making use of (7.19)–(7.22), we obtain:

$$| \lambda_k - \lambda_{k_*-1} | c_k^2(t_{N+1}^*) + \dot{c}_k^2(t_{N+1}^*) = | \lambda_k - \lambda_{k_*-1} | c_k^2(t_N) + \dot{c}_k^2(t_N), \quad k > k_*. \tag{7.46}$$

Then, similar to the argument in (7.30)–(7.31), on the interval (t_{N+1}^*, T),

$$\begin{aligned}
| \lambda_k - \lambda_{k_*} | c_k^2(T) + \dot{c}_k^2(T) &= | \lambda_k - \lambda_{k_*} | c_k^2(t_{N+1}^*) + \dot{c}_k^2(t_{N+1}^*) \\
&\leq | \lambda_k - \lambda_{k_*-1} | c_k^2(t_{N+1}^*) + \dot{c}_k^2(t_{N+1}^*) \\
&= | \lambda_k - \lambda_{k_*-1} | c_k^2(t_N) + \dot{c}_k^2(t_N), \quad k > k_*.
\end{aligned} \tag{7.47}$$

These estimates allow us to extend Proposition 7.2 with respect to $k = k_* + 1, \ldots, N$ to the moment $t = T$ in place of $t = t_N$.

Case $k = 1, \ldots, k_* - 1$: In turn, Remark 7.3 indicates that our control actions on (t_N, T), described in Step 8, are not affected by the choice of the parameters in (7.28)–(7.29). This (along the argument leading to Proposition 7.1) allows us to extend Proposition 7.2 with respect to $k = 1, \ldots, k_* - 1$ to the moment $t = T$ in place of $t = t_N$.

Thus, we established the following extension of Proposition 7.2.

Proposition 7.3. *For any given positive integers k_*, N, $k_* < N$ in (7.19)–(7.22), the value of*

$$\sum_{k=1, k \neq k_*}^{N} (c_k^2(T) + \dot{c}_k^2(T))$$

can be made as small as we wish by suitable selection of $\mu_{k0}, \mu_{k1}, k = 1, \ldots, N, k \neq k_*$ in (7.39).

Evaluation of $c_k(T), \dot{c}_k(T), k > N$. To this ends, note that (7.46) and (7.38) provide us with the following estimates:

$$\begin{aligned}
| \lambda_k - \lambda_{k_*-1} | c_k^2(t_{N+1}^*) + \dot{c}_k^2(t_{N+1}^*) &= | \lambda_k - \lambda_N + \lambda_N - \lambda_{k_*-1} | c_k^2(t_N) + \dot{c}_k^2(t_N) \\
&= (| \lambda_k - \lambda_N | c_k^2(t_N) + \dot{c}_k^2(t_N)) \\
&\quad + | \lambda_N - \lambda_{k_*-1} | c_k^2(t_N) \\
&\leq \left(1 + \frac{\lambda_{k_*-1} - \lambda_N}{| \lambda_k - \lambda_N |} \right) r_{k0}^2 \\
&\leq \left(1 + \frac{\lambda_{k_*-1} - \lambda_N}{\lambda_N - \lambda_{N+1}} \right) r_{k0}^2, \quad k > N.
\end{aligned}$$

Furthermore, combining this with (7.47), we obtain:

$$
| \lambda_k - \lambda_{k_*} | \, c_k^2(T) + \dot{c}_k^2(T) \leq | \lambda_k - \lambda_{k_*-1} | \, c_k^2(t_{N+1}^*) + \dot{c}_k^2(t_{N+1}^*)
$$

$$
\leq \left(1 + \frac{\lambda_{k_*-1} - \lambda_N}{\lambda_N - \lambda_{N+1}} \right) r_{k0}^2, \quad k > N. \tag{7.48}
$$

Now, due to (7.7), (7.14), (7.42), (7.45), (7.48), and (7.27), for some constants $C_1, C_2 > 0$ we have:

$$
\| (y_d, 0) - (y(T, \cdot), y_t(T, \cdot) \|_{H_0^1(0,1) \times L^2(0,1)}^2
$$

$$
\leq C_1 \| y_d - y(T, \cdot) \|_{\lambda_0}^2 + \| y_t(T, \cdot) \|_{L^2(0,1)}^2
$$

$$
= v_{k_*1}^2 + \sum_{k=1, k \neq k_*}^{\infty} \left(C_1 \, | \lambda_k - \lambda_0 | \, c_k^2(T) + \dot{c}_k^2(T) \right)
$$

$$
= v_{k_*1}^2 + \sum_{k=1, k \neq k_*}^{N} \left(C_1 \, | \lambda_k - \lambda_0 | \, c_k^2(T) + \dot{c}_k^2(T) \right)
$$

$$
+ \sum_{k=N+1}^{\infty} \left(C_1 \, | \lambda_k - \lambda_0 | \, c_k^2(T) + \dot{c}_k^2(T) \right)
$$

$$
\leq v_{k_*1}^2 + \sum_{k=1, k \neq k_*}^{N} \left(C_1 \, | \lambda_k - \lambda_0 | \, c_k^2(T) + \dot{c}_k^2(T) \right)
$$

$$
+ C_2 \sum_{k=N+1}^{\infty} \left(| \lambda_k - \lambda_{k_*} | \, c_k^2(T) + \dot{c}_k^2(T) \right)
$$

$$
\leq v_{k_*1}^2 + \sum_{k=1, k \neq k_*}^{N} \left(C_1 \, | \lambda_k - \lambda_0 | \, c_k^2(T) + \dot{c}_k^2(T) \right)
$$

$$
+ C_2 \sum_{k=N+1}^{\infty} \frac{1}{\lambda_0 - \lambda_k} \left(1 + \frac{\lambda_{k_*-1} - \lambda_N}{\lambda_N - \lambda_{N+1}} \right)
$$

$$
\times \left(| \lambda_k - \lambda_0 |^2 \, c_k^2(0) + | \lambda_k - \lambda_0 | \, \dot{c}_k^2(0) \right). \tag{7.49}
$$

To establish (7.3), we need to show that all the terms on the right in (7.49) can be made arbitrarily small by

(A) selecting suitable N in (7.18) - to make the last term in the right-hand side of (7.49) "small," and then

(B) (given N) selecting suitable $\mu_{k0}, \mu_{k1}, k = 1, \ldots, N, k \neq k_*$ in (7.39) to make the second term in the right-hand side of (7.49) "small," and

(C) selecting a "sufficiently small" v_{k_*1} in (7.43).

First of all, recall that

$$
\lim_{k \to \infty} \frac{-\lambda_k}{k^2} = \text{a positive constant.}
$$

This means, in particular, that there exists a sequence $\{N_i\}_{i=1}^{\infty}$ such that

$$\frac{1}{\lambda_0 - \lambda_{N_i}} \leq 1, \quad \frac{1}{\lambda_{N_i} - \lambda_{N_i+1}} \leq 1, \quad i = 1, \ldots$$

Therefore, we have

$$\frac{1}{\lambda_0 - \lambda_k}\left(1 + \frac{\lambda_{k_*-1} - \lambda_{N_i}}{\lambda_{N_i} - \lambda_{N_i+1}}\right) \leq \frac{1}{\lambda_0 - \lambda_{N_i}}(1 + \lambda_{k_*-1} - \lambda_{N_i}) \leq 2 \qquad (7.50)$$

for all i's and $k > N_i > k_*$.
Combining (7.50) with (7.49) and (7.18) yields:

$$\| (y_d, 0) - (y(T, \cdot), y_t(T, \cdot)) \|_{H_0^1(0,1) \times L^2(0,1)}^2$$

$$\leq v_{k_*1}^2 + \sum_{k=1, k \neq k_*}^{N_i} \left(C_1 \mid \lambda_k - \lambda_0 \mid c_k^2(T) + \dot{c}_k^2(T)\right) + 2C_2\gamma^2(N_i), \quad i > i_*, \ (7.51)$$

which, since v_{k_*1} can be selected arbitrarily small, in view of (7.18) and Proposition 7.3, yields (7.3) in the case of Option #1.

Step 10. Option #2: $\| y_d \|_{L^2 0,1} > A$. The difference between this case and that in Option #1 is that now we have to increase the value of $c_{k_*}(t)$ to make it equal to $\| y_d \|_{L^2(0,1)}$ at some moment T, while, again, making the value of $\dot{c}_{k_*}(T)$ "small". Accordingly, the argument of Option #1 applies in the case of Option #2 as well, with the only difference that in (7.43)–(7.45) we select the moment t_{N+1}^* such that

$$c_{k_*}(t_{N+1}^*) = v_{k_*0} \in (A, \| y_d \|_{L^2 0,1}), \ \dot{c}_{k_*}(t_{N+1}^*)) = v_{k_*1} > 0.$$

(Again, $v_{k_*1} > 0$ can be made arbitrarily small.)
    Then, at time t_{N+1}^* we "switch" according to (7.19)–(7.22) to the control $h_{N+1} = -\lambda_{k_*}$. Making use of the solution formula (7.9), at time

$$T = t_{N+1}^* + \frac{\| y_d \|_{L^2 0,1} - v_{k_*0}}{v_{k_*1}}$$

we will have $c_{k_*}(T) = \| y_d \|_{L^2 0,1}, \ \dot{c}_1(t_1)) = v_{k_*1}$. This completes the proof of Theorem 7.1.   □

Chapter 8
Reachability of Nonnegative Equilibrium States for the Semilinear Vibrating String

Abstract We show that the set of nonnegative equilibrium-like states, namely, like $(y_d, 0)$ of the *semilinear* vibrating string that can be reached from any nonzero initial state $(y_0, y_1) \in H_0^1(0,1) \times L^2(0,1)$, by varying its axial load and the gain of damping, is dense in the *"nonnegative"* part of the subspace $L^2(0,1) \times \{0\}$ of $L^2(0,1) \times H^{-1}(0,1)$. Our main results deal with nonlinear terms which admit at most the linear growth at infinity in y and satisfy certain restriction on their total impact on $(0, \infty)$ with respect to the time-variable.

8.1 Introduction

8.1.1 Problem Setting

In this chapter we consider the following initial and boundary-value problem for the *semilinear* one dimensional wave equation:

$$y_{tt} = y_{xx} + v(x,t)y - \gamma(t)y_t - F(x,t,y), \quad x \in (0,1), \ t > 0, \tag{8.1}$$
$$y(0,t) = y(1,t) = 0, \quad y(x,0) = y_0(x), \ y_t(x,0) = y_1(x).$$

In (8.1) the functions $v \in L^\infty((0,1) \times (0,\infty))$ and $\gamma \in L^\infty(0,\infty)$ are regarded again as *bilinear* or *multiplicative controls* (which we "can choose"). The nonlinear term F is assumed to be fixed but not necessarily known. We regard it as a "disturbance" from a certain class of functions whose properties are described below. Our *goal* is to investigate the reachability (or controllability) properties of model (8.1). Namely:

Given any nonzero initial state $(y_0, y_1) \in H_0^1(0,1) \times L^2(0,1)$, we would like to know what states $(y(\cdot,t), y_t(\cdot,t))$ can be reached by system (8.1) at times $t > 0$ by applying various aforementioned bilinear controls v and γ.

Remark 8.1. If $F(x,t,0) = 0$, then the zero state $(y_0, y_1) = (0,0)$ is a fixed point for the solution mapping of system (8.1), regardless of the choice of controls v and γ. Hence, it cannot be steered anywhere from this state.

A.Y. Khapalov, *Controllability of Partial Differential Equations Governed by Multiplicative Controls*, Lecture Notes in Mathematics 1995, DOI 10.1007/978-3-642-12413-6_8, © Springer-Verlag Berlin Heidelberg 2010

8.1.2   Main Result

As before in Chapter 6 and 7, our main results deal only with the steering properties of the semilinear system (8.1) to "non-negative equilibrium states" like

$$(y_d, 0), \quad y_d(x) \geq 0 \text{ a.e. in } (0,1) \tag{8.2}$$

and with the following classes functions v, γ, and F in (8.1).

Structural and regularity assumptions on v, γ, and F:

$$v(x,t) = v_*(x) + h(t), \quad v_* \in L^\infty(0,1), \quad v_*(x) \leq \mu < 0 \text{ a.e. in } (0,1); \tag{8.3}$$

$$h(t) = \begin{cases} h_1 & \text{for } t \in (0,t_1), \\ h_2 & \text{for } t > t_1, \end{cases} \tag{8.4}$$

where $\mu < 0, h_1 < 0, t_1 > 0$, and $h_2 > 0$ are some real numbers;

$$\gamma(t) = \begin{cases} 0 & \text{for } t \in (0,t_1), \\ \gamma & \text{for } t > t_1, \end{cases} \tag{8.5}$$

where γ is some *positive* number.

We assume that $F(x,t,u)$ is measurable in $x,t \in (0,1) \times (0,T)$ for all $T > 0$, continuous in $u \in R$ and

$$|F(x,t,u)| \leq G(t)(1+|u|) \text{ for a.a. } x,t \in (0,1) \times (0,\infty), \forall u \in R, \tag{8.6}$$

where $G \in L^\infty(0,\infty) \bigcap L^1(0,\infty)$ is some positive-valued function.

We use below (8.6) in a different form, namely, we assume that G can be represented as

$$G(t) = q(t)g(t), \quad g(t), q(t) > 0 \text{ for a.a. } t \in (0,\infty),$$

(for example, $q = g = G^{1/2}$) and

$$q,g \in L^\infty(0,\infty) \bigcap L^2(0,\infty).$$

In other words, we assume that F admits the following representation:

$$F(x,t,y(x,t)) = q(t)f(x,t,y(x,t)) \quad \left(f(x,t,y(x,t)) = \frac{F(x,t,y(x,t))}{q(t)} \right),$$

where

$$|f(x,t,u)| \leq g(t)(1+|u|) \text{ for a.a. } x,t \in (0,1) \times (0,\infty), \forall u \in R. \tag{8.7}$$

Denote:

$$\| g \|_{L^\infty(0,\infty)} = K, \ \| g \|^2_{L^2(0,\infty)} = L, \ \| q \|_{L^\infty(0,\infty)} = B, \ \| q \|^2_{L^2(0,\infty)} = D. \tag{8.8}$$

Conditions (8.3)–(8.8) ensure that system (8.1) has a unique solution in the space $C([0,T];H_0^1(0,1)) \cap C^1([0,T];L^2(0,1))$ for every pair $(y_0,y_1) \in H_0^1(0,1) \times L^2(0,1)$ and every $T > 0$.

Remark 8.2. As before, denote by λ_k and ω_k, $k = 1,\dots$ respectively the eigenvalues and the orthonormal in $L^2(0,1)$ eigenfunctions associated with the spectral problem

$$\omega_{xx} + v_*(x)\omega = \lambda\omega, \quad \omega \in H_0^1(0,1). \tag{8.9}$$

Then (8.3) implies that

$$0 > \lambda_1 > \lambda_2 > \dots . \tag{8.10}$$

Our main result is as follows.

Theorem 8.1. *Let (8.3)–(8.8) hold. Consider an arbitrary nonzero initial state $(y_0,y_1) \in H_0^1(0,1) \times L^2(0,1)$, $(y_0,y_1) \neq (0,0)$, and any non-negative target equilibrium-like state $(y_d,0)$, $y_d(x) \geq 0$, $y_d \in L^2(0,1)$. Then for every $\varepsilon > 0$ there are a time $T = T(\varepsilon,y_0,y_1,y_d,L,K,B,D) > 0$ and piecewise constant-in-time controls (v,γ) of the form (8.3)–(8.5) such that*

$$\| y_d - y(\cdot,T) \|_{L^2(0,1)} + \| y_t(\cdot,T) \|_{H^{-1}(0,1)} \leq \varepsilon. \tag{8.11}$$

(In the above $H^{-1}(0,1) = (H_0^1(0,1))'$.)

As the reader noticed, in Theorem 8.1 we require the initial datum in (8.1) be of a higher regularity compared to that of the target state. This "discrepancy" arises within the framework of our method due to the presence of nonlinear term F in (8.1).

Our approach to prove Theorem 8.1 is to try *first* to find bilinear controls v and γ of the form (8.3)–(8.5), which steer the homogeneous truncated version of (8.1), namely,

$$u_{tt} = u_{xx} + v(x,t)u - \gamma(t)u_t, \quad x \in (0,1), \ t > 0, \tag{8.12}$$

$$u(0,t) = u(1,t) = 0, \quad u(x,0) = y_0(x), \ u_t(x,0) = y_1(x),$$

from the same initial state (y_0,y_1) toward the same target $(y_d,0)$. While doing this, we intend to select the aforementioned pair (v,γ) in such a way that they would generate a trajectory of the original system (8.1) that follows "closely" to that of (8.12). We will use the following auxiliary result.

Theorem 8.2. *Consider an arbitrary nonzero initial state* $(y_0, y_1) \in H_0^1(0, 1) \times L^2(0, 1)$ *and any* <u>non-negative</u> *target equilibrium-like state* $(y_d, 0)$, $y_d(x) \geq 0$, $y_d \in L^2(0, 1)$. *Then for every* $\varepsilon > 0$ *there are a time* $T = T(\varepsilon, y_0, y_1, y_d) > 0$ *and piecewise constant-in-time controls* (v, γ) *of the form (8.3)–(8.5) such that*

$$\| y_d - u(\cdot, T) \|_{L^2(0,1)} + \| u_t(\cdot, T) \|_{H^{-1}(0,1)} \leq \varepsilon. \tag{8.13}$$

As the reader may recall, in Chapter 6 we already established the controllability result of Theorem 8.2 in the stronger energy space $H_0^1(0, 1) \times L^2(0, 1)$, which yields (8.13), even for the case of not-necessarily non-negative equilibrium target states. (In Chapter 7 we obtained a similar result by making use of a *single*, more complex, multiplicative control v for almost all initial data from $H_0^1(0, 1) \times L^2(0, 1)$.) However, the methods of Chapters 6 and 7 do not directly apply to the semilinear model (8.1). In particular, in the aforementioned sections the corresponding controllability results were achieved by applying "relatively small" (i.e., "*more practical*") multiplicative controls for "relatively long" periods of time.

In the present Chapters 8 such approach will not work due to the presence of the nonlinear term, because the latter would "corrupt" the dynamics of the system at hand over a "long" period of time. Therefore, making use of the qualitative ideas of Chapter 6, we will now try to achieve the desirable controllability results by applying "relatively large" controls either for "very short" or for suitably selected "large" periods of time, which will allow us to overcome the presence of the *additive* nonlinear term. However, this will result in the change to a weaker energy space $L^2(0, 1) \times H^{-1}(0, 1)$ in (8.11).

Below we give a quite different proof of Theorem 8.2, compared to that of Chapter 6, specifically tailored as an auxiliary step to prove Theorem 8.1 (to ensure, e.g., (8.65), (8.66) in the proof below).

8.1.3 Some Useful Observations and Notations

Assuming that (8.3) holds, introduce the following norm equivalent to the standard norm one in $H_0^1(0, 1)$:

$$\| \phi \|_{v_*} = \left(\sum_{k=1}^{\infty} (-\lambda_k) \left(\int_0^1 \phi \omega_k dx \right)^2 \right)^{1/2} = \left(\int_0^1 (\phi_x^2 - v_*(x)\phi^2) dx \right)^{1/2}.$$

Accordingly, we will use the following norm in $H^{-1}(0, 1)$ dual of $H_0^1(0, 1)$:

$$\| \phi \|_{H^{-1}(0,1)} = \left(\sum_{k=1}^{\infty} (-\lambda_k)^{-1} \left(\int_0^1 \phi \omega_k dx \right)^2 \right)^{1/2}.$$

We introduce next the following energy norm:

$$E^{1/2}(\phi(\cdot,t)) = \left(\frac{1}{2} \| \phi \|_{v_*}^2 + \frac{1}{2} \| \phi \|_{L^2(0,1)}^2\right)^{1/2}.$$

Note that (8.11) deals with a weaker energy norm, namely, as follows:

$$E_*^{1/2}(\phi(\cdot,t)) = \left(\frac{1}{2} \| \phi \|_{L^2(0,1)}^2 + \frac{1}{2} \| \phi \|_{H^{-1}(0,1)}^2\right)^{1/2}.$$

8.2   Solution Formulas for (8.1)

Consider any $T > 0$, $(y_0, y_1) \in H_0^1(0,1) \times L^2(0,1)$ and any v, γ, q, and f satisfying conditions (8.3)–(8.8).
    Denote

$$p(x,t) = F(x,t,y(x,t)) = q(t)f(x,t,y(x,t)). \tag{8.14}$$

Note that, due to (8.6)–(8.8), we have $p \in L^2(Q_t)$ and for any $t > 0$:

$$\int\int_{Q_t} p^2 dxdt \le B^2 \int\int_{Q_t} f^2 dxdt \le 2LB^2 \left(1 + \max_{\tau \in [0,t]} \left\{\int_0^1 y^2(x,\tau)dx\right\}\right), \quad (8.15)$$

$$\int\int_{Q_t} p^2 dxdt \le B^2 \int\int_{Q_t} f^2 dxdt \le 2K^2 B^2 t \left(1 + \max_{\tau \in [0,t]} \left\{\int_0^1 y^2(x,\tau)dx\right\}\right), (8.16)$$

$$\sup_{\tau \in (0,t)} \int_0^1 p^2 dx = \sup_{\tau \in (0,t)} \int_0^1 q^2(\tau)f^2(x,\tau,y(x,\tau))dx$$

$$\le 2K^2 B^2 \left(1 + \max_{\tau \in [0,t]} \left\{\int_0^1 y^2(x,\tau)dx\right\}\right). \tag{8.17}$$

Consider any $v(x) = v_*(x) + h$, where v_* is as in (8.3) and h is any real number. Then the eigenvalues and the orthonormalized (in $L^2(0,1)$) eigenfunctions, associated with the spectral problem

$$\omega_{xx} + v(x)\omega = \lambda\omega, \quad \omega \in H_0^1(0,1),$$

are respectively $\lambda_1 + h > \lambda_2 + h > \dots$ and $\omega_1, \omega_2, \dots$ (see (8.9)–(8.10)).
    We have the following classical formula for solutions to the boundary problem (8.1) with this $v(x)$:

$$y(x,t) = \sum_{k=1}^{\infty} c_k(t)\omega_k(x), \quad x \in (0,1), \ t > 0, \tag{8.18}$$

where

$$\frac{d^2 c_k}{dt^2} = (\lambda_k + h)c_k - \gamma\frac{dc_k}{dt} + \int_0^1 (-p(x,t))\omega_k(x)dx, \quad t > 0,$$

$$c_k(0) = \int_0^1 y_0 \omega_k dx, \quad \dot{c}_k(0) = \int_0^1 y_1 \omega_k dx. \tag{8.19}$$

Solving the system of ordinary differential equations (8.18)–(8.19) yields:

For the index k such that $\gamma^2 + 4(\lambda_k + h) = 0$:

$$c_k(t) = e^{-\frac{\gamma}{2}t} c_k(0) + t\left(\dot{c}_k(0) + \frac{c_k(0)\gamma}{2}\right)e^{-\frac{\gamma}{2}t}$$

$$+ \int_0^t e^{\gamma(t-\tau)/2}(t-\tau)\int_0^1 (-p(x,\tau))\omega_k(x)dx. \tag{8.20}$$

For all k's such that $\gamma^2 + 4(\lambda_k + h) > 0$:

$$c_k(t) = e^{(-\frac{\gamma}{2}+\sqrt{\gamma^2/4+\lambda_k+h})t} \frac{\dot{c}_k(0) - c_k(0)(-\frac{\gamma}{2} - \sqrt{\gamma^2/4+\lambda_k+h})}{\sqrt{\gamma^2+4(\lambda_k+h)}}$$

$$+ e^{(-\frac{\gamma}{2}-\sqrt{\gamma^2/4+\lambda_k+h})t} \frac{-\dot{c}_k(0) + c_k(0)(-\frac{\gamma}{2} + \sqrt{\gamma^2/4+\lambda_k+h})}{\sqrt{\gamma^2+4(\lambda_k+h)}}$$

$$+ \int_0^t [e^{(-\frac{\gamma}{2}+\sqrt{\gamma^2/4+\lambda_k+h})(t-\tau)} \frac{1}{\sqrt{\gamma^2+4(\lambda_k+h)}}$$

$$- e^{(-\frac{\gamma}{2}-\sqrt{\gamma^2/4+\lambda_k+h})(t-\tau)} \frac{1}{\sqrt{\gamma^2+4(\lambda_k+h)}}]$$

$$\times \int_0^1 (-p(x,\tau))\omega_k(x)dxd\tau; \tag{8.21}$$

$$\dot{c}_k(t) = e^{(-\frac{\gamma}{2}+\sqrt{\gamma^2/4+\lambda_k+h})t}\left(-\frac{\gamma}{2} + \sqrt{\gamma^2/4+\lambda_k+h}\right)$$

$$\times \frac{\dot{c}_k(0) - c_k(0)(-\frac{\gamma}{2} - \sqrt{\gamma^2/4+\lambda_k+h})}{\sqrt{\gamma^2+4(\lambda_k+h)}}$$

$$+ e^{(-\frac{\gamma}{2}-\sqrt{\gamma^2/4+\lambda_k+h})t}\left(-\frac{\gamma}{2} - \sqrt{\gamma^2/4+\lambda_k+h}\right)$$

$$\times \frac{-\dot{c}_k(0) + c_k(0)(-\frac{\gamma}{2} + \sqrt{\gamma^2/4+\lambda_k+h})}{\sqrt{\gamma^2+4(\lambda_k+h)}}$$

$$+ \int_0^t [((-\frac{\gamma}{2} + \sqrt{\gamma^2/4 + \lambda_k + h})e^{(-\frac{\gamma}{2} + \sqrt{\gamma^2/4 + \lambda_k + h})(t-\tau)} \frac{1}{\sqrt{\gamma^2 + 4(\lambda_k + h)}}$$

$$+ (\frac{\gamma}{2} + \sqrt{\gamma^2/4 + \lambda_k + h})$$

$$\times e^{(-\frac{\gamma}{2} - \sqrt{\gamma^2/4 + \lambda_k + h})(t-\tau)} \frac{1}{\sqrt{\gamma^2 + 4(\lambda_k + h)}}]$$

$$\times \int_0^1 (-p(x,\tau))\omega_k(x)dxd\tau. \tag{8.22}$$

For all k's such that $\gamma^2 + 4(\lambda_k + h) < 0$:

$$c_k(t) = c_k(0)e^{-\frac{\gamma}{2}t} \cos\left(t\sqrt{-\gamma^2/4 - \lambda_k - h}\right)$$

$$+ e^{-\frac{\gamma}{2}t} \sin\left(t\sqrt{-\gamma^2/4 - \lambda_k - h}\right) \frac{\dot{c}_k(0) + \gamma c_k(0)/2}{\sqrt{-\gamma^2/4 - \lambda_k - h}}$$

$$+ \int_0^t \left[e^{-\frac{\gamma(t-\tau)}{2}} \frac{\sin\left((t-\tau)\sqrt{-\gamma^2/4 - \lambda_k - h}\right)}{\sqrt{-\gamma^2/4 - \lambda_k - h}} \right]$$

$$\times \int_0^1 (-p(x,\tau))\omega_k(x)dx; \tag{8.23}$$

$$\dot{c}_k(t) = -c_k(0)e^{-\frac{\gamma}{2}t}\sqrt{-\gamma^2/4 - \lambda_k - h}\,\sin\left(t\sqrt{-\gamma^2/4 - \lambda_k - h}\right)$$

$$+ \frac{-\gamma}{2}e^{-\frac{\gamma}{2}t}\sin\left(t\sqrt{-\gamma^2/4 - \lambda_k - h}\right)\frac{\dot{c}_k(0) + \gamma c_k(0)/2}{\sqrt{-\gamma^2/4 - \lambda_k - h}}$$

$$+ e^{-\frac{\gamma}{2}t}\dot{c}_k(0)\cos\left(t\sqrt{-\gamma^2/4 - \lambda_k - h}\right)$$

$$+ \int_0^t [-\frac{\gamma}{2}e^{-\frac{\gamma(t-\tau)}{2}} \frac{\sin\left((t-\tau)\sqrt{-\gamma^2/4 - \lambda_k - h}\right)}{\sqrt{-\gamma^2/4 - \lambda_k - h}}$$

$$+ e^{-\frac{\gamma(t-\tau)}{2}} \cos\left((t-\tau)\sqrt{-\gamma^2/4 - \lambda_k - h}\right)]$$

$$\times \int_0^1 (-p(x,\tau))\omega_k(x)dxd\tau, \tag{8.24}$$

where $p(x,t) = F(x,t,y(x,t))$.

The series (8.18)–(8.24) converges in $C([0,T]; H_0^1(0,1)) \cap C^1([0,T]; L^2(0,1))$ for every $T > 0$.

8.3 Proof of Theorem 8.2: Steering to Nonnegative States Revisited

Step 1. The set of target states. Note that, to prove Theorem 8.2, it is sufficient to do it for any set of non-negative target state y_d which is dense in the non-negative part of the space $H_0^1(0,1)$. To this end, we will further consider only:

(A) non-negative continuously differentiable functions $y_d = y_d(x), x \in [0,1]$ that

(B) vanish at $x = 0, x = 1$, i.e., $y_d(0) = y_d(1) = 0$, and

(C) whose second derivatives are piecewise continuous with finitely many discontinuities of the first kind (hence, $y_d \in H^2(0,1) \cap H_0^1(0,1) \subset C^1[0,1]$) and are such that

(D)

$$y_d(x) > 0 \text{ in } (0,1) \quad \text{and} \quad \frac{y_{dxx}}{y_d} \in L^\infty(0,1). \tag{8.25}$$

Further control strategy. We intend to apply the following *strategy to prove Theorem 8.2*:

- First of all, given the target state y_d (selected arbitrarily within the set outlined in Step 1), in Step 2 we intend to choose v_* as in (8.3) such that the first eigenfunction ω_1, associated with the spectral problem (8.9)–(8.10), will be co-linear with y_d and the first eigenvalue λ_1 will be negative.
- Then, in Step 3, we intend to apply control $v_* + h_1$ of the form (8.3)–(8.4) with *any* $h_1 < 0$ and to show that for some $t_1 = t_1(h_1)$ we can have

$$u_1(t_1) \in (0, \| y_d \|_{L^2(0,1)}),$$

where $u_1(t)\omega_1$ stands for the first term in the solution expansion of the form (8.18) for system (8.12). The key observation here, allowing us to achieve this goal, is the fact that the solution formulas (8.23)–(8.24) with $F = 0$ describe the *ellipsoidal motion* of the point $(u_1(t), \dot{u}_1(t))$ in the associated two dimensional space when the aforementioned constant controls are used.
- After that, given the selection of v_*, h_1, and t_1, we will simply evaluate the difference between $(y_d, 0)$ and $(u(t), u_t(t))$ in the space $L^2(0,1) \times H^{-1}(0,1)$ for $t > t_1$. Our goal here is,
  (a) making use of a *specific choice of the value for h_2* as given in (8.3)–(8.5),
  (b) to apply the effect of *carefully designed damping gain γ* to ensure that
  (c) the above difference can be made as small as we wish at some carefully selected moments of time.
  The choice of the above parameters is rather technical and is tailored specifically to do it so that the proof of Theorem 8.2 can later be extended to prove

Theorem 8.1. Namely, we need to make sure that the corresponding terms in the expansion (8.18)–(8.24) containing F can be suppressed (see, e.g., the key estimates (8.65) and (8.66)).

Step 2. Selection of $v_*(x)$. Consider any initial state $(y_0, y_1) \in H_0^1(0,1) \times L^2(0,1)$, $(y_0, y_1) \neq (0,0)$ and any <u>non-negative</u> target equilibrium state $(y_d, 0)$, $y_d(x) \geq 0, y_d \in H_0^1(0,1)$, satisfying the aforementioned conditions (A)–(D).

Select

$$v_*(x) = -\frac{y_{dxx}(x)}{y_d(x)} - 2 \left\| \frac{y_{dxx}}{y_d} \right\|_{L^\infty(0,1)}, \quad x \in (0,1). \tag{8.26}$$

(Note that the inequality in (8.3) is satisfied for our choice.)

Since $y_d(x) > 0$ in $(0,1)$ (see (8.25)), condition (8.26) implies that

$$\frac{y_d}{\| y_d \|_{L^2(0,1)}} = \omega_1 \tag{8.27}$$

(see also, e.g., Chapter 2 for more details) and that

$$\lambda_1 = \int_0^1 (-\omega_{1x}^2 + v_* \omega_1^2) dx = -2 \left\| \frac{y_{dxx}}{y_d} \right\|_{L^\infty(0,1)}, < 0. \tag{8.28}$$

Remark 8.3. Without loss of generality, we further assume that

$$\left(\int_0^1 y_d y_0 dx \right)^2 + \left(\int_0^1 y_d y_1 dx \right)^2 \neq 0.$$

(Indeed, otherwise, we just replace y_d with an alternative "arbitrarily close" to it target function for which this inequality holds.)

Step 3. Selection of t_1 for any $h(t) = h_1 < 0$ in (8.4). Given the initial datum (y_0, y_1), our goal is to select a control pair $(h_1 < 0, \gamma = 0)$ in such way that at some time $t = t_1$ the solution expansion for system (8.12) will look like (see (8.18)):

$$u(x, t_1) = u_1(t_1) \omega_1(x)$$
$$+ \sum_{k=2}^\infty u_k(t_1) \omega_k(x), \text{ where } u_1(t_1) \in (0, \| y_d \|_{L^2(0,1)}), \tag{8.29}$$

$$u_t(x, t_1) = \sum_{k=1}^\infty \dot{u}_k(t_1) \omega_k(x). \tag{8.30}$$

(Recall that $\| y_d \|_{L^2(0,1)} \omega_1(x) = y_d$.)

Consider any $h_1 < 0$. Then the eigenvalues of the corresponding spectral problem $\omega_{xx} + v\omega = \lambda\omega$, associated with $v(x) = v_*(x) + h_1$, are as follows:

$$0 > \lambda_1 + h_1 > \lambda_2 + h_1 > \dots,$$

while the respective eigenfunctions ω_k's will remain the same as in the case when $v = v_*$ (see (8.9)–(8.10), (8.27), (8.28)).

Then (8.23)–(8.24), applied to the solution of (8.12) with $v(x) = v_*(x) + h_1$ and $\gamma = 0$, yield that the point $(u_1(t), \dot{u}_1(t))$ moves with period $2\pi/(-h_1 - \lambda_1)^{1/2}$ in the associated two dimensional space around the origin along the ellipse

$$|\lambda_1 + h_1| \, u_1^2(t) + \dot{u}_1^2(t) = |\lambda_1 + h_1| \, u_1^2(0) + \dot{u}_1^2(0). \tag{8.31}$$

Hence, we can find a t_1 such that $u_1(t_1) \in (0, \| \, y_d \, \|_{L^2(0,1)})$ as in (8.29), *within one rotation around the origin*, that is,

$$t_1 \in [0, \frac{2\pi}{\sqrt{-h_1 - \lambda_1}}]. \tag{8.32}$$

Step 4. Steering for $t > t_1$. In this step we intend to steer the system (8.12) from the state (8.29)–(8.30), (8.32) "arbitrarily close" in $L^2(0,1) \times H^{-1}(0,1)$ to our desirable target state

$$(y_d(x), 0), \tag{8.33}$$

by making use of the following pair of controls of the form (8.3)–(8.4), (8.5):

$$h_2 = -\lambda_1 + r, \quad \gamma > 0, \quad \text{where } r \in (0, 0.5\min\{1, \lambda_1 - \lambda_2\}) \tag{8.34}$$

and γ is "large enough." In other words, we intend to apply damping to all the terms in the expansion (8.29)–(8.30), except for the first term in (8.29), which we intend to steer to our target y_d. Our argument is split below into several steps.

3.4.1. Selection of $\gamma > 0$. We will now select a special sequence $\{\gamma_i\}_{i=1}^\infty, \gamma_i > 0$ for the values of damping γ.

Recall that

$$\lim_{k \to \infty} \frac{-\lambda_k}{k^2} = \text{a positive constant.}$$

This means, in particular, that there exists a sequence $\{k_i\}_{i=1}^\infty$ such that

$$\lambda_{k_i} - \lambda_{k_i+1} \to \infty \quad \text{as } i \to \infty. \tag{8.35}$$

This allows us to select $\gamma_i^2 \in (-4(\lambda_{k_i} - \lambda_1), -4(\lambda_{k_i+1} - \lambda_1))$ such that

$$|\gamma_i^2 + 4(\lambda_{k_i} - \lambda_1)|, |\gamma_i^2 + 4(\lambda_{k_i+1} - \lambda_1)| \to \infty \tag{8.36}$$

as $\gamma_i \to \infty$ by choosing γ_i^2's as the "midpoints" of the aforementioned intervals:

$$\gamma_i = \sqrt{-4(\lambda_{k_i} - \lambda_1) + \frac{1}{2}\left(-4(\lambda_{k_i+1} - \lambda_1) - (-4(\lambda_{k_i} - \lambda_1))\right)}$$
$$= \sqrt{4\lambda_1 - 2(\lambda_{k_i+1} + \lambda_{k_i})}, \quad i = 1,\ldots \tag{8.37}$$

Denote

$$b(\gamma) = \inf_{k=1,\ldots; \, r \in (0, 0.5\min\{1, \lambda_1 - \lambda_2\})} |\gamma^2 + 4(\lambda_k - \lambda_1 + r)|. \tag{8.38}$$

Conditions (8.35)–(8.36) ensure that

$$b(\gamma_i) \to \infty \quad \text{as } i \to \infty. \tag{8.39}$$

Below we deal only with γ's from the sequence in (8.37), denoting them just by γ.

3.4.2. Evaluation of $\| u(\cdot,t) - y_d \|_{L^2(0,1)}^2$. We often use below the following formulas:

$$-\frac{\gamma}{2} + \sqrt{\gamma^2/4 + (\lambda_k - \lambda_1 + r)} = \frac{\lambda_k - \lambda_1 + r}{\gamma/2 + \sqrt{\gamma^2/4 + (\lambda_k - \lambda_1 + r)}}, \tag{8.40}$$
$$e^{(-\frac{\gamma}{2} + \sqrt{\gamma^2/4 + (\lambda_k - \lambda_1 + r)})(t-t_1)} \le e^{(\lambda_k - \lambda_1 + r)\frac{t-t_1}{\gamma}}, \tag{8.41}$$

for all $t > t_1$ whenever $\gamma^2/4 + (\lambda_k - \lambda_1 + r) > 0$ and $\lambda_k - \lambda_1 + r < 0$.

Denote $Q_2 = (0,1) \times (t_1, t)$ and let $N(\gamma, r)$ be the largest positive integer k such that $\gamma^2 + 4(\lambda_k - \lambda_1 + r) > 0$ (when this set is nonempty).

Then, the solution formulas (8.21), (8.23) (applied for $t \ge t_1$ with the initial datum (8.29)–(8.30) and the aforementioned pair of controls $(v = v_* + (-\lambda_1 + r), \gamma > 0)$ along with (8.6)–(8.8), (8.34) and (8.37) yield as $\gamma \to \infty$:

$$\| u(\cdot,t) - y_d \|_{L^2(0,1)}^2 = \| u(\cdot,t) - \| y_d \|_{L^2(0,1)} \omega_1 \|_{L^2(0,1)}^2$$

$$\le M \left(e^{0.5 + \sqrt{0.25 + r/\gamma^2} \frac{t-t_1}{\gamma}} \frac{u_1(t_1)}{2} \left(\frac{\gamma}{\sqrt{\gamma^2 + 4r}} + 1 \right) - \| y_d \|_{L^2(0,1)} \right)^2$$

$$+ M e^{0.5 + \sqrt{0.25 + r/\gamma^2} \frac{2r}{\gamma}} \frac{t-t_1}{\gamma} \frac{(\int_0^1 u_t(s,t_1)\omega_1(s)ds)^2}{\gamma^2 + 4r}$$

$$+ M e^{-\gamma(t-t_1)} \frac{r^2(\int_0^1 u(s,t_1)\omega_1(s)ds)^2}{(\gamma^2 + 4r)(\gamma/2 + \sqrt{\gamma^2/4 + r})^2}$$

$$+ M \sum_{k=2}^{N(\gamma,r)} e^{2(\lambda_k - \lambda_1 + r)\frac{t-t_1}{\gamma}} \left(\frac{\gamma^2}{\gamma^2 + 4(\lambda_k - \lambda_1 + r)} + 1 \right)$$

$$\times(\int_0^1 u(s,t_1)\omega_k(s)ds)^2$$

$$+ \sum_{k>N(\gamma,r)} e^{-\gamma(t-t_1)}(\frac{\gamma^2}{b(\gamma)}+1)(\int_0^1 u(s,t_1)\omega_k(s)ds)^2$$

$$+M\sum_{k=2}^\infty \frac{1}{b(\gamma)}(\int_0^1 u_t(s,t_1)\omega_k(s)ds)^2 \tag{8.42}$$

where $M>0$ is some constant.

Here we used (8.21) and (8.40)–(8.41) to evaluate the first $N(\gamma,r)$ terms in the expansion like in (8.29) for the solution to (8.12) for $t>t_1$ (i.e., when $\gamma^2+4(\lambda_k-\lambda_1+r)>0$), and (8.23) for the remaining terms (the case (8.20) is excluded due to (8.34), (8.36), (8.37), and (8.39)).

The difference in the appearance of the first three terms in (8.42), evaluating the first term in this expansion, reflects the fact that only for $k=1$ the expression $\lambda_k-\lambda_1+r$ is positive (due to (8.34)) and equal to r, while it is negative for $k\geq 2$.

We want to show that with γ as in (8.37) we can select $t>t_1$ and r in (8.34) which will make the right-hand side of (8.42) as small as we wish.

<u>Selection of the duration for control action.</u> From now on, given the sequence of values for γ as in (8.37), we will deal with the following specially selected *sequence of values for the duration of the control time-intervals* (t_1,t) (we will explain why later):

$$t=t_1+\frac{1}{C^*(1+KB)(B+(-\lambda_1+0.5)^2+BK)}\gamma\ln\gamma. \tag{8.43}$$

Then for any r as in (8.34) the *3-rd, 5-th, and the 6-th terms* in (8.42) converge to zero as $\gamma\to\infty$ along (8.37), (8.39), and (8.43) uniformly over r as in (8.34). For the 5-th term we also use the fact that

$$e^{-\gamma(t-t_1)}\gamma^2 \to 0 \text{ as } \gamma\to\infty$$

and $t-t_1\to\infty$ along (8.37) and (8.43).

Next *we select* $r=r(\gamma)$ to make *the 1-st term* on the right in (8.42) to be zero, namely, from the condition

$$e^{\frac{r}{0.5+\sqrt{0.25+r/\gamma^2}}\frac{t-t_1}{\gamma}}\frac{c_1(t_1)}{2}(\frac{\gamma}{\sqrt{\gamma^2+4r}}+1) = \gamma^{0.5+\sqrt{0.25+r/\gamma^2}}\frac{r}{C^*(1+KB)(B+(-\lambda_1+0.5)^2+BK)}\frac{c_1(t_1)}{2}$$

$$\times(\frac{1}{\sqrt{1+4r/\gamma^2}}+1) = \|y_d\|_{L^2(0,1)}. \tag{8.44}$$

Such $r(\gamma)$ exists for any $\gamma > 1$ (recall here (8.37)), because the left-hand side of (8.44) for $r = 0$ is equal to $c_1(t_1)$ lying in the interval $(0, \| y_d \|_{L^2(0,1)})$ (as in (8.29)) and it tends to ∞ as $r \to \infty$. Moreover, for any $r > 0$ the left-hand side of (8.44) tends to ∞ as $\gamma \to \infty$. Hence, this $r = r(\gamma)$ can be selected to satisfy (8.34), more precisely,

$$r = r(\gamma) \to 0 \text{ as } \gamma \to \infty \tag{8.45}$$

along (8.37) and t calculated as in (8.43).

We intend now to show that

$$\sup_{k=2,\ldots,N(\gamma,r)} e^{2(\lambda_k - \lambda_1 + r)\frac{t-t_1}{\gamma}} \frac{\gamma^2}{\gamma^2 + 4(\lambda_k - \lambda_1 + r)} \to 0, \tag{8.46}$$

$$\frac{1}{\gamma^2} e^{0.5 + \sqrt{0.25 + r/\gamma^2} \,\frac{2r}{\gamma}\,\frac{t-t_1}{\gamma}} \to 0 \tag{8.47}$$

under the conditions (8.34), (8.37), (8.43), and (8.44)–(8.45), which will allow us to make the *remaining 2-nd and 4-th terms* in (8.42) as small as we wish.

3.4.3. Derivation of (8.47). Note that for any $\beta \in (0,1)$ we have:

$$e^{0.5 + \sqrt{0.25 + r/\gamma^2} \,\frac{2r}{\gamma}\,\frac{t-t_1}{\gamma}} = \gamma^{0.5 + \sqrt{0.25 + r/\gamma^2} \,\frac{2r}{\gamma}\,\frac{1}{0.5 + C^*(1+K)(B+(-\lambda_1+0.5)^2 + KB)}} \leq \gamma^{\beta} \text{ as } \gamma \to \infty \tag{8.48}$$

for large γ's in (8.37), due to (8.43),(8.45), which yields (8.47).

3.4.4. Derivation of (8.46). To show that the right-hand side of (8.46) tend to zero along the conditions (8.34), (8.37), (8.43), and (8.44)–(8.45), note first that for $k = 2, \ldots, N(\gamma, r)$:

$$\gamma^2 + 4(\lambda_k - \lambda_1 + r) > 0,$$

and

$$\sup_{k=2,\ldots,N(\gamma,r)} e^{2(\lambda_k - \lambda_1 + r)\frac{t-t_1}{\gamma}} \frac{\gamma^2}{\gamma^2 + 4(\lambda_k - \lambda_1 + r)}$$

$$= \sup_{k=2,\ldots,N(\gamma,r)} \gamma^{\frac{2(\lambda_k - \lambda_1 + r)}{C^*(1+KB)(B+(-\lambda_1+0.5)^2 + BK)}} \frac{\gamma^2}{\gamma^2 + 4(\lambda_k - \lambda_1 + r)}.$$

Furthermore, for all the aforementioned k's such that

$$\frac{2(\lambda_k - \lambda_1 + r)}{C^*(1+KB)(B+(-\lambda_1+0.5)^2 + BK)} + 2 \leq 0,$$

we have:

$$\gamma^{\frac{2(\lambda_k-\lambda_1+r)}{C^*(1+KB)(B+(-\lambda_1+0.5)^2+BK)}}\frac{\gamma^2}{\gamma^2+4(\lambda_k-\lambda_1+r)} \le \frac{1}{b(\gamma)}. \tag{8.49}$$

For the remaining no more than finitely many k's, say, $k = 2,\ldots,k_* \le N(\gamma,r)$ such that

$$\frac{2(\lambda_k-\lambda_1+r)}{C^*(1+KB)(B+(-\lambda_1+0.5)^2+BK)}+2>0$$

(note that k_*, if such exists, is bounded above by a number which does not depend on r and γ), we have

$$\gamma^{\frac{2(\lambda_k-\lambda_1+r)}{C^*(1+KB)(B+(-\lambda_1+0.5)^2+BK)}}\frac{\gamma^2}{\gamma^2+4(\lambda_k-\lambda_1+r)}$$

$$\le \frac{\gamma^{l(r)}}{\gamma^2+4(\lambda_{k_*}-\lambda_1+r)} \to 0 \text{ as } \gamma \to \infty \ \forall k = 2,\ldots,k_*, \tag{8.50}$$

where

$$l(r) = \max_{k=2,\ldots,k_*}\left(\frac{2(\lambda_k-\lambda_1+r)}{C^*(1+KB)(B+(-\lambda_1+0.5)^2+BK)}+2\right)$$

$$< \left(\frac{2(\lambda_2-\lambda_1+0.5\min\{1,\lambda_1-\lambda_2\})}{C^*(1+KB)(B+(-\lambda_1+0.5)^2+BK)}+2\right) < 2,$$

because $\lambda_2 - \lambda_1 + 0.5\min\{1,\lambda_1-\lambda_2\} < 0$. Combining (8.49)–(8.50) yields (8.46).

This completes the proof that the right-hand side of (8.42) tends to zero along the relations (8.37), (8.43), (8.34), and (8.44)–(8.45).   □

3.4.5. Evaluation of $\| u_t(\cdot,t)) \|_{H^{-1}(0,1)}$. Analogously, using the formulas (8.22), (8.24), we obtain for $t \ge t_1$:

$$\| u_t(\cdot,t) \|^2_{H^{-1}(0,1)}$$

$$\le M\frac{1}{-\lambda_1}\frac{r^2}{\gamma^2+4r}e^{0.5+\sqrt{0.25+r/\gamma^2}\frac{2r}{\gamma}\frac{t-t_1}{\gamma}}(\int_0^1 u(s,t_1)\omega_1(s)ds)^2$$

$$+M\sum_{k=2}^{N(\gamma,r)}\frac{1}{b(\gamma)}\frac{|\lambda_k-\lambda_1+r|^2}{-\lambda_k}(\int_0^1 u(s,t_1)\omega_k(s)ds)^2$$

$$+M\frac{1}{-\lambda_1}\frac{1}{\gamma^2+4r}e^{0.5+\sqrt{0.25+r/\gamma^2}\frac{2r}{\gamma}\frac{t-t_1}{\gamma}}\left(\frac{r}{0.5\gamma+\sqrt{0.25\gamma+r}}\right)^2(\int_0^1 u_t(s,t_1)\omega_1(s)ds)^2$$

$$+M\sum_{k=2}^{N(\gamma,r)}\frac{1}{-\lambda_k}\frac{1}{b(\gamma)}\left(\frac{|\lambda_k-\lambda_1+r|}{0.5\gamma+\sqrt{0.25\gamma+\lambda_k-\lambda_1+r}}\right)^2\left(\int_0^1 u_t(s,t_1)\omega_k(s)ds\right)^2$$

$$+M\frac{1}{-\lambda_1}\left(e^{(-\frac{\gamma}{2}-\sqrt{\frac{\gamma^2}{4}+r})(t-t_1)}(\frac{\gamma}{2}+\sqrt{\frac{\gamma^2}{4}}+r)\right)^2\frac{1}{\gamma^2+4r}\left(\int_0^1 u_t(s,t_1)\omega_1(s)ds\right)^2$$

$$+M\sum_{k=2}^{N(\gamma,r)}\frac{1}{-\lambda_k}\frac{1}{b(\gamma)}\left(e^{(-\frac{\gamma}{2}-\sqrt{\frac{\gamma^2}{4}+\lambda_k-\lambda_1+r})(t-t_1)}(\frac{\gamma}{2}+\sqrt{\frac{\gamma^2}{4}+\lambda_k-\lambda_1+r})\right)^2$$

$$\times\left(\int_0^1 u_t(s,t_1)\omega_k(s)ds\right)^2$$

$$+M\sum_{k=N(\gamma,r)+1}^{\infty}\frac{1}{-\lambda_k}e^{-\gamma(t-t_1)}\left(\frac{\gamma^2}{4}+|\lambda_k-\lambda_1+r|+\frac{\gamma^4}{b(\gamma)}\right)\left(\int_0^1 u(s,t_1)\omega_k(s)ds\right)^2$$

$$+M\sum_{k=N(\gamma,r)+1}^{\infty}\frac{1}{-\lambda_k}e^{-\gamma(t-t_1)}\left(1+\frac{\gamma^2}{b(\gamma)}\right)\left(\int_0^1 u_t(s,t_2)\omega_1(s)ds\right)^2, \qquad (8.51)$$

where $M>0$ is some positive constant.

An argument similar to that we applied to evaluate (8.42) implies that the right-hand side in (8.51) tends to zero as $\gamma\to\infty$ under the conditions (8.37), (8.43), (8.34), and (8.44)–(8.45).

In particular, (8.47) can be used to evaluate the *1-st and the 3-rd terms*.

To evaluate the *2-nd and the 4-th terms* we can use (8.39) along with the following observations.

Recalling that $u(\cdot,t_1)\in H_0^1(0,1)$, we have:

$$\sum_{k=1}^{\infty}|\lambda_k|\left(\int_0^1 u(s,t_1)\omega_k(s)ds\right)^2<\infty,$$

and, hence, we can evaluate the 2-nd term as

$$M\sum_{k=2}^{N(\gamma,r)}\frac{1}{b(\gamma)}\frac{|\lambda_k-\lambda_1+r|^2}{-\lambda_k}\left(\int_0^1 u(s,t_1)\omega_k(s)ds\right)^2$$

$$\leq M\sum_{k=2}^{N(\gamma,r)}\frac{1}{b(\gamma)}\frac{|\lambda_k-\lambda_1+r|^2}{-\lambda_k^2}\left(-\lambda_k(\int_0^1 u(s,t_1)\omega_k(s)ds)^2\right).$$

To evaluate the *4-th term*, note that, uniformly over $k = 2, \ldots, N(\gamma, r)$,

$$\frac{1}{-\lambda_k} \frac{1}{b(\gamma)} \left(\frac{|\lambda_k - \lambda_1 + r|}{0.5\gamma + \sqrt{0.25\gamma^2 + \lambda_k - \lambda_1 + r}} \right)^2$$

$$\leq \frac{1}{b(\gamma)} \left(\frac{(-\lambda_k)^{1/2} + |-\lambda_1 + r| (-\lambda_k)^{-1/2}}{0.5\gamma} \right)^2$$

$$\leq \frac{1}{b(\gamma)} \left(\frac{(0.25\gamma^2 - \lambda_1 + r)^{1/2} + |-\lambda_1 + r| (-\lambda_k)^{-1/2}}{0.5\gamma} \right)^2 \to 0$$

as $\gamma \to \infty$, because

$$0 < -\lambda_k < 0.25\gamma^2 - \lambda_1 + r$$

for our $k = 2, \ldots, N(\gamma, r)$.

To evaluate the remaining *5-th, 6-th, 7-th, and 8-th terms*, we can use the fact that

$$e^{-\gamma(t - t_1)} \gamma^\kappa \to 0 \quad \text{as } \gamma \to \infty \text{ and } t - t_1 \to \infty$$

along (8.37) and (8.43) for any $\kappa \geq 0$ (and also the fact that $\lambda_k - \lambda_1 + r < 0$ for the 6-th term).

Therefore, there is a pair of controls (v, γ) and time $T > 0$ described in Theorem 8.2 such that (8.13) holds. This ends the proof of Theorem 8.2. $\qquad\Box$

8.4 Proof of Theorem 8.1: Controllability of the Semilinear Equation

Our plan is simply to show that we can further modify the control pair (v, γ) within the restrictions imposed on it in the proof of Theorem 8.2 in such a way that they would generate a trajectory of the original system (8.1) that follows "closely" to that of (8.12), and, thus, system (8.1) will arrive to the same desirable target $(y_d, 0)$ as (8.12) as well.

The key observation here is that the evaluation of the difference between (y, y_t) and (u, u_t) can be reduced to the evaluation of the terms in (8.18)–(8.24) containing F.

8.4.1 Some Preliminary Estimates

Lemma 8.1. *Let $h \geq 0$ and $\gamma > 0$ be given real numbers. Then the following estimate holds for system (8.1), (8.6)–(8.7), (8.8) and (8.3) with constant $h(t) = h$ for all $t > 0$:*

$$\int_0^1 y^2(x, t) dx \leq 2C^* E(y(\cdot, t)) \leq 2C^* (E(y(\cdot, 0)) + \frac{\rho}{2} LB) e^{\rho C^* (1 + KB)t} \qquad (8.52)$$

$\forall \rho \geq \frac{1}{2\gamma}(B + h^2 + KB)$, where C^* is a positive constant from Poincare's inequality (see (8.54) below) and K, L, and B are from (8.8).

Proof. Note first that, due to the inequality $ab \leq \rho a^2/2 + b^2/(2\rho)$ ($\forall \rho > 0$),

$$hyy_t \leq \frac{\rho}{2}y^2 + \frac{h^2}{2\rho}y_t^2,$$

and, due to (8.8),

$$\int\int_{Q_t}(1 + |y|)g|y_t|\,dxd\tau \leq \frac{\rho}{2}L + \int\int_{Q_t}(\frac{1}{2\rho}y_t^2 + K\frac{\rho}{2}y^2 + \frac{K}{2\rho}y_t^2)dxd\tau,$$

where $Q_t = (0, 1) \times (0, t)$.

Multiply the equation in (8.1) by y_t and integrate the resulting expression over Q_t. Then, after standard calculations and making use of the above inequalities, we obtain the following:

$$E(y(\cdot, t)) \leq E(y(\cdot, 0)) + h\int\int_{Q_t}yy_t\,dxd\tau - \gamma\int\int_{Q_t}y_t^2\,dxd\tau$$

$$+ \int\int_{Q_t}B(1 + |y|)g|y_t|\,dxd\tau$$

$$\leq E(y(\cdot, 0)) + \frac{\rho}{2}LB + \frac{\rho}{2}(1 + KB)\int\int_{Q_t}y^2\,dxd\tau$$

$$+ \left(\frac{1}{2\rho}(B + h^2 + KB) - \gamma\right)\int\int_{Q_t}y_t^2\,dxd\tau$$

$$\leq (E(y(\cdot, 0)) + \frac{\rho}{2}LB) + C^*\rho(1 + KB)\int_0^t E(y(\cdot, \tau))d\tau \quad \forall \gamma$$

$$\geq \frac{1}{2\rho}(B + h^2 + KB), \tag{8.53}$$

where C^* is the constant in Poincare's inequality:

$$\int_0^1 \phi^2(x)dx \leq C^*\int_0^1 \phi_x^2(x)dx \quad \forall \phi \in H_0^1(0, 1). \tag{8.54}$$

Applying Gronwall-Bellman inequality to (8.53) yields the 2-nd inequality in (8.52). To complete the proof of Lemma 8.1, note that the 1-st inequality in (8.52) is just an immediate consequence of (8.54).

Lemma 8.2. *Let $t > 0$ be given. Then the following estimate holds for system (8.1), (8.6)–(8.8) and (8.3) with any constant $h(t) = h < 0$ and $\gamma = 0$:*

$$\int_0^1 y^2(x,\tau)dx \le \frac{1}{-h}E(y(\cdot,0)) + \frac{LB}{-2h} + \int_0^1 y_0^2(x)dx$$

$$+B(1+K+C^*K)t\left(\frac{1}{-h}E(y(\cdot,0)) + \frac{LB}{-2h} + \int_0^1 y_0^2(x)dx\right)$$

$$\times e^{B(1+K+C^*K)t} \quad \forall \tau \in [0,t], \tag{8.55}$$

where C^* is from (8.54) and L, K, and B are from (8.8).

Proof. Indeed, similar to (8.53) (but with $\rho = 1$) for any $\tau \in [0,t]$:

$$E(y(\cdot,\tau)) - h\int_0^1 y^2(x,\tau)dx \le (E(y(\cdot,0)) + \frac{LB}{2}) - h\int_0^1 y_0^2(x)dx$$

$$+B(1+K+C^*K)\int_0^\tau E(y(\cdot,s))ds. \tag{8.56}$$

Applying Gronwall-Bellman inequality to (8.56) yields:

$$E(y(\cdot,\tau)) \le \left(E(y(\cdot,0)) + \frac{LB}{2} - h\int_0^1 y_0^2(x)dx\right)e^{B(1+K+C^*K)t} \quad \forall \tau \in [0,t] \tag{8.57}$$

Dividing (8.56) by $-h > 0$ and making use of (8.57) yields (8.55). This ends the proof of Lemma 8.53.   □

8.4.2 Proof of Theorem 8.1

Step 1. Steering on $(0,t_1)$ and selection of h_1. In the Step 3 in the proof of Theorem 8.2 we showed that, given y_d and the initial datum (y_0, y_1), for any control pair $(h_1, \gamma = 0)$, where h_1 is an arbitrary negative number, we can find the moment $t = t_1$ as in (8.32) such that (8.29)–(8.30) holds.

Noting now that the solution expansion (8.18) for (8.1) at time $t = t_1$ looks like

$$y(x,t_1) = c_1(t_1)\omega_1(x) + \sum_{k=2}^{\infty} c_k(t_1)\omega_k(x), \quad y_t(x,t_1) = \sum_{k=1}^{\infty} \dot{c}_k(t_1)\omega_k(x), \quad (8.58)$$

we will now try to select the value for h_1 such that the solution of (8.1) would follow "closely" the corresponding solution to (8.12) when the same controls $v = v_* + h_1, \gamma = 0$ are applied.

To this end, we will evaluate the difference between $(u_1(t), \dot{u}_1(t))$ and $(c_1(t), \dot{c}_1(t))$ on $[0, 2t_1^*]$, where

$$t_1^* = \frac{2\pi}{\sqrt{-h_1 - \lambda_1}}$$

is the period of one rotation of the point $(u_1(t), \dot{u}_1(t))$ about the origin as in (8.31).

First of all, note that the function

$$z(x,t) = y(x,t) - u(x,t)$$

satisfies the following mixed problem:

$$z_{tt} = z_{xx} + (v_*(x) + h_1)z - \gamma(t)z_t - q(t)f(x,t,y(x,t)),$$

$$(x,t) \in Q_{2t_1^*} = (0,1) \times (0, 2t_1^*),$$

$$z(0,t) = z(1,t) = 0, \quad z(x,0) = 0, \quad z_t(x,0) = 0. \quad (8.59)$$

Applying (8.23)–(8.24) now to (8.59), we obtain the following chain of inequalities for every $t \in [0, 2t_1^*]$:

$$\frac{1}{2}\left(|\lambda_1|(c_1(t) - u_1(t))^2 + (\dot{c}_1(t) - \dot{u}_1(t))^2\right)$$

$$\leq \frac{1}{2}\sum_{k=1}^{\infty}(-\lambda_k)$$

$$\times \left(\int_0^{2t_1^*}\left[\frac{\sin\left((t-\tau)\sqrt{-h_1-\lambda_k}\right)}{\sqrt{-h_1-\lambda_k}}\int_0^1(-q(\tau)f(x,t,y(x,\tau)))\omega_k(x)dx\right]d\tau\right)^2$$

$$+\frac{1}{2}\sum_{k=1}^{\infty}\left(\int_0^{2t_1^*}\left[\cos\left((t-\tau)\sqrt{-h_1-\lambda_k}\right)\int_0^1(-q(\tau)f(x,t,y(x,\tau)))\omega_k(x)dx\right]d\tau\right)^2$$

$$\leq 2t_1^*B^2\int\int_{Q_{2t_1^*}} f^2 dxdt. \quad (8.60)$$

In turn, (8.6)–(8.7), (8.8), (8.16), and (8.55) yield:

$$\iint\limits_{Q_{2t_1^*}} f^2 dx dt \leq 4K^2 t_1^* \{1 + \frac{1}{-h_1} E(y(\cdot,0)) + \frac{LB}{-2h_1} + \int_0^1 y_0^2(x) dx$$

$$+B(1+K+C^*K)2t_1^* \left(\frac{1}{-h_1} E(y(\cdot,0)) + \frac{LB}{-2h_1} + \int_0^1 y_0^2(x) dx \right)$$

$$\times e^{B(1+K+C^*K)2t_1^*} \}. \tag{8.61}$$

Due to (8.61), with $h_1 \to -\infty$, the right-hand side of (8.60) tends to zero uniformly over $t \in [0, 2t_1^*]$. In other words, the point $(c_1(t), \dot{c}_1(t))$ follows "arbitrarily closely" the two subsequent periodic elliptic motions of the point $(u_1(t), \dot{u}_1(t))$ around the origin in the associated two dimensional space. Hence, there exists a pair $(h_1 < 0, t_1 > 0)$ such that (8.29)–(8.30) holds for the corresponding solution to (8.1) as well, namely,

$$c_1(t_1) \in (0, \| y_d \|_{L^2(0,1)}). \tag{8.62}$$

Step 2. Steering for $t > t_1$.

4.2.1. Evaluation of $\| y(\cdot,t) - y_d \|_{L^2(0,1)}^2$. Denote by $u(x,t,y(x,t_1))$ solution to (8.12) with the initial datum the same as for (8.1) at $t = t_1$, namely, (8.58), (8.62).

Remark 8.4. From now on when we will be referring to the results in section 8.3 for $t > t_1$ we will assume that we deal with the just-introduced initial conditions for u at time $t = t_1$.

The solution formulas (8.21), (8.23), applied for $t \geq t_1$ with the initial datum (8.58), (8.62) and the aforementioned pair of controls $v = v_* + (-\lambda_1 + r), \gamma > 0$ as in the proof of Theorem 8.2 (see (8.34)), along with (8.6)–(8.7), (8.8), and (8.37) yield as $\gamma \to \infty$:

$$\| y(\cdot,t) - y_d \|_{L^2(0,1)}^2 = \| y(\cdot,t) - \| y_d \|_{L^2(0,1)} \omega_1 \|_{L^2(0,1)}^2$$

$$\leq \| y(\cdot,t) - u(x,t,y(\cdot,t_1)) \|_{L^2(0,1)}^2$$

$$+ \| u(x,t,y(\cdot,t_1)) - \| y_d \|_{L^2(0,1)} \omega_1 \|_{L^2(0,1)}^2$$

$$\leq MD \frac{1}{\gamma^2} e^{0.5 + \sqrt{0.25 + r/\gamma^2} \frac{t-t_1}{\gamma}} \iint\limits_{Q_2} f^2(x,\tau,y(x,\tau)) dx d\tau$$

$$+ M \frac{D}{b(\gamma)} \iint\limits_{Q_2} f^2(x,\tau,y(x,\tau)) dx d\tau$$

$$+ \| u(x,t,y(\cdot,t_1)) - \| y_d \|_{L^2(0,1)} \omega_1 \|_{L^2(0,1)}^2, \tag{8.63}$$

where $M > 0$ is some constant.

Indeed, to evaluate

$$\| y(\cdot,t) - u(x,t,y(\cdot,t_1)) \|^2_{L^2(0,1)}$$

in (8.63) we just need to evaluate the contribution of the terms containing F in (8.21), (8.23).

For example, for $k = 1$, (8.21) yields with our $h = -\lambda_1 + r$ from (8.34):

$$\Big(\int_{t_1}^{t} [e^{(-\frac{\gamma}{2}+\sqrt{\gamma^2/4+\lambda_k+h})(t-\tau)} \frac{1}{\sqrt{\gamma^2+4(\lambda_k+h)}}$$

$$-e^{(-\frac{\gamma}{2}-\sqrt{\gamma^2/4+\lambda_k+h})(t-\tau))} \frac{1}{\sqrt{\gamma^2+4(\lambda_k+h)}}] \int_0^1 (-p(x,\tau))\omega_k(x)dxd\tau)^2$$

$$\leq 2 \left(\frac{1}{\sqrt{\gamma^2+4r}} e^{(-\frac{\gamma}{2}+\sqrt{\gamma^2/4+r})(t-t_1)} \int_{t_1}^{t} q(\tau) \left(\int_0^1 f(x,\tau,y(x,\tau))\omega_1(x)dx \right) d\tau \right)^2$$

$$+2 \left(\frac{1}{\sqrt{\gamma^2+4r}} \int_{t_1}^{t} q(\tau) \left(\int_0^1 f(x,\tau,y(x,\tau))\omega_1(x)dx \right) d\tau \right)^2$$

$$\leq 2D \left(\frac{1}{\gamma^2} e^{0.5+\sqrt{0.25+r/\gamma^2} \frac{2r}{\gamma} \frac{t-t_1}{\gamma}} + \min\{\frac{1}{\gamma^2}, \frac{1}{b(\gamma)}\} \right)$$

$$\times \iint_{Q_2} f^2(x,\tau,y(x,\tau))dxd\tau, \tag{8.64}$$

which, in view of (8.40)–(8.41) and (8.38), provides the 1-st term on the right in (8.63).

Analogously, making use of (8.21) for $k = 2,\ldots,N(\gamma,r)$ when $\gamma^2 + 4(\lambda_k - \lambda_1 + r) > 0$, and of (8.23), we can obtain the 2-nd term on the right in (8.63).

Then, exactly as in the proof of Theorem 8.2 beginning from subsection 3.4.2, but now with the initial condition as in (8.58), (8.62) at time $t = t_1$, we obtain that the last term on the right of (8.63) tends to zero along the relations (8.37), (8.43), (8.34), and (8.44)–(8.45).

To show that the remaining two terms containing f also converge to zero, it is sufficient to show that along the relations (8.37), (8.43), (8.34), and (8.44)–(8.45) the value of $\| f \|_{L^2(Q_2)}$ is uniformly bounded, e.g., as follows:

$$\iint_{Q_2} f^2(x,\tau,y(x,\tau))dxd\tau$$

$$\leq 2L \left(1 + 8 \| y_d \|_{L^2(0,1)} + \frac{4}{-\lambda_2+\lambda_1 - r} \left(E(y(\cdot,t_1)+\frac{1}{2}) \right) \right), \tag{8.65}$$

and that

$$\frac{1}{\gamma^2} e^{0.5+\sqrt{0.25+r/\gamma^2}\,\frac{2r}{\gamma}}\, ^{\frac{t-t_1}{\gamma}} \int\!\!\int_{Q_2} f^2(x,\tau,y(x,\tau))dxd\tau \;\to\; 0 \qquad (8.66)$$

along the same relations.

4.2.2. Derivation of (8.66).

We make use of (8.52) from Lemma 8.1 and of (8.15), (8.6)–(8.7), (8.8):

$$\int\!\!\int_{Q_2} f^2(x,\tau,y(x,\tau))dxd\tau \le 2 \int\!\!\int_{Q_2} g^2(\tau)dxd\tau + 2\max_{\tau\in[t_1,t]}\left(\int_0^1 y^2 dx\right)\int_{t_1}^t g^2(\tau)d\tau$$

$$\le 2L + 4LC^*(E(y(\cdot,t_1))$$
$$+\frac{\rho}{2}LB))e^{\rho C^*(1+KB)(t-t_1)} \quad \forall t > t_1$$

and for $\rho = \frac{1}{2\gamma}(B+(-\lambda_1+r)^2+KB)$. Hence,

$$\int\!\!\int_{Q_2} f^2(x,\tau,y(x,\tau))dxd\tau$$

$$\le 2L + 4LC^*\left(E(y(\cdot,t_1))+\frac{1}{4\gamma}LB(B+(-\lambda_1+r)^2+KB)\right)$$

$$\times e^{C^*(1+KB)(B+(\lambda_1+0.5)^2+KB)\frac{t-t_1}{2\gamma}}$$

$$\le 2L + 4LC^*\left(E(y(\cdot,t_1))+\frac{1}{4\gamma}LB(B+(-\lambda_1+r)^2+KB)\right)\gamma^{1/2}$$

as $\gamma \to \infty$ along (8.37), due to the formula (recall (8.43))

$$e^{\frac{t-t_1}{\gamma}} = \gamma^{\frac{1}{C^*(1+KB)(B+(-\lambda_1+0.5)^2+BK)}}.$$

This yields (8.66), by making use of (8.48).

4.2.3. Derivation of (8.65).

Note first that (8.44)–(8.45), (8.47) and already established (8.66) mean, as it follows from the first three terms in (8.42) (rewritten for our new initial datum (8.58), (8.62)) and the estimates (8.64) (combined with (8.66)), evaluating $|c_1(t)- \| y_d \|_{L^2(0,1)}|$ in (8.63), that the value of $|c_1(t)|$ will be bounded as long as we select r,t and $\gamma \to \infty$ from (8.34), (8.37), (8.43) and (8.44)–(8.45). Without loss of generality, we can assume that starting with some γ_*,

$$|c_1(t)| \le 2\| y_d \|_{L^2(0,1)}. \qquad (8.67)$$

Multiplying equations (8.19) by $\dot{c}_k(t)$'s, then integrating them over (t_1, t) and summing the results, we obtain:

$$\frac{1}{2}\sum_{k=1}^{\infty}\dot{c}_k^2(t) + \frac{1}{2}\sum_{k=2}^{\infty}(-\lambda_k+\lambda_1-r)c_k^2(t) \leq \frac{1}{2}\sum_{k=1}^{\infty}\dot{c}_k^2(t_1) + \frac{1}{2}\sum_{k=2}^{\infty}(-\lambda_k)c_k^2(t_1)$$
$$- \gamma\iint_{Q_2} y_t^2\,dx\,d\tau + \iint_{Q_2} Bg(\tau)(1+|y|)\,|\,y_t\,|\,dx\,d\tau. \tag{8.68}$$

Note that, for any $v > 0$,

$$\iint_{Q_2} Bg(\tau)(1+|y|)\,|\,y_t\,|\,dx\,d\tau \leq \frac{v}{2}\iint_{Q_2}(Bg(\tau)(1+|y|))^2\,dx\,d\tau + \frac{1}{2v}\iint_{Q_2} y_t^2\,dx\,d\tau$$
$$\leq \frac{v}{2}\iint_{Q_2} 2B^2g^2(\tau)\,dx\,d\tau + \frac{v}{2}\iint_{Q_2} 2B^2g^2(\tau)y^2\,dx\,d\tau$$
$$+ \frac{1}{2v}\iint_{Q_2} y_t^2\,dx\,d\tau$$
$$\leq vLB^2 + vLB^2 \max_{\tau\in[t_1,t]}\int_0^1 y^2(x,\tau)\,dx$$
$$+ \frac{1}{2v}\iint_{Q_2} y_t^2\,dx\,d\tau \tag{8.69}$$

Hence, making use of (8.67) and (8.69), and selecting

$$v = \frac{-\lambda_2+\lambda_1-r}{4LB^2} \quad \text{(we will need this next)}$$

and $\gamma \geq \frac{1}{2v}$, to get rid of the terms containing y_t in (8.68)–(8.69), we obtain from (8.68) that:

$$c_k^2(t) + \sum_{k=2}^{\infty} c_k^2(t) = \int_0^1 y^2(x,t)\,dx$$
$$\leq 4\,\|\,y_d\,\|_{L^2(0,1)}^2$$
$$+ \frac{2}{-\lambda_2+\lambda_1-r}\frac{1}{2}\sum_{k=2}^{\infty}(-\lambda_k+\lambda_1-r)c_k^2(t)$$
$$\leq 4\,\|\,y_d\,\|_{L^2(0,1)}^2$$
$$+ \frac{2}{-\lambda_2+\lambda_1-r}E(y(\cdot,t_1))$$
$$+ \frac{2}{-\lambda_2+\lambda_1-r}vLB^2\left(1+\max_{\tau\in[t_1,t]}\int_0^1 y^2(x,\tau)\,dx\right).$$

Therefore, for $\gamma \geq \frac{1}{2\nu}$ and our selected ν we will have

$$\int_0^1 y^2(x,t)dx \leq 8 \parallel y_d \parallel_{L^2(0,1)}^2 + \frac{4}{-\lambda_2 + \lambda_1 - r}\left(E(y(\cdot,t_1)) + \frac{1}{2}\right). \qquad (8.70)$$

This, similar to (8.15), yields (8.65).

4.2.4. Evaluation of $\parallel y_t(\cdot,t)) \parallel_{H^{-1}(0,1)}$. Making use of the formulas (8.22),(8.24), similar to the derivation of (8.63), we obtain for $t \geq t_1$:

$$\parallel y_t(\cdot,t) \parallel_{H^{-1}(0,1)}^2 \leq \parallel y_t(\cdot,t) - u_t(\cdot,t,y(\cdot,t_1)) \parallel_{H^{-1}(0,1)}^2 + \parallel u_t(\cdot,t,y(\cdot,t_1)) \parallel_{H^{-1}(0,1)}^2$$

$$\leq \frac{M}{b(\gamma)} \sup_{\tau \in (t_1,t)} \parallel f(\cdot,\tau,y(\cdot,\tau)) \parallel_{L^2(0,1)}^2 \sum_{k=2}^{\infty} \frac{B^2}{-\lambda_k}$$

$$+ \frac{MB^2}{\gamma^2 + 4r} e^{0.5 + \sqrt{0.25 + r/\gamma^2}} \frac{2r}{\gamma} \frac{t - t_1}{\gamma} \sup_{\tau \in (t_1,t)} \parallel f(\cdot,\tau,y(\cdot,\tau)) \parallel_{L^2(0,1)}^2$$

$$+ M\left(\frac{1}{b(\gamma)} + \frac{1}{\gamma^2}\right) \sup_{\tau \in (t_1,t)} \parallel f(\cdot,\tau,y(\cdot,\tau)) \parallel_{L^2(0,1)}^2$$

$$\times \sum_{k > N(\gamma,r)}^{\infty} \frac{B^2}{-\lambda_k} + \parallel u_t(\cdot,t,y(\cdot,t_1)) \parallel_{H^{-1}(0,1)}^2 \qquad (8.71)$$

where $M > 0$ is some positive constant.

In (8.71), to evaluate the terms in (8.22), (8.24) containing f, we also employed the following estimates:

$$\left| \int_{t_1}^t ae^{a(t-\tau)}d\tau \right| = 1 - e^{a(t-t_1)} \leq 1 \ \forall t \geq t_1, \ a < 0,$$

$$0 < \int_{t_1}^t ae^{a(t-\tau)}d\tau = e^{a(t-t_1)} - 1 \ \forall t \geq t_1, \ a > 0,$$

$$\left| \int_0^1 f(x,t)\omega_k(x)dx \right| \leq \left(\int_0^1 f^2(x,t)dx\right)^{1/2}\left(\int_0^1 \omega_k^2(x)dx\right)^{1/2} = \left(\int_0^1 f^2(x,t)dx\right)^{1/2}.$$

For example (see (8.24) on (t_1,t)):

$$-\frac{1}{-\lambda_k}(\int_{t_1}^{t}[-\frac{\gamma}{2}e^{-\frac{\gamma(t-\tau)}{2}}\frac{\sin\left((t-\tau)\sqrt{-\gamma^2/4-\lambda_k-h}\right)}{\sqrt{-\gamma^2/4-\lambda_k-h}}$$

$$+e^{-\frac{\gamma(t-\tau)}{2}}\cos\left((t-\tau)\sqrt{-\gamma^2/4-\lambda_k-h}\right)]\int_0^1(-p(x,\tau))\omega_k(x)dxd\tau)^2$$

$$\leq \frac{2}{-\lambda_k}\frac{B^2}{-\gamma^2/4-\lambda_k+\lambda_1-r}\sup_{\tau\in(t_1,t)}\|f(\cdot,\tau,y(\cdot,\tau))\|^2_{L^2(0,1)}$$

$$\times\left(\int_{t_1}^{t}(-\frac{\gamma}{2}e^{-\frac{\gamma(t-\tau)}{2}})d\tau\right)^2$$

$$+\frac{2B^2}{-\lambda_k}\sup_{\tau\in(t_1,t)}\|f(\cdot,\tau,y(\cdot,\tau))\|^2_{L^2(0,1)}\left(\int_{t_1}^{t}e^{-\frac{\gamma(t-\tau)}{2}}d\tau\right)^2$$

$$\leq\frac{8}{-\lambda_k}B^2\sup_{\tau\in(t_1,t)}\|f(y(\cdot,\tau))\|^2_{L^2(0,1)}\left(\frac{1}{b(\gamma)}+\frac{1}{\gamma^2}\right).$$

Again, as in the proof of Theorem 8.2 in subsection 3.4.5, but now with the initial condition as in (8.58), (8.62) at time $t=t_1$, we obtain that *the last term* on the right of (8.71) tends to zero along the relations (8.37), (8.43), (8.34), and (8.44)–(8.45).

It remains to notice that, due to (8.70), (8.6)–(8.7) and (8.8) (similar to (8.17)),

$$\sup_{\tau\in(t_1,t)}\|f(\cdot,\tau,y(\cdot,\tau))\|^2_{L^2(0,1)}$$

is uniformly bounded for $t>0$ as in (8.43), and to use (8.47) and (8.39) to show that *the first three terms* in the right-hand side of (8.70) converge to zero as $\gamma\to\infty$ under the conditions (8.37), (8.43), (8.34), and (8.44)–(8.45).

Combining this with the above evaluation of (8.63) yields that there is a pair of controls (v,γ) and time $T>0$ described in Theorem 8.1 such that (8.11) holds. This completes the proof of Theorem 8.1. $\qquad\square$

Chapter 9
The 1-*D* Wave and Rod Equations Governed by Controls That Are Time-Dependent Only

Abstract We discuss the early results from [8] on global approximate reachability of the 1-*D* wave equation with Dirichlet boundary conditions and the rod equation with hinged ends in the case when the multiplicative control is time-dependent only. The methods of [8] make use of the inverse function theorem and involve dealing with the associated Riesz bases of exponential time-dependent functions under the additional assumption that all the modes in the initial datum are present.

9.1 Abstract Setting of Controllability Problem

In this chapter we discuss multiplicative controllability of hyperbolic equations along the approach due to J.M. Ball, J.E. Marsden, and M. Slemrod. We describe the main ideas of this approach and the principal results relevant to the global reachability properties of these equations. For the complete account of all details we refer the reader to the original paper [8].

Chapter 9 is organized as follows. In section 9.1 a class of abstract evolution equations, governed in a Banach space X by multiplicative time-dependent controls, is introduced. However, it turns out that the local controllability of these equations (i.e., *anywhere* within some neighborhood of a drifting trajectory) is out of question when X is infinite dimensional. In section 9.2 we distinguish a more specific subclass within the aforementioned class of equations – the abstract hyperbolic equations, and also remind the reader the concept of Riesz bases. In section 9.3 we show how the global approximate reachability of the latter equations can be established under a number of additional assumptions. In particular, it is assumed that the *eigenvalues of the linear part of the equation at hand must all be integers that are multiplied by the same number and that all modes in the initial datum must be present* (see Assumptions 1-4 below). In section 9.4 this abstract theory is applied to the 1-*D* wave and rod equations. In section 9.5 we discuss two results on local controllability for these equations due to K. Beauchard.

A.Y. Khapalov, *Controllability of Partial Differential Equations Governed*
by Multiplicative Controls, Lecture Notes in Mathematics 1995,
DOI 10.1007/978-3-642-12413-6_9, © Springer-Verlag Berlin Heidelberg 2010

In [8] the following abstract evolution equation was considered:

$$\frac{dw(t)}{dt} = \mathbf{A}w(t) + p(t)\mathbf{B}(w(t)), \quad t > 0, \tag{9.1}$$

$$w(0) = w_0,$$

where $\mathbf{A}$ generates a C^o semigroup of bounded linear operators on a (possibly complex) Banach space X, $\mathbf{B} : X \to X$ is a C^1 map, and $p \in L^1([0,T];R)$ is a multiplicative control. We assume further that $w_0 \neq 0$. This setup was motivated by the wave and rod equations.

Let $Z(T)$ be a Banach space continuously and densely included in $L^1([0,T];R)$, where $T > 0$ is given. It can be shown [8] that for each $w_0 \in X, p \in Z(T)$, there exists a $t_0 \in (0,T]$ such that (9.1) has a unique mild solution in $C([0,t_0];X)$, namely, when (9.1) is understood in the following integral form:

$$w(t) = e^{\mathbf{A}t}w_0 + \int_0^t e^{\mathbf{A}(t-s)}p(s)\mathbf{B}(w(s))ds.$$

Furthermore, it can also be shown that for $p = 0$, the Fréchet derivative of $w = w(t; p, w_0)$ with respect to p in the direction of p is defined by the formula:

$$D_p(w(t;0,w_0)) \cdot p = \int_0^t e^{\mathbf{A}(t-s)}p(s)\mathbf{B}(e^{\mathbf{A}t}w_0)ds. \tag{9.2}$$

Define the linear operator $L_T : Z(T) \to X$ (for the given w_0),

$$L_T p = D_p(w(T;0,w_0)) \cdot p.$$

By employing the generalized inverse function theorem (e.g., [114], p. 240), the authors of [8] established that, *if Range* $(L_T) = X$, then (9.1) is locally controllable near the drifting (i.e., uncontrolled) solution $e^{\mathbf{A}T}w_0$ of (9.1) at time $t = T$. The respective rigorous formulation of this result is as follows (Theorem 3.1 from [8]).

Theorem 9.1. *Let $\mathbf{A}$ be the infinitisimal generator of a C^0 semigroup of bounded linear operators on the Banach space X, and let $\mathbf{B} : X \to X$ be a C^k map, $k \geq 1$, which satisfies*

$$\| \mathbf{B}x \|_X \leq C + K \| x \|_X \quad \forall x \in X,$$

where C and K are constants. Suppose that Range $(L_T) = X$. Then there is an $\varepsilon > 0$ such that $w(T; p, w_0) = h$ for some $p \in Z(T)$, provided $\| h - e^{\mathbf{A}t}w_0 \|_X < \varepsilon$.

A major difficulty with Theorem 9.1 is that the surjectivity of L_T is not something that one normally encounters when dealing with distributed parameter systems. In fact, if X is infinite dimensional, this operator *will not* in general be surjective, though it may have dense range, as implied by Theorem 9.2 below (or Theorem 3.6 in [8]). This prevents us from applying Theorem 9.1 to PDE's.

Theorem 9.2. *Let X be a Banach space with dim $X = \infty$. Let $\mathbf{A}$ generate a semigroup of bounded linear operators on X, and let $\mathbf{B} : X \rightarrow X$ be bounded. Let $w_0 \in X$ be fixed and $w(t; p, w_0)$ denote the unique solution to (9.1) with $p \in L^1_{loc}([0, \infty); R)$. Then the set of all states of (9.1) accessible from w_0,*

$$S(w_0) = \bigcup_{t \geq 0,\, p \in L^1_{loc}([0,\infty);R),\, r>1} w(t; p, w_0)$$

is contained in a countable union of compact subsets of X, and, in particular, has a nonempty compliment.

In other words, Bair's Category Theorem yields that $S(w_0)$ is of the 1-st category and is incomplete in the infinite dimensional space X. On the other hand, if $S(w_0)$ turns out to be dense in X, this will make the respective system (9.1) approximately controllable (reachable) in X. We intend to explore this idea below.

9.2 Abstract Hyperbolic Equations

Let us consider the following 2-nd order abstract evolution equation:

$$\frac{d^2u}{dt^2} + Au + p(t)Bu = 0, \quad t > 0, \tag{9.3}$$

$$u(0) = u_0 \in D(A^{1/2}), \quad \frac{du}{dt} = u_1 \in H,$$

where A is a positive definite self-adjoint operator with dense domain $D(A)$ in the real Hilbert space H, B is a bounded operator from $D(A^{1/2})$ to H, and p is a real-valued control. We suppose that A^{-1} is compact, and that A has simple eigenvalues $\lambda_n^2, n = 1, 2, \ldots$, where $0 < \lambda_1 < \lambda_2 < \ldots$. Then there exists a corresponding complete orthonormal (in H) basis $\{\phi_n\}_{n=1}^{\infty}$ of eigenfunctions:

$$A\phi_n = \lambda_n^2 \phi_n, \quad n = 1, \ldots$$

Note that (9.3) can be re-written in the 1-st order form described by (9.1) with

$$\mathbf{A} = \begin{pmatrix} 0 & I \\ -A & 0 \end{pmatrix}, \quad \mathbf{B} = \begin{pmatrix} 0 & 0 \\ -B & 0 \end{pmatrix}, \quad w = \begin{pmatrix} u \\ \frac{du}{dt} \end{pmatrix}$$

and $X = D(A^{1/2}) \times H$.

Let $\mathbf{H}$ denote the complexified Hilbert space $H \oplus iH$ with the inner product defined by

$$< x_1 + iy_1, x_2 + iy_2 >_{\mathbf{H}} = (x_1, x_2) + (y_1, y_2) + i[(y_1, x_2) - (x_1, y_2)]$$

for $x_1, x_2, y_1, y_2 \in H$, where $(\cdot, \cdot)$ stands for the inner product in H. The map $\psi : X \to \mathbf{H}$ defined by

$$\psi(u_1, u_2) = A^{1/2} u_1 + i u_2$$

is an isometry. Let $z = A^{1/2} u + i \frac{du}{dt}$, in which case (9.3) becomes

$$i\frac{dz}{dt} = A^{1/2} z + p(t) B A^{-1/2} \mathrm{Re}\, z, \tag{9.4}$$

$$z(0) = z_0 = A^{1/2} u_0 + i u_1 \in \mathbf{H}.$$

We also have

$$z(t) = \sum_{n=1}^{\infty} z_n(t)\phi_n, \quad z_0 = \sum_{n=1}^{\infty} z_{0n}\phi_n,$$

where $(z_1, \ldots) \in l_2$ are complex components of z relative to the basis $\{\phi_n\}_{n=1}^{\infty}$.

Riesz bases. A sequence of elements $\{\zeta_j\}_{j=1}^{\infty}$ of (real or complex) Hilbert space Z is called a Riesz basis of Z if every $\eta \in Z$ has a unique expansion

$$\eta = \sum_{j=1}^{\infty} a_j \zeta_j$$

converging in Z and

$$C_1 \sum_{j=1}^{\infty} |a_j|^2 \le \| \eta \|_Z \le C_2 \sum_{j=1}^{\infty} |a_j|^2$$

for some positive constants C_1 and C_2.

The well-known (e.g., [8, 52]) important feature of Riesz basis is that, *given any sequence $(a_1, \ldots) \in l_2$, there exists a unique solution $\eta \in Z$ of the equations*

$$[\eta, \zeta_j] = a_j, \quad j = 1, \ldots, \tag{9.5}$$

where $[\cdot, \cdot]$ stands for the inner product in Z.

The following criterion is useful for the construction of a Riesz basis of the space $L^2([0, T]; C)$ (see [8, 132]).

Theorem 9.3. *Let $0 = \mu_0 < \mu_1 < \ldots, \mu_k = -\mu_{-k}, k = 1, \ldots,$ and suppose that*

$$\lim_{k \to \infty} (\mu_k - \mu_{k-1}) \ge \gamma > 0.$$

Then, for any $T > 2\pi/\gamma$ the functions $\{e^{i\mu_k t}\}_{k=-\infty}^{k=\infty}$ may be extended to a Riesz basis of $L^2([0, T]; C)$.

9.3 Approximate Controllability for System (9.4)

Assumption 1. *We will further assume that the operator B is diagonal:*

$$B(\sum_{n=1}^{\infty} \alpha_n \phi_n) \; = \; \sum_{n=1}^{\infty} b_n \alpha_n \phi_n,$$ (9.6)

for some $b_n, n = 1, \ldots$

In this case, (9.4) reduces to the infinite system of uncoupled ode's:

$$\frac{dz}{dt} = -i\lambda_n z_n - ip(t)\frac{b_n}{\lambda_n} \operatorname{Re} z_n, \quad z_n(0) = z_{0n}, \quad n = 1, \ldots$$ (9.7)

Let us note that the fact that $BA^{1/2}$ is a bounded operator from H to H is equivalent to the condition that

$$(\frac{b_1}{\lambda_1}, \ldots) \in l_{\infty}.$$

Assumption 2. *We will also further assume that*

$$\frac{b_n}{\lambda_n} = c + \gamma_n$$ (9.8)

for some $c \in R$ and $(\gamma_1, \ldots) \in l_2$.

Denote

$$P(t) = \int_0^t p(s)ds$$ (9.9)

and introduce the following change of variables:

$$\xi_n(t) = \frac{\lambda_n}{b_n}\left[\frac{z_n(t)}{z_{0n}}e^{i(\lambda_n t + \frac{b_n}{2\lambda_n}P(t))} - 1\right].$$

Then (9.4) yields:

$$\frac{d\xi_n}{dt} = -i\frac{p(t)}{2}\frac{\bar{z}_{0n}}{z_{0n}}\left(\frac{b_n}{\lambda_n}\xi_n(t) + 1\right)e^{2i(\lambda_n t + \frac{b_n}{2\lambda_n}P(t))},$$ (9.10)
$$\xi_n(0) = 0,$$

where $\bar{a}$ stands for the complex conjugate of a.

The following result establishes the local controllability of equations (9.10) in l_2 (*not of the original system (9.4)*) under the following condition:

Assumption 3. *Let $(z_{o1}, \ldots) \in l_2$, $z_{0n} \neq 0$, $b_n \neq 0$, $n = 1, \ldots$.*

Theorem 9.4. *Suppose that Assumptions 1-3 hold, and that* $\{1, e^{\pm 2i\lambda_n t}, n = 1, \ldots\}$ *can be extended to a Riesz basis of* $L^2([0, l]; C)$ *for some* $l > 0$. *Then there exists* $\varepsilon_l > 0$ *such that if* $\| h \|_{l_2} + | \theta | < \varepsilon_l$, *where* $h = (h_1, \ldots) \in l_2$ *and* $\theta \in R$, *then*

$$\xi_n(l) = \frac{\lambda_n}{b_n} \left[\frac{z_n(l)}{z_{0n}} e^{i(\lambda_n l + \frac{b_n}{2\lambda_n} \theta)} - 1 \right] = h_n, \quad n = 1, \ldots \tag{9.11}$$

for some $p \in L^2([0, l]; R)$ *with* $\int_0^l p(s)ds = \theta$.

Proof. Denote solutions to (9.10) on $[0, l]$ by $\xi_n(l; p), n = 1, \ldots$ Consider the map $Q : L^2([0, l]; R) \rightarrow l_2 \times R$ defined by

$$Q(p) = \left((\xi_1(l; p), \ldots), \int_0^l p(t)dt \right).$$

It can be shown ([8], Theorem 5.7) that for $p = 0$, the Fréchet derivative of Q with respect to p in the direction of p is defined by the formula:

$$D_p Q(0) \cdot p = \left(\left(-\frac{i}{2} \frac{\bar{z}_{01}}{z_{01}} \int_0^l p(t)e^{2i\lambda_1 t} dt, \ldots \right), \int_0^l p(t)dt \right).$$

The desirable result of Theorem 9.4 then follows by employing the generalized inverse function theorem (compare to Theorem 9.1 in the above), if we will show that $D_p Q(0)$ is surjective.

Let $(a_1, \ldots) \in l_2, \alpha \in R$. Write $c_n = 2i(z_{0n}/\bar{z}_{0n})a_n$. Then, making use of the fact that $\{1, e^{\pm 2i\lambda_n t}, n = 1, \ldots\}$ can be extended to a Riesz basis of $L^2([0, l], C)$, we conclude from (9.5) that we can solve the equations

$$\int_0^l q(t)e^{2i\lambda_n t} dt = c_n, \quad \int_0^l q(t)e^{-2i\lambda_n t} dt = \bar{c}_n, \quad n = 1, \ldots,$$

$$\int_0^l q(t)dt = \alpha$$

for $q \in L^2([0, l]; C)$. Setting $p(t) = \text{Re} \, q(t)$, we derive that $D_p Q(0)$ is surjective. This ends the proof of Theorem 9.4. □

Remark 9.1. The set of $z(l) = \sum_{n=1}^{\infty} z_n(l)\phi_n$ in **H** described by (9.4) such that for some $\theta \in R$ the respective $\xi(l) = (\xi_1(l), \ldots)$ in (9.11) belongs to the ball of radius ε_l is compact and hence *nowhere dense (or rare)* in **H**. (Compactness follows from the estimate $| z_n | \leq (| \frac{b_n \xi_n(l)}{\lambda_n} | + 1) | z_{0n} | \leq (C\varepsilon_l^{1/2} + 1) | z_{0n} |$ for some $C > 0$.) Therefore the result of Theorem 9.4 does not imply that we can steer (9.4) to a dense subset of some neighborhood in **H** in *finite time*. In other words, Theorem 9.4 does not imply the approximate controllability of (9.4). To achieve it, we will extend Theorem 9.4 to the case when ε_l can be selected arbitrarily, namely, under the following additional condition:

Assumption 4. *Let* λ_n/σ *be an integer for all* $n = 1, \ldots$ *for some* $\sigma > 0$.

This conditions ensures that $e^{i\lambda_n \frac{2m\pi}{\sigma}} = 1, m = 1,\ldots$ used below to derive (9.12) from (9.11). Also, due to Theorem 9.3, $\{1, e^{\pm 2i\lambda_n t}, n = 1,\ldots\}$ can be extended to a Riesz basis of $L^2([0,T],C)$ for some $T > 2\pi/\sigma$, which allows application of Theorem 9.4.

Theorem 9.5. *Let Assumptions 1-4 hold. Then for any $h = (h_1,\ldots) \in l_2$ with $1 + (b_n/\lambda_n)h_n \neq 0$ for all n, and any $\theta \in R$, there exists a positive integer m and a control $p \in L^2([0,2m\pi/\sigma];R)$ such that*

$$z_n(2m\pi/\sigma) = z_{0n}e^{-\frac{ib_n\theta}{2\lambda_n}}\left(1 + \frac{b_n}{\lambda_n}h_n\right), \quad n = 1,\ldots \qquad (9.12)$$

The proof of this result consists of establishing the fact that the following two sets:

$$\{(h,\theta) \in l_2 \times R \mid z_n(2m\pi/\sigma)$$
$$= z_{0n}e^{-\frac{ib_n\theta}{2\lambda_n}}\left(1 + \frac{b_n}{\lambda_n}h_n\right), n = 1,\ldots, \quad m = 1,\ldots, p \in L^2([0,2m\pi/\sigma];R)\}$$

and

$$\{(h,\theta) \in l_2 \times R \mid 1 + (b_n/\lambda_n)h_n \neq 0, \ n = 1,\ldots\}$$

are the same. For further details we refer to [8].

Since the set

$$\{h \in l_2 \mid 1 + (b_n/\lambda_n)h_n \neq 0, \ n = 1,\ldots\}$$

is dense in l_2, the equality of the aforementioned sets implies the global approximate controllability of system (9.3) or, more precisely, reachability, since the control time is not fixed. We can formulate this as follows.

Theorem 9.6. *Let Assumptions 1-4 hold. Then the attainable set*

$$s(z_0) = \bigcup_{t \geq 0, \ p \in L^2_{loc}([0,\infty);R)} z(t;p,z_0)$$

is dense in **H**, *where $z(t;p,z_0)$ stands for solution to (9.3).*

9.4 Applications to PDE's

In this section we provide two examples taken from [8].

1-D wave equation with Dirichlet boundary conditions. Consider the following mixed problem:

$$u_{tt} - u_{xx} + p(t)u = 0, \quad x \in (0,1), \ t > 0, \qquad (9.13)$$
$$u(0,t) = u(1,t) = 0, \quad u(x,0) = u_0(x), \ u_t(x,0) = u_1(x).$$

In the notations of (9.3) we have:

$$A = -\frac{\partial^2}{\partial x^2}, \quad B = I, \quad H = L^2(0,1),$$

$$D(A) = H^2(0,1) \bigcap H_0^1(0,1), \quad D(A^{1/2}) = H_0^1(0,1),$$

$$\lambda_n = n\pi, \quad \phi_n = \sqrt{2}\sin \pi nx, \quad b_n = 1, \quad n = 1,\dots$$

Assumptions 1-2, 4 hold. If we additionally assume that *all modes in the initial datum are present*:

$$\left(\int_0^1 u_0 \sin \pi nx dx\right)^2 + \left(\int_0^1 u_1 \sin \pi nx dx\right)^2 \neq 0 \quad \forall n = 1,\dots, \qquad (9.14)$$

we will satisfy Assumption 3 as well. Then from Theorems 9.4-9.6 we conclude that *system (9.13), (9.14) is approximately globally reachable in* $H_0^1(0,1) \bigcap L^2(0,1)$.

1-D rod equation with hinged ends. Consider the system:

$$u_{tt} + u_{xxxx} + p(t)u_{xx} = 0, \quad x \in (0,1), \quad t > 0, \qquad (9.15)$$

$$u(0,t) = u(1,t) = u_{xx}(0,t) = u_{xx}(1,t) = 0, \quad u(x,0) = u_0(x), \quad u_t(x,0) = u_1(x).$$

In the notations of (9.3) we have:

$$A = \frac{\partial^4}{\partial x^4}, \quad B = \frac{\partial^2}{\partial x^2}, \quad H = L^2(0,1),$$

$$D(A) = \{u \in H^4(0,1) \mid u, u_{xx} \in H_0^1(0,1)\}, \quad D(A^{1/2}) = H^2(0,1) \bigcap H_0^1(0,1),$$

$$\lambda_n = (n\pi)^2, \quad \phi_n = \sqrt{2}\sin \pi nx, \quad b_n = -n^2\pi^2, \quad n = 1,\dots$$

Assumptions 1-2, 4 hold. To satisfy Assumption 3 we will need to assume (9.14). Then Theorems 9.4-9.6 yield that *system (9.15), (9.14) is approximately globally reachable in* $(H^2(0,1) \bigcap H_0^1(0,1)) \bigcap L^2(0,1)$.

9.5   Local Controllability Properties for the Beam and Wave Equations in Subspaces of X

Let us remind the reader that Theorem 9.2 states that the set $S(w_0)$ of all solutions to equation (9.1) is incomplete in its natural phase-space X where its solutions lie, namely, on which operator A generates a C^o semigroup of bounded linear operators. This means that both global and local (exact) controllability of (9.1) in X is out of question.

However, (for the time-reversible equations) one may try to study such controllability in suitable subspaces of X. In fact, for the abstract hyperbolic system (9.3) Theorem 9.4 actually establishes exact reachability of the states described in Remark 9.1. Respectively, one may ask a question - *"Can these states cover a full neighborhood in some Banach space which is other than X?"*

In this section we discuss two examples due to K. Beauchard [13,15] who showed that in suitably selected phase-spaces (on which the respective operator A does *not* generate a C^o semigroup of bounded linear operators) one can actually steer the solutions of the rod and wave equations, governed by multiplicative controls, to a full neighborhood. (In Chapter 16 we discuss more results of this type for the Schrödinger equation.)

1-*D* rod equation with clamped ends. Consider the system:

$$u_{tt} + u_{xxxx} + p(t)u_{xx} = 0, \quad x \in (0,1), \ t > 0, \qquad (9.16)$$

$$u(0,t) = u(1,t) = u_x(0,t) = u_x(1,t) = 0, \quad u(x,0) = u_0(x), \ u_t(x,0) = u_1(x).$$

We have:

$$A = \frac{\partial^4}{\partial x^4}, \ B = \frac{\partial^2}{\partial x^2}, \ H = L^2(0,1),$$

$$D(A) = H^4(0,1) \bigcap H_0^2(0,1), \ D(A^{1/2}) = H_0^2(0,1).$$

Denote

$$H_{(0)}^s(0,1) = \{\varphi \mid \varphi \in H^s(0,1), \varphi = \varphi_x = 0 \text{ at } x = 0,1\}, \ s = 2,3,4,$$

$$H_{(0)}^5(0,1) = \{\varphi \mid \varphi \in H^5(0,1), \varphi = \varphi_x = \varphi_{xxxx} = 0 \text{ at } x = 0,1\}.$$

As usual, let $\{\lambda_n, \varphi_n\}_{n=1}^\infty$ be the eigenelements associated with (9.16).

In [13] the author studied solutions to (9.16) near its uncontrolled trajectory emitted from the initial state

$$(u_{0r}, u_{1r}) = (0, \sqrt{\lambda_2}\varphi_2 + \sqrt{\lambda_3}\varphi_3),$$

namely,

$$u_r(x,t) = \varphi_2(x)\sin(\sqrt{\lambda_2}t) + \varphi_3(x)\sin(\sqrt{\lambda_3}t).$$

The main result of [13] is that at time $T = 8/\pi$ one can steer system (9.16) from any state in some neighborhood of (u_{0r}, u_{1r}) to some *full* neighborhood of $(u_r(\cdot,T), u_{rt}(\cdot,T))$ in the space $H_{(0)}^{5+\varepsilon}(0,1) \times H_{(0)}^{3+\varepsilon}(0,1), \ \varepsilon > 0$, with $p \in H_0^1(0,T)$.

The argument of [13] employs the classical approach, based on the use of suitable inverse mapping or implicit function theorems (e.g., as Theorems 9.1 and 9.4 in the above). In this case the Nash-Moser implicit function theorem is employed. This strategy was used in [11,12,14] and [16] to study controllability of the Schrödinger equation (see Chapter 16 below for more details) and in the next example.

The Neumann problem for the 1-D wave equation. In [15] a result similar to the just-described was obtained for the Neumann boundary problem:

$$u_{tt} = u_{xx} + p(t)\mu(x)u, \quad x \in (0,1), \ t > 0, \tag{9.17}$$
$$u_x(0,t) = u_x(1,t) = 0.$$

In this case the reference trajectory u_r was the one emitted from

$$u(x,0) = u_0(x) = 1, \ u_t(x,0) = u_1(x) = 0. \tag{9.18}$$

The main result of [15] states that, if $\mu(x)$ is properly chosen, the solutions to (9.17), (9.18) at time $T > 2$ will cover a full neighborhood in the space $\{\varphi \in H^3(0,1) \mid \varphi_x(0) = \varphi_x(1) = 0\} \times \{\varphi \in H^2(0,1) \mid \varphi_x(0) = \varphi_x(1) = 0\}$ with $p \in L^2(0,T)$.

Part III
Controllability for Swimming Phenomenon

The subject of our interest in Chapters 10-15 is the swimming phenomenon from the viewpoint of controllability theory for pde's. We intend to address this problem by investigating the controllability properties of an abstract object (a "swimmer") which applies *fishlike or rowing motion to propel itself in a fluid* (as opposed to bodies that are drifting, or being pushed/pulled in fluid by external forces). This object can be viewed as a simplified model of a swimming living organism or an artificial "mechanical device".

Chapter 10
Introduction

The swimming phenomenon has been a source of great interest and inspiration for many researchers for a long time, with formal publications traced as far back as to the works of G. Borelli in 1680–1681. It is impossible to give a complete account of all findings in this area. However, we will attempt to classify some of them as it seems relevant to the motivation of the content of this chapter.

Historically, the first efforts were originated in the natural sciences and were aimed at understanding of biomechanics of swimming of specific types of species, with principal contributions made by Gray [53] (1932), Gray and Hancock [54] (1951), Taylor [139] (1951), [140] (1952), Wu [145] (1971), Lighthill [108] (1975), and others. This research resulted in the derivation of a number of mathematical models for swimming motion, see, e.g., Childress [27] (1981) and the references therein. Based on the size of Reynolds number, it was suggested to divide swimming models into three groups: microswimmers (such as flagella, spermatozoa, etc.) with "insignificant" inertia; "regular" swimmers (fish, dolphins, humans, etc.), whose motion takes into account both viscosity of fluid and inertia; and Euler's swimmers, in which case viscosity is to be "ignored". Let us elaborate on the reasoning behind this division along the argument given in [27].

The aforementioned models are based on the assumption that the incompressible medium for the swimmer is governed by the Navier-Stokes equation:

$$\frac{\partial u}{\partial t} + u \cdot \nabla u + \frac{1}{\rho}\nabla p - v\nabla^2 u = 0, \ \nabla \cdot u = 0. \tag{10.1}$$

This equation is to be satisfied in the (unbounded) region in R^3 (or R^2, if the model it two dimensional) exterior to the swimmer. In the above $u(x,t)$ is the fluid velocity at point $x = (x_1, x_2, x_3)$ at time t, ρ is a constant density of the fluid and v is a constant kinematic viscosity.

A swimmer moves in a fluid due to its shape change, resulting from the actions of its internal forces. In turn, the boundary of the swimmer, denote it by S, will also react to the response of the fluid. As a result, S may have a very complicated time dependent form, which creates a principal difficulty for the study of swimming motion. This difficulty can be avoided if one will tackle a simpler problem, namely, when S is assumed to be *known*, while the propulsion force acting upon S is *unknown*.

A.Y. Khapalov, *Controllability of Partial Differential Equations Governed by Multiplicative Controls*, Lecture Notes in Mathematics 1995, DOI 10.1007/978-3-642-12413-6_10, © Springer-Verlag Berlin Heidelberg 2010

Typically, each swimmer is associated with some characteristic length L for the boundary S and a characteristic frequency ω for the motions of S. These motions will result in a motion of the fluid about the swimmer with a characteristic speed U. Respectively, equations (10.1) can be re-written in dimensionless variables as follows ([27], p. 2):

$$\sigma \frac{\partial u^*}{\partial t^*} + u^* \cdot \nabla^* u^* + \nabla^* p^* - \text{Re}^{-1} \nabla^{*2} u^* = 0, \quad \nabla^* \cdot u^* = 0, \tag{10.2}$$

where $u^* = U^{-1}u, p^* = (\rho U^2)^{-1}p, \nabla^* = L\nabla, t^* = \omega t, \text{Re} = \frac{UL}{\nu}$ is Reynolds number, and $\sigma = \frac{\omega L}{U}$ is the frequency parameter.

The motion of the microorganism in the above-cited literature is associated with the case when σ is a quantity of order unity and $\text{Re} \ll 1$. In this case if one formally takes Re to zero, (10.1) will be reduced to the *stationary* Stokes equation of type ([27], p. 13):

$$\nabla p - \nabla^2 u = 0, \quad \nabla \cdot u = 0. \tag{10.3}$$

In respect to the motion of a "miscroscopic" scallop in the fluid described by this equation (i.e., time-reversible when it is excited by time-dependent forces), E.M. Purcell in his famous lecture [127] (1977) made the following observation (also known as the "scallop" theorem): "... *the scallop at low Reynolds number [is no good. It] can't swim because it only has one hinge, and if you have only one degree of freedom in configuration space, you are bound to make a reciprocal motion. There is nothing else you can do. The simplest animal that can swim that way is an animal with two hinges.....*" Of course, the cited "scallop" should be viewed as an abstract swimming object with one hinge and no inertia:

An "abstract scallop" applying folding forces.

In turn, for Euler's swimmers, one neglects the viscosity and sets $\nu = 0$, in which case (10.1) becomes Euler's equation for incompressible, inviscid fluid ([27], p. 77):

$$\frac{\partial u}{\partial t} + u \cdot \nabla u + \frac{1}{\rho} \nabla p = 0, \quad \nabla \cdot u = 0.$$

It should be noted that the above-described models deal with the whole 2-*D* or 3-*D* spaces for swimmer and use the latter as the reference frame, with fluid moving about it. *Such approach makes it difficult to track the actual position of swimmer in a fluid which is our primary goal in this chapter.* Also, we consider a swimmer in a *bounded* domain.

A different modeling approach was proposed by Peskin in the computational mathematical biology (see Peskin [125] (1975), Fauci and Peskin [40] (1988), Fauci

[41] (1993), Peskin and McQueen [126] (1994) and the references therein), where a "thin" swimmer is modeled as an immaterial "immersed" boundary. In this case a fluid equation is to be complemented by a coupled infinite dimensional differential equation for the aforementioned immersed boundary. The following equation and figure illustrate this modeling approach in the case when the fluid density $\rho = 1$ (e.g., [41], p. 93):

$$\frac{\partial u}{\partial t} + u \cdot \nabla u + \nabla p = \nu \Delta u + F(x,t), \ \nabla \cdot u = 0, \tag{10.4}$$

$$F(x,t) = \int f(s,t)\delta[x - X(s,t)]ds. \tag{10.5}$$

In the above the state of the (massless) organism is given by the configuration of points $X(s,t)$, where s is an arc-length parameter; $f(s,t)$ is the boundary force per unit length at each point of the immersed organism, determined by its configuration at time t. The integration is over the immersed organism, and δ denotes the two dimensional delta function. The velocity of each point of the organism is equal to the fluid velocity evaluated at that point:

$$\frac{\partial X(s,t)}{\partial t} = \int u(x,t)\delta[x - X(s,t)]dx, \tag{10.6}$$

where the integration is over the entire fluid domain.

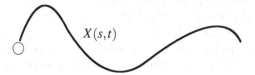

"Immersed" worm − like swimmer (e.g. spermatozoon).

The "immersed" boundary approach served us as a source and motivation for the models we study below.

In recent years a number of significant efforts, both theoretical and experimental, were made to study models of possible bio-mimetic mechanical devices, which employ the change of their geometry, inflicted by internal forces, as the means for self-propulsion, e.g., S. Hirose [58] (1993), Mason and Burdick [116] (2000); McIsaac and Ostrowski [117] (2000); Martinez and J. Cortes [118] (2001); Trintafyllou et al. [143] (2000); Morgansen et al. [121] (2001); Fakuda et al. [37] (2002); Guo et al. [55] (2002); Hawthornee et al. [56] (2004), and the references therein. It should be noted that in the cited references the swimming models typically employ the systems of ODE's only.

In terms of modeling, a direction, which can be associated with this approach, deals with attempts to apply certain reduction techniques to convert swimming model equations into ODE's, making use of applicable analytical considerations, empiric observations and experimental data, e.g., Becker et al. [19] (2003); Kanso et al. [59] (2005); San Martin et al. [134] (2007); Alogues et al. [2] (2008), and the references therein. For example, in the case of fluid governed by stationary Stokes equation, its linearity with respect to forcing terms suggests that the swimming model equations can be sought in the form of linear algebraic equations for the velocities defining the motion of the swimmer, with matrices whose entries depend on viscosity, size and instantaneous swimmer's shape and can be computed using suitable approximation or analytically (in some special cases), see, e.g., [2, 19] and the references therein. The dependence on the geometry of swimmer appears to be the crux of the problem here. This approach does not easily apply if the space domain is bounded (when one also needs to take into account the position of swimmer relative to domain's boundary) or if the fluid is nonstationary.

In this chapter we intend to deal with the swimming phenomenon in its original intrinsic realm of pde's. We consider a fluid governed in bounded domain by the nonstationary Stokes equation in which case the solutions to modeling equations depend both on the forcing terms and instantaneous fluid velocity at every point of the space domain (as the initial condition), as well as on the instantaneous shape *and* position of swimmer in the latter. In other words, *our respective governing model equations pose a distributed-parameter infinite dimensional problem.* (We discuss the geometric aspects of our approach in Chapter 13 below.)

It was recognized that sophistication and complexity of design of bio-mimetic robots gives rise to control-theoretic methods. Available results in this area deal with control problems in the framework of ODE's, e.g., Koiller et al. [91] (1996); McIsaac and Ostrowski [117] (2000); Martinez and Cortes [118] (2001); Trintafyllou et al. [143] (2000); San Martin et al. [133] (2007), Alouges et al. [2] (2008), Sigalotti and Vivalda [135] (2009), and the references therein.

For example, in [2] the authors studied controllability properties of a simple yet "unconventional" swimmer due to [123]. It consists of three identical balls, which are assumed to be located on the same straight line in R^3 at all times, while acting upon each other in a push/pull fashion:

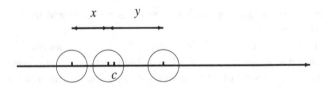

Swimmer consisting of three balls on the same line.

The state of this swimmer is given by (x, y, c), where c is the center of mass of swimmer, x and y are the distances between the centers of balls as shown. Each ball is acted upon by a force $f_i, i = 1, 2, 3$ (coming from the other balls) which is transmitted to the surrounding fluid, governed by the *stationary Stokes equation* in R^3. *This choice of fluid equation allowed* the authors of [2] to conclude that the relation between forces and velocities of the balls, denoted by $u^{(i)}, i = 1, 2, 3$ is linear and thus can be described as follows:

$$\begin{pmatrix} u^{(1)} \\ u^{(2)} \\ u^{(3)} \end{pmatrix} = S(x,y) \begin{pmatrix} f_1 \\ f_2 \\ f_2 \end{pmatrix}, \tag{10.7}$$

where $S(x,y)$ is the Oseen matrix [123]. The authors of [2] showed that the center of mass c satisfies the one dimensional autonomous ode,

$$\frac{dc}{dt} = V_x(x,y)\frac{dx}{dt} + V_y(x,y)\frac{dy}{dt},$$

where $V_x(x,y)$ and $V_x(x,y)$ *are determined by* $S(x,y)$ *and* $\frac{dx}{dt}$ *and* $\frac{dy}{dt}$ *serve as multiplicative controls.* Making use of (implicit) Lie algebra technique it was established that the above finite dimensional system is globally controllable in the sense that it can reach any state (x_1, y_1, c_1) from any initial state (x_0, y_0, c_0) on the given straight line by selecting suitable controls.

These results from [2] are very interesting in the sense that they demonstrate a principal possibility of self-propulsion in a fluid by applying a suitable push/pull swimming technique. However, the finite-dimensional model (10.7), derived in [2] making use of the stationary Stokes equation in the whole space, is quite different from what one would expect to happen to the above-described 3-ball swimmer in reality. One, of course, should not expect that in the "real-world" *nonstationary* 3-*D* fluid this swimmer will be traveling along the same line all the time (namely, due to the motion of the fluid and, in particular, due to the distortions introduced to it by the swimmer itself). We can interpret this remark as that a more accurate model of this swimmer (in this respect) has to be infinite dimensional.

In [133] and [135] the authors followed the same approach, namely, first they reduced the original pde model for the motion of cilia to a system of nonlinear ODE's, governed by *additive control*, and then studied its controllability.

Contrary to the above-cited results, in this chapter (as well as in [76–82] (2005–2009) *we approach the controllability problem for swimming models as that for a coupled (non-stationary) pde system, thus leaving it in its original infinite dimensional realm.*

It should be noted that in the above-outlined modeling results the classical mathematical issues of wellposedness (such as existence, uniqueness and regularity of solutions) were not addressed (in the context of pde's). Apparently, for the first time they were studied by Galdi [51] (1999) for a model of swimming micromotions in R^3 (with swimmer as reference frame).

In [76] (2005) (see also [77–82]) we introduced a swimming model, governed in bounded domain by a fluid equation coupled with a system of ODE's describing the spatial position of swimmer. We established wellposedness of this model (in respective function spaces) up to the contact of swimmer either with the space boundary or with itself (see Chapters 11 and 12 below). Our model was derived from the above-mentioned modeling approach due to Peskin in the case when swimmer is "small" and is identified with the fluid it occupies. The need of such type of models was motivated by the intention to investigate their controllability properties. In Chapters 14 and 15 below we give rigorous formulation of the controllability problems of our interest in this chapter.

In the case when a swimmer is a solid body, performing certain prescribed undulatory self-deformations, a similar model was considered in the recent paper by San Martin et al. [134] (2008), namely, with the model equations including an ode for a swimmer, coupled with a fluid equation. The authors established (in a different technical framework) a similar wellposedness result – "up to contact" (with the boundary). We will discuss [134] in the end of Chapter 12 in more detail.

Chapter 11
A "Basic" 2-*D* Swimming Model

Abstract We introduce a "basic" mathematical model of a swimmer in a fluid, governed within a bounded 2-*D* domain by the nonstationary Stokes equation. Its body consists of finitely many subsequently connected small sets each of which can act upon any adjacent set in a rotation fashion with the purpose to generate its fishlike or rowing motion. The shape of the object is maintained by respective elastic forces.

11.1 Modeling Philosophy

Our main concern with respect to the choice of modeling approach is that even the "simplest" swimming models tend to be highly nonlinear, which makes it difficult to approach them from the controllability theory viewpoint. Indeed, in its current state, the mathematical controllability theory is very comprehensive and thorough when dealing with linear pde's. The developed linear controllability methods were quite successfully extended to the semilinear pde's in the case when the nonlinear terms (as well as controls) are *additive* and are to be viewed and dealt with as though they are *generic nonlinear disturbances (to be "overwhelmed")*. Clearly, this is not the case of highly nonlinear swimming models where the specifics of nonlinear mapping *control → system's state* is the crux of the problem. From this viewpoint, a suitable modeling approach can be a critical tool to understanding the controllability side of swimming phenomenon.

In this chapter we introduce a simplified swimming model with the intention to try to describe the swimming process on, what we call, an "elemental level". Our goal is to attempt to distinguish principal elements which play most crucial role in the process of how a "swimmer" can propel itself in a fluid. The latter means that this must be achieved by internal forces, in which case the swimming motion can occur *only* if swimmer interacts with surrounding medium. Respectively, the goal of our models is to isolate and expose certain (but, of course, not all) aspects of this interaction, which, as we believe, are most critical for any self-propelling motion in a fluid.

A.Y. Khapalov, *Controllability of Partial Differential Equations Governed by Multiplicative Controls*, Lecture Notes in Mathematics 1995, DOI 10.1007/978-3-642-12413-6_11, © Springer-Verlag Berlin Heidelberg 2010

More precisely, we would like to explore an empiric observation that a body will move in a fluid in the direction where it meets least resistance. To this end:

- Our swimmer is described below as a collection of small separate sets of given shape, which, to achieve the desirable motion, can change their spatial orientation. (The objective of controllability theory is exactly to find out if this is possible.)
- We identify the swimmer with fluid it occupies. This seems to be a reasonable assumption if swimmer is "small" or "narrow". (Recall along these lines that in many theoretical works a solid body is viewed as a limit of a sequence of fluids of increasing density occupying its volume.)

As we mentioned before, our modeling approach can be viewed as one derived from the approach developed by C.S. Peskin in computational mathematical biology, where an object in a fluid is modeled as an immaterial curve (immersed boundary), identified with the fluid, further discretized for computational purposes on some grid. In turn, our swimmer in model (11.1)–(11.3) can be viewed as such already discretized curve supported on the respective cells of the aforementioned grid, see Figs. 1–3.

The equations (11.1)–(11.3) below resemble, in particular, (10.4)–(10.6) in the above and the equations (2.9) in [125], p. 223, where an object in a fluid is modeled as a collection of countably many points linked by internal forces instead of our finitely many "thick" points, which allows us to "replaces" the δ-functions in the limit description of the forcing term in [125] with the integral terms in (11.2) and a finite sum in (11.3) below involving "more analysis-friendly" characteristic functions.

11.2 Description of a "Basic" Swimming Model

We consider the following model, consisting of a fluid pde:

$$\frac{\partial y}{\partial t} = \nu \Delta y + F(z, v) - \nabla p \quad \text{in} \quad Q_T = \Omega \times (0, T), \tag{11.1}$$

$$\text{div}\, y = 0 \ \text{in} \ Q_T, \quad y = 0 \ \text{in} \ \Sigma_T = \partial\Omega \times (0, T), \quad y|_{t=0} = y_0 \ \text{in} \ \Omega,$$

coupled with a system of ode's for the position of swimmer in it:

$$\frac{dz_i}{dt} = \frac{1}{\text{mes}\,\{S(0)\}} \int_{S(z_i(t))} y(x, t)dx, \quad z_i(0) = z_{i0}, \quad i = 1, \dots, n, \quad n > 2, \tag{11.2}$$

where for $t \in [0, T]$

$$z(t) = (z_1(t), \ldots, z_n(t)), \ z_i(t) \in R^2, \ i = 1, \ldots, n,$$
$$v(t) = (v_1(t), \ldots, v_{n-1}(t)) \in R^{n-1},$$

$$(F(z,v))(x,t) = \sum_{i=2}^{n+1} \xi_{i-1}(x,t)[k_{i-1}\frac{(\| z_i(t) - z_{i-1}(t) \|_{R^2} - l_{i-1})}{\| z_i(t) - z_{i-1}(t) \|_{R^2}}(z_i(t) - z_{i-1}(t))$$
$$+ k_{i-2}\frac{(\| z_{i-2}(t) - z_{i-1}(t) \|_{R^2} - l_{i-2})}{\| z_{i-2}(t) - z_{i-1}(t) \|_{R^2}}(z_{i-2}(t) - z_{i-1}(t))]$$
$$+ \sum_{i=2}^{n+1} \xi_{i-1}(x,t)$$
$$\times \left(v_{i-1}(t) \begin{pmatrix} 0 & 1 \\ -1 & 0 \end{pmatrix} (z_i(t) - z_{i-1}(t)) \right.$$
$$\left. + v_{i-2}(t) \begin{pmatrix} 0 & 1 \\ -1 & 0 \end{pmatrix} (z_{i-2}(t) - z_{i-1}(t)) \right). \tag{11.3}$$

In the above, Ω is a bounded domain in R^2 with boundary $\partial \Omega$ of class C^2, $y = (y_1(x,t), y_2(x,t))$ and $p(x,t)$ are respectively the velocity and the pressure of the fluid at point $x = (x_1, x_2) \in \Omega$ at time t, while v is a kinematic viscosity constant. Also, to simplify the Σ-notations in (1.3) and below, *throughout Part III*, we use two auxiliary fictitious points z_0 and z_{n+1} as $z_0(t) = z_1(t), z_n(t) = z_{n+1}(t)$ and setting accordingly $v_0 = v_n = k_0 = k_n = l_0 = l_n = 0$ (see below for more details).

Let us explain the terms in (11.1)–(11.3) in more detail.

Swimmer's body. The swimmer in (11.1)–(11.3) is modeled as a collection of finitely many points subsequently connected by flexible immaterial (or, say, which have a "negligible affect" on the swimming process) internal links. Each of these points is surrounded by *"small"* support, identified with the fluid its occupies, see Fig. 1.

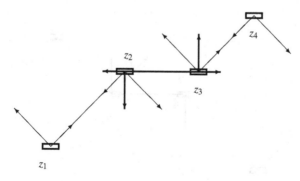

Fig. 11.1 The case n = 4

At any given moment of time the swimming object is represented by a "broken-line" structure, formed by an ordered sequence of "thick points" $S(z_1(t)), \ldots,$ $S(z_n(t))$ (on Figures 1–3 they are rectangles), where $z_i(t), i = 1, \ldots, n$ are points in Ω ("swimmer's skeleton"). Respectively, ξ_i's denote the characteristic functions of $S(z_i(t))$'s:

$$\xi_i(x,t) = \begin{cases} 1, & \text{if } x \in S(z_i(t)), \\ 0, & \text{if } x \in \Omega \backslash S(z_i(t)), \end{cases} \quad i = 1, \ldots, n. \tag{11.4}$$

Throughout Part III we assume that $S(0)$ is the given set of nonzero measure, lying in a "small" neighborhood of the origin of radius $r > 0$ with its center of mass at the origin (i.e., if $S(0)$ is treated as a thin plate). Some of our results below also require that $S(0)$ is open, which the reader can always assume for simplicity. $S(a)$ denotes the set $S(0)$ shifted to point a without changing its orientation in space.

Forces. The term $F(z,v)$ in (11.3) represents the *internal forces* (their sum is zero) generated by the swimmer, acting in turn as *external forces* upon the fluid in the fluid equation (11.1). We assume that all swimmer's forces act through the immaterial links attached to the centers of mass of sets $S(z_i(t))$'s, i.e., to the points $z_i(t)$'s, and then transmitted as such to all points in their respective supports. The latter will create a pressure upon the surrounding fluid, thus acting as external forces upon it.

Each of the points $z_i(t)$ can force any of the adjacent points to "rotate" about it. In turn, by the 3-rd Newton's Law, the affected point will act back on $z_i(t)$ with the opposite force. For example, $z_1(t)$ can act upon $z_2(t)$ with the force perpendicular to the vector $z_2(t) - z_1(t)$ and $z_2(t)$ will act back with the opposite force. These two forces, being transmitted to their respective supports provide two terms in the last line in (11.3), namely,

$$\xi_1(x,t)v_1(t) \begin{pmatrix} 0 & 1 \\ -1 & 0 \end{pmatrix} (z_2(t) - z_1(t)) + \xi_2(x,t)v_1(t) \begin{pmatrix} 0 & 1 \\ -1 & 0 \end{pmatrix} (z_1(t) - z_2(t)).$$

The magnitudes and directions of the applied rotation forces (shown on Fig. 2) are determined by the coefficients $v_i, i = 1, \ldots, n-1$, which we regard as *multiplicative* controls.

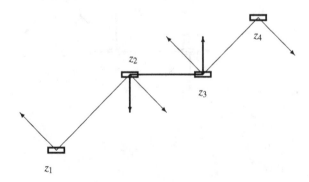

Fig. 11.2 Controlling rotation forces, n=4

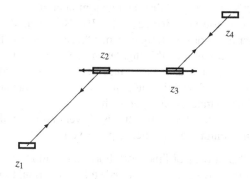

Fig. 11.3 Elastic forces, n = 4

The shape of the swimmer is preserved by the elastic forces (shown on Fig. 3 which, in general, act according to Hooke's Law when the distances between any two adjacent points $z_i(t)$ and $z_{i-1}(t), i = 2,\ldots,n$ deviate from the respective given values

$$l_{i-1} > 0, \quad i = 2,\ldots,n, \tag{11.5}$$

as described in the first two lines in (11.3), where the given parameters $k_i > 0$, $i = 1,\ldots,n-1$ characterize the rigidity of the links $z_{i-1}(t)z_i(t)$, $i = 2,\ldots,n$. (For the auxiliary points/links we set $k_0 = k_n = l_0 = l_n = 0$.) In Chapters 11–14 we assume that the coefficients k_i's are constant, while in Chapter 15 we also consider the case when they are variable and serve as *additional multiplicative controls*.

Remark 11.1. Note that, when the adjacent points in the swimming object share the same position in space, the forcing term F in (11.3) and hence the model (11.1)–(11.3) become undefined. While such development seems physically plausible, even if, prior to this, the solution to this system exists on some time-interval, it does not necessarily have to happen. The former issue, namely, the existence on some time-interval $(0,T)$, is addressed below in the next section. The latter issue can be viewed as the issue of *controllability*, namely, when one tries to select available controls with the purpose to ensure that swimmer avoids the aforementioned ill-defined situation.

Position of swimmer in the fluid. The motion of swimmer in (11.1)–(11.3) is described through the motion of the points $z_i(t), i = 1,\ldots,n$, whose velocities are determined by the average motion of the fluid within the respective supports $S(z_i(t))$'s as described in (11.2).

Internal forces and the conservation of momenta:

- We want to emphasize again that all forces in (11.3) satisfy the 3-rd Newton's Law and their sum is equal to zero. Thus, they are internal with respect to the swimmer and cannot move its center of mass without interaction with the fluid. This is the principal feature of an object which is "swimming-by-itself".

- The 3-rd Newton's Law ensures that the linear momentums, generated by swimmer's forces, are conserved (see, e.g., [137]). However, the rotation forces produce, in general, a nonzero torque in our "basic model (11.1)–(11.3). This means that the conservation of the angulaɪ momentums should hold in a more general framework, which also takes into account some "additional control forces" (from an "engine" with its mutually counter-rotating parts), also internal with respect to the apparatus, that generate the corresponding "negating" torques. Nonetheless, in Chapter 12 we also discuss a version of model (11.1)–(11.3) whose angular momentums are explicitly preserved.

Comments on the choice of fluid equation. In our models in this chapter we chose the fluid governed by the *nonstationary* Stokes equation. Let us elaborate on this choice:

In the publications discussed in the beginning of this chapter, a typical choice of fluid for microswimmers (as in our case) is the fluid governed by the *stationary* Stokes equation. The empiric reasoning behind this is that, due to the small size of swimmer, the inertia terms in the Navier-Stokes equation, containing the 1-st order derivatives in t and x, can be omitted, provided that the frequency parameter σ is a quantity of order unity (see the discussion in Chapter 10 around (10.1)–(10.3)). However, it was noted that a microswimmer (e.g., a nano-size robot) may use a rather high frequency of motion, which may justify at least in some cases the need for the term y_t in the model equations. In general, it seems reasonable to suggest that the presence of this term in the equation (in a number of cases) can provide a better approximation of the Navier-Stokes equation than the lack of it. We also point out that in [40, 41, 126] the full-size Navier-Stokes equation is used for microswimmers. It also seems that the methods we use in this chapter, i.e., developed for the nonstationary Stokes equation (as opposed to stationary Stokes equation), may serve as a natural step toward the swimming models, based on the Navier-Stokes equation.

The models and results below are formulated for swimming objects that consist of identical small sets. They can be extended, of course, to the case when these sets are different. However, in the latter case the respective sets must be of the same measure to preserve the 3-rd Newton's Law as given in (11.3) (otherwise, one needs to make respective adjustments).

Chapter 12
The Well-Posedness of a 2-D Swimming Model

Abstract We introduce a "more complex" version of the swimming model from Chapter 11, namely, for which both the linear and angular momenta are conserved. Then we discuss the issues of existence, uniqueness and continuity for its solutions.

12.1 A "Basic" Swimming Model with Conserved Linear and Angular Momenta

In this chapter we consider a different, somewhat "more complex" version of the swimming model from Chapter 11, namely, as follows:

$$\frac{\partial y}{\partial t} = v \Delta y + F(z,v) - \nabla p \text{ in } Q_T = \Omega \times (0,T), \tag{12.1}$$

$$\text{div } y = 0 \text{ in } Q_T, \quad y = 0 \text{ in } \Sigma_T = \partial \Omega \times (0,T), \quad y|_{t=0} = y_0,$$

$$\frac{dz_i}{dt} = \frac{1}{\text{mes}(S(0))} \int_{S(z_i(t))} y(x,t)dx, \quad z_i(0) = z_{i,0}, \quad i = 1,\ldots,n, \tag{12.2}$$

where

$$z(t) = (z_1(t),\ldots,z_n(t)) \in [R^2]^n, \quad v(t) = (v_1(t),\ldots,v_{n-2}(t)) \in R^{n-2},$$

$$(F(z,v))(x,t) = \sum_{i=2}^{n} \left[\xi_{i-1}(x,t)k_{i-1} \frac{(\|z_i(t) - z_{i-1}(t)\|_{R^2} - l_{i-1})}{\|z_i(t) - z_{i-1}(t)\|_{R^2}} (z_i(t) - z_{i-1}(t)) \right.$$

$$\left. + \xi_i(x,t)k_{i-1} \frac{(\|z_i(t) - z_{i-1}(t)\|_{R^2} - l_{i-1})}{\|z_i(t) - z_{i-1}(t)\|_{R^2}} (z_{i-1}(t) - z_i(t)) \right]$$

$$+ \sum_{i=2}^{n-1} v_{i-1}(t) \times \left[\xi_{i-1}(x,t)A(z_{i-1}(t) - z_i(t)) \right.$$

$$\left. - \xi_{i+1}(x,t) \frac{\|z_{i-1}(t) - z_i(t)\|^2}{\|z_{i+1}(t) - z_i(t)\|^2} A(z_{i+1}(t) - z_i(t)) \right]$$

A.Y. Khapalov, *Controllability of Partial Differential Equations Governed by Multiplicative Controls*, Lecture Notes in Mathematics 1995, DOI 10.1007/978-3-642-12413-6_12, © Springer-Verlag Berlin Heidelberg 2010

$$+ \sum_{i=2}^{n-1} \xi_i(x,t) v_{i-1}(t) \times \left[A(z_i(t) - z_{i-1}(t)) \right.$$

$$\left. - \frac{\| z_{i-1}(t) - z_i(t) \|^2}{\| z_{i+1}(t) - z_i(t) \|^2} A(z_i(t) - z_{i+1}(t)) \right], \qquad (12.3)$$

where

$$A = \begin{pmatrix} 0 & 1 \\ -1 & 0 \end{pmatrix}.$$

Mathematics-wise, the only difference between this model and that of Chapter 11 is the shape of the forcing term $F(z,v)$. Note, however, that $F(z,v)$ from (12.3) is a particular example of $F(z,v)$ from (11.3) in Chapter 11.

Once again, as in Chapter 11, the 1-st sum on the right in (12.3) describes the elastic forces intended to preserve the structure of the swimming object, see Fig. 3 in Chapter 11. Respectively, the 2-nd sum in (12.3) (where matrix A is the same as in Chapter 11) describes the fact that each of the "middle points" $z_i(t)$ for $i = 2, \ldots, n-1$ can force the two adjacent points to rotate about it. The magnitudes and directions of the rotation forces are determined by the coefficients $v_i(t)$, $i = 1, \ldots, n-2$. The 3-rd sum on the right in (12.3) represents the effect of the above-described rotation forces on the respective "middle points" generating them, in accordance with Newton's 3rd Law, see Figures 1–3.

Due to Newton's 3rd Law, the sum of the internal forces acting on the swimming object is zero and cannot move its center of mass without interacting with the surrounding fluid.

Conservation of momenta. As before in Chapter 11, Newton's 3-rd Law is sufficient for the conservation of linear momentum. Furthermore, since the elastic forces described in the first sum in (12.3) act along the lines connecting the centers of mass of adjacent thick points, the angular momentum due to these forces is conserved as well.

We will now show that angular momentum, which is due to the rotation forces described in the second sum in (12.3), is conserved. We begin by computing the

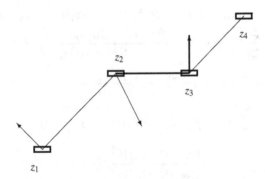

Fig. 12.1 Controlling rotation forces, generated by the point z_2 only, n$=$4

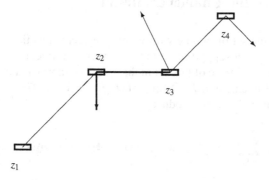

Fig. 12.2 Controlling rotation forces, generated by the point z_3 only, n$=$4

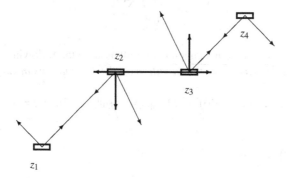

Fig. 12.3 All forces, n$=$4

torque T_2 generated by the pair of rotation forces acting adjacent to the 1-st "middle point" $z_2(t)$ (see Fig. 1). Making use of (3.45) on p. 83 in [137], we obtain:

$$T_2 = -\|v_1(t)A(z_1(t) - z_2(t))\|_{R^2}\|z_1(t) - z_2(t)\|_{R^2}$$
$$+ \left\|v_1(t)\frac{\|z_1(t) - z_2(t)\|_{R^2}^2}{\|z_3(t) - z_2(t)\|_{R^2}^2}(-A)(z_3(t) - z_2(t))\right\|_{R^2}\|z_3(t) - z_2(t)\|_{R^2}$$
$$= -|v_1(t)|\|z_1(t) - z_2(t)\|_{R^2}^2 + |v_1(t)|\|z_1(t) - z_2(t)\|_{R^2}^2 = 0.$$

We can compute the torques $T_3,\ldots,T_{n-1}$ generated by the pair of respective rotation forces as described in the second term of (12.3) about the thick points $S(z_3(t)),\ldots,$ $S(z_{n-1}(t))$ in a similar manner. Thus, the angular momentum resulting from the swimming object's internal forces is conserved.

12.2 Existence and Uniqueness Result

Let $J(\Omega)$ denote the set of infinitely differentiable vector functions with values in R^2 which have compact support in Ω and are divergence-free, i.e., $\mathrm{div}\phi = 0$ in Ω. Denote by $J_o(\Omega)$ the closure of this set in the $(L^2(\Omega))^2$-norm and by $G(\Omega)$ denote the orthogonal complement of $J_o(\Omega)$ in $(L^2(\Omega))^2$ (see, e.g., [97]).

In $J(\Omega)$ introduce the scalar product

$$[\phi_1, \phi_2] = \int_\Omega \sum_{j=1}^{2} \sum_{i=1}^{2} \phi_{1\,j x_i} \phi_{2\,j x_i}\, dx, \quad \phi_1(x) = (\phi_{11}, \phi_{12}), \quad \phi_2(x) = (\phi_{21}, \phi_{22}).$$

Denote by $H(\Omega)$ the Hilbert space which is the completion of $J(\Omega)$ in the norm

$$\|\phi_1\|_{H(\Omega)} = \sqrt{\int_\Omega \sum_{j=1}^{2} \sum_{i=1}^{2} \phi_{1\,j x_i}^2\, dx}.$$

Henceforth throughout this chapter, we will assume the following:
Assumption P *Assume that (in addition on the assumptions in Chapter 11)*

$$l_{i-1} > 2r, \ i = 2,\dots,n; \ \ \overline{S}(z_i(0)) \subset \Omega, \ \ \|z_{i,0} - z_{j,0}\|_{R^2} > 2r, \ i,j = 1,\dots,n, \ i \neq j;$$

$$(12.4)$$

and the set $S(0)$ is such that

$$\int_{(S(0)\cup S(h))\setminus(S(0)\cap S(h))} dx = \int_\Omega |\xi(x) - \xi(x - h)|dx \leq C\|h\|_{R^2} \quad \forall h \in B_{h_0}(0)$$

$$(12.5)$$

for some positive constants h_0 and C, where $\xi(x)$ is the characteristic function of $S(0)$ and $B_{h_0}(0) = \{x \,|\, \|x\|_{R^2} < h_0\} \subset R^2$.

In the above, conditions (12.4) mean that at time $t = 0$, any two thick points $S(z_i(0))$'s (not only adjacent) forming our swimming object do not overlap, and that the apparatus as a whole lies strictly in Ω. Also, condition (12.5) is essentially a regularity assumption of Lipschitz type regarding the shift of the set $S(0)$. It is satisfied, for instance, for circles and rectangles.

Throughout Chapters 11–15 we assume that $y_0 \in H(\Omega)$.

We now proceed to the main result of this section.

Theorem 12.1. *Let $y_0 \in H(\Omega)$; $T > 0$; $k_i > 0$, $i = 1,\dots,n-1$; $v_i \in L^\infty(0,T)$, $i = 1,\dots,n-2$; and $z_{i,0} \in \Omega$, $i = 1,\dots,n$, and let Assumption P hold. Then there exists a $T^* = T^*(z_{1,0},\dots,z_{n,0}, \|v_1\|_{L^\infty(0,T)}, \dots, \|v_{n-2}\|_{L^\infty(0,T)}, \Omega) \in (0,T)$ such that system (12.1) - (12.3), (12.4), (12.5) admits a unique solution $\{y,p,z\}$ on $(0,T^*)$, $\{y, \nabla p, z\} \in L^2(0,T^*; J_o(\Omega)) \times L^2(0,T^*; G(\Omega)) \times [C([0,T^*]; R^2)]^n$. Moreover, $y \in$*

$C([0,T^*];H(\Omega))$, $y_t, y_{x_i x_j}, p_{x_i} \in (L^2(Q_{T^*}))^2$, *where* $i, j = 1, 2$, *and equations (12.1)*
and (12.2) are satisfied almost everywhere, while

$$\|z_i(t) - z_j(t)\|_{R^2} > 2r, \ i, j = 1, \ldots, n, i \neq j; \ \overline{S}(z_i(t)) \subset \Omega, \ t \in [0, T^*], \ i = 1, \ldots, n. \tag{12.6}$$

Remark 12.1. • The conditions contained in (12.6) imply that for the solution of
(12.1) - (12.3), (12.4), (12.5), whose existence is established in Theorem 12.1, we
are able to guarantee for some interval $[0, T^*]$ (see (12.41)) that no parts of the
swimming object will "collide", and simultaneously, that it stays strictly inside
of Ω. These conditions allow us to maintain the mathematical and physical wellposedness of the model (12.1) - (12.3).
• As it will follow from the proof below, the existence result of Theorem 12.1 is
not necessarily of "local" nature in time. Namely, conditions (12.6), treated as
new initial conditions, allow further extension in time of the solutions to (12.1) -
(12.3). This depends on the choice of "controls" $v_1(t), \ldots, v_{n-2}(t)$, and can be
viewed as the goal of the respective *multiplicative controllability problem*.

Our plan to prove Theorem 12.1 is to proceed stepwise as follows:
In Section 12.3, we will prove the existence and uniqueness of the solutions to
the decoupled version of (1.2). Then, in Section 12.4, based in part on the properties
of these solutions, we will construct three continuous mappings for the decoupled
version of the system (12.1) - (12.3). In Section 12.5, we will apply a fixed point
argument to prove that the system (12.1) - (12.3), (12.4), (12.5) has a solution with
required regularity properties, and prove that this solution is unique. In addition,
in order to avoid the collision of any part of the swimming object with itself or
the boundary of Ω, we will derive suitable a priori estimates for the "decoupled"
solutions of (12.2) and for (12.1) so that we can carry such estimates over to the
solution of (12.1) - (12.3), (12.4), (12.5) via the fixed point argument, while at the
same time ensuring (12.6) to avoid such collision.

Corollary 12.1. *Since the forcing term (11.3) in Chapter 11 is a particular "simpler" version of the term (12.3), Theorem 12.1 implies the same well-posedness
result for the model (11.1) - (11.3) in Chapter 11 as well.*

12.3 Preliminary Results

Consider the following decoupled version of system (12.2):

$$\frac{dw_i}{dt} = \frac{1}{\mathrm{mes}(S(0))} \int_{S(w_i(t))} u(x,t)dx, \ \ w_i(0) = z_{i,0}, \ \ i = 1, \ldots, n, \tag{12.7}$$

where $u(x,t)$ is some given function.

Lemma 12.1. *Let $T > 0$ and $u \in (L^2(0,T;L^\infty(\Omega)))^2$ be given. Then there is a $T^* \in (0,T)$ such that system (12.7) has a unique solution in $C([0,T^*];R^2)$ satisfying (12.6).*

Proof. We will use the contraction principle to prove existence and uniqueness.

For h_0, C from (12.5), select:

$$0 < T_0 < \min\left\{\frac{\text{mes}(S(0))h_0^2}{4\|u\|_{(L^2(Q_T))^2}^2}, \frac{(\text{mes}(S(0)))^2}{C^2\|u\|_{(L^2(0,T;L^\infty(\Omega)))^2}^2}, T\right\}. \tag{12.8}$$

Denote

$$B_{h_0/2}(0) = \left\{z \in C([0,T_0];R^2) \mid \|z\|_{C([0,T_0];R^2)} \le \frac{h_0}{2}\right\} \subset C([0,T_0];R^2).$$

For each $i = 1,\ldots,n$, define a mapping $D_i : B_{h_0/2}(z_{i,0}) \longrightarrow C([0,T_0];R^2)$ by

$$D_i(w_i(t)) = z_{i,0} + \frac{1}{\text{mes}(S(0))}\int_0^t \int_{S(w_i(\tau))} u(x,\tau)dxd\tau.$$

We have the following estimate:

$$\|D_i(w_i(t))\|_{R^2} = \left\|z_{i,0} + \frac{1}{\text{mes}(S(0))}\int_0^t \int_{S(w_i(\tau))} u(x,\tau)dxd\tau\right\|_{R^2}$$

$$\le \|z_{i,0}\|_{R^2}$$

$$+ \frac{\sqrt{T_0}}{\sqrt{\text{mes}(S(0))}}\|u\|_{(L^2(Q_T))^2} \quad \forall t \in [0,T_0], \ i = 1,\ldots,n. \tag{12.9}$$

Similarly, in view of (12.8):

$$\|D_i(w_i(t)) - z_{i,0}\|_{C([0,T_0];R^2)} \le \frac{\sqrt{T_0}}{\sqrt{\text{mes}(S(0))}}\|u\|_{(L^2(Q_T))^2} < \frac{h_0}{2}. \tag{12.10}$$

Thus, D_i maps $B_{h_0/2}(z_{i,0})$ into itself for each $i = 1,\ldots,n$.

Let $w_i^{(1)}(t), w_i^{(2)}(t) \in B_{h_0/2}(z_{i,0})$. Denote the characteristic function of the set $S \subset R^2$ by $\xi(x,S)$. Note that $\|w_i^{(1)}(t) - w_i^{(2)}(t)\|_{R^2} \le h_0$ for all $t \in [0,T_0]$, and so using (12.5) we obtain for any $t \in [0,T_0], i = 1,\ldots,n$:

$$\|D_i(w_i^{(1)}(t)) - D_i(w_i^{(2)}(t))\|_{R^2}$$

$$= \frac{1}{\text{mes}(S(0))}\left\|\int_0^t \int_{S(w_i^{(1)}(t))} u(x,\tau)dxd\tau - \int_0^t \int_{S(w_i^{(2)}(\tau))} u(x,\tau)dxd\tau\right\|_{R^2}$$

$$= \frac{1}{\text{mes}(S(0))}\left\|\int_0^t \int_\Omega u(x,\tau)\left(\xi(x,S(w_i^{(1)}(t))) - \xi(x,S(w_i^{(2)}(\tau)))\right)dxd\tau\right\|_{R^2}$$

$$\leq \frac{1}{\text{mes}(S(0))} \left\| \int_0^t \int_\Omega u(x,\tau) |\xi(x) - \xi(x - (w_i^{(1)}(\tau) - w_i^{(2)}(\tau)))| dx d\tau \right\|_{R^2}$$

$$\leq \frac{1}{\text{mes}(S(0))}$$

$$\times \sqrt{\sum_{j=1}^2 \left(\int_0^t \|u_j(\cdot,\tau)\|_{L^\infty(\Omega)} \int_\Omega |\xi(x) - \xi(x - (w_i^{(1)}(\tau) - w_i^{(2)}(\tau)))| dx d\tau \right)^2}$$

$$\leq \frac{1}{\text{mes}(S(0))}$$

$$\times \sqrt{\int_0^t \left(\int_\Omega |\xi(x) - \xi(x - (w_i^{(1)}(\tau) - w_i^{(2)}(\tau)))| dx \right)^2 d\tau \sum_{j=1}^2 \int_0^t \|u_j(\cdot,\tau)\|_{L^\infty(\Omega)}^2 d\tau}$$

$$\leq \frac{C\sqrt{T_0}}{\text{mes}(S(0))} \|u\|_{(L^2(0,T;L^\infty(\Omega)))^2} \|w_i^{(1)}(t) - w_i^{(2)}(t)\|_{C([0,T_0];R^2)}. \tag{12.11}$$

Therefore, after maximizing the left-hand side of (12.11) over $[0,T_0]$, we conclude:

$$\|D_i(w_i^{(1)}(t)) - D_i(w_i^{(2)}(t))\|_{C([0,T_0];R^2)}$$

$$\leq \frac{C\sqrt{T_0}}{\text{mes}(S(0))} \|u\|_{(L^2(0,T;L^\infty(\Omega)))^2} \|w_i^{(1)} - w_i^{(2)}\|_{C([0,T_0];R^2)}. \tag{12.12}$$

Hence, in view of (12.8),

$$\frac{C\sqrt{T_0}}{\text{mes}(S(0))} \|u\|_{(L^2(0,T;L^\infty(\Omega)))^2} < 1,$$

it follows from (12.12) that D_i is a contraction mapping on $B_{h_0/2}(z_{i,0})$ for each $i = 1, \ldots, n$. Therefore, there exists a unique $w_i(t) \in C([0,T_0];R^2)$ such that $D_i(w_i(t)) = w_i(t)$, i.e.,

$$w_i(t) = z_{i,0} + \frac{1}{\text{mes}(S(0))} \int_0^t \int_{S(w_i(\tau))} u(x,\tau) dx d\tau, \quad i = 1, \ldots, n, \tag{12.13}$$

which yields (12.7).

Remark 12.2. The selection of T_0 in (12.8) is time invariant. Hence, we can extend the above proof of existence to $[T_0, 2T_0], [2T_0, 3T_0], \ldots$, until we reach T.

Condition (12.6). Estimates like (12.9)–(12.10) (with the solution $w_i(t)$ in place of $D_i(w_i(t))$) imply that we may select a $T^* \in (0, T_0)$ such that for any $t \in [0, T^*]$, all $w_i(t)$'s will stay "close enough" to their initial values $z_{i,0}$'s in order that (12.6) holds for $w_i(t)$, $i = 1, \ldots, n$. In particular, let

$$d_{R^2}(S(z_{i,0}), \partial\Omega) = \inf\{\|x - y\|_{R^2} \mid x \in S(z_{i,0}), y \in \partial\Omega\}$$

denote the distance in R^2 between $S(z_{i,0})$ and the boundary of Ω. In view of (12.4), select $T^* \in (0, T_0)$ such that:

$$0 < T^* < \min\left\{ \frac{\text{mes}(S(0))}{4\|u\|^2_{(L^2(Q_T))^2}} \left(\min_{i=1,\ldots,n} \{d_{R^2}(S(z_i(0)), \partial\Omega)\} \right)^2, \right.$$

$$\left. \frac{\text{mes}(S(0))}{4\|u\|^2_{(L^2(Q_T))^2}} \left(\min_{\substack{i,j=1,\ldots,n \\ i\neq j}} \{\|z_{i,0} - z_{j,0}\|_{R^2}\} - 2r \right)^2, T_0 \right\}. \tag{12.14}$$

Hence, applying (12.10) to the solution $w_i(t)$, for $t \in [0, T^*]$ and each $i = 1, \ldots, n$ we have:

$$\|w_i(t) - z_{i,0}\|_{R^2} \leq \frac{\sqrt{T^*}}{\sqrt{\text{mes}(S(0))}} \|u\|_{(L^2(Q_T))^2} < \frac{1}{2} \left(\min_{\substack{i,j=1,\ldots,n \\ i\neq j}} \{\|z_{i,0} - z_{j,0}\|_{R^2}\} - 2r \right).$$

This inequality and (12.4) imply that for $t \in [0, T^*]$ and each $i, j = 1, \ldots, n$, $i \neq j$:

$$0 < \|z_{i,0} - z_{j,0}\|_{R^2} < \min_{\substack{i,j=1,\ldots,n \\ i\neq j}} \{\|z_{i,0} - z_{j,0}\|_{R^2}\} - 2r + \|w_i(t) - w_j(t)\|_{R^2}.$$

So,

$$2r < \|w_i(t) - w_j(t)\|_{R^2}, \ \forall\, t \in [0, T^*], \forall\, i, j = 1, \ldots, n, \ i \neq j. \tag{12.15}$$

Additionally, (12.10) and (12.14) imply that for $t \in [0, T^*]$, and each $i = 1, \ldots, n$:

$$\|w_i(t) - z_{i,0}\|_{R^2} < \frac{1}{2} \min_{i=1,\ldots,n} \{d_{R^2}(S(z_i(0)), \partial\Omega)\} < \frac{1}{2} d_{R^2}(S(z_i(0)), \partial\Omega).$$

Hence $\bar{S}(w_i(t)) \subset \Omega$ for $t \in [0, T^*]$ and each $i = 1, \ldots, n$. Thus, (12.6) holds for (12.13) with the choice of T^* in (12.14). This ends the proof of Lemma 12.1.

12.4 Decoupled Solution Mappings

Let $\mathscr{B}_q(0)$ denote a closed ball of radius q (which will be selected in subsection 12.5) with center at the origin in the space $L^2(0, T; J_o(\Omega)) \cap L^2(0, T; (H^2(\Omega))^2)$ endowed with the norm of $L^2(0, T; (H^2(\Omega))^2)$:

$$\mathscr{B}_q(0) = \left\{ \phi \in L^2(0, T; J_o(\Omega)) \cap L^2(0, T; (H^2(\Omega))^2) \mid \|\phi\|_{L^2(0,T;(H^2(\Omega))^2)} \leq q \right\}.$$

Note that the space $(H^2(\Omega))^2$ is continuously embedded into $C(\bar{\Omega})$, and thus the space $L^2(0,T;(H^2(\Omega))^2)$ is continuously embedded into $(L^2(0,T;L^\infty(\Omega)))^2$. This yields the estimate

$$\|\phi\|_{(L^2(0,T;L^\infty(\Omega)))^2} \leq K\|\phi\|_{L^2(0,T;(H^2(\Omega))^2)} \quad \text{for some } K > 0. \tag{12.16}$$

This implies that Lemma 12.7 holds for any $u \in \mathscr{B}_q(0)$ if T is sufficiently small.

Solution mapping for $z_i(t), i = 1, \ldots, n$. We now intend to show that the operator

$$\mathbf{A} : \mathscr{B}_q(0) \longrightarrow [C([0,T];R^2)]^n, \quad \mathbf{A}u = w = (w_1, \ldots, w_n),$$

where the w_i's solve (12.7), is continuous and compact if $T > 0$ is sufficiently small.

Continuity. Let $u^{(1)}, u^{(2)} \in \mathscr{B}_q(0)$ with T_1 in place of T, where $T_1 > 0$ satisfies (12.14) with T_1 in place of T^*. Define $\mathbf{A}u^{(j)} = w^{(j)} = (w_1^{(j)}, \ldots, w_n^{(j)})$ for $j = 1, 2$. To show $\mathbf{A}$ is continuous, we will evaluate

$$\|\mathbf{A}u^{(1)} - \mathbf{A}u^{(2)}\|_{[C([0,T];R^2)]^n}$$

term-by-term. To this end, similar to (12.11), we have the following estimate:

$$\|w_i^{(1)}(t) - w_i^{(2)}(t)\|_{R^2}$$
$$= \left\| \frac{1}{\text{mes}(S(0))} \int_0^t \int_{S(w_i^{(1)}(\tau))} u^{(1)}(x,\tau)dxd\tau \right.$$
$$\left. - \frac{1}{\text{mes}(S(0))} \int_0^t \int_{S(w_i^{(2)}(\tau))} u^{(2)}(x,\tau)dxd\tau \right\|_{R^2}$$
$$= \frac{1}{\text{mes}(S(0))} \left\| \int_0^t \int_{S(w_i^{(1)}(\tau))} u^{(1)}(x,\tau)dxd\tau \right.$$
$$- \int_0^t \int_{S(w_i^{(1)}(\tau))} u^{(2)}(x,\tau)dxd\tau$$
$$\left. + \int_0^t \int_{S(w_i^{(1)}(\tau))} u^{(2)}(x,\tau)dxd\tau - \int_0^t \int_{S(w_i^{(2)}(\tau))} u^{(2)}(x,\tau)dxd\tau \right\|_{R^2}$$
$$\leq \frac{1}{\text{mes}(S(0))} \left(\left\| \int_0^t \int_{\Omega} (u^{(1)}(x,\tau) - u^{(2)}(x,\tau))\xi(x,S(w_i^{(1)}(\tau)))dxd\tau \right\|_{R^2} \right.$$
$$\left. + \left\| \int_0^t \int_{\Omega} u^{(2)}(x,\tau)(\xi(x,S(w_i^{(1)}(\tau))) - \xi(x,S(w_i^{(2)}(\tau))))dxd\tau \right\|_{R^2} \right)$$
$$\leq \frac{\sqrt{T_1}}{\sqrt{\text{mes}(S(0))}} \|u^{(1)} - u^{(2)}\|_{(L^2(Q_{T_1}))^2}$$
$$+ \frac{C\sqrt{T_1}}{\text{mes}(S(0))} \|u^{(2)}\|_{(L^2(0,T_1;L^\infty(\Omega)))^2} \|w_i^{(1)} - w_i^{(2)}\|_{C([0,T_1];R^2)}. \tag{12.17}$$

Recall from (12.16) that

$$\|u^{(2)}\|_{(L^2(0,T_1;L^\infty(\Omega)))^2} \le Kq.$$

So, select a $T > 0$ as follows:

$$0 < T < \min\left\{ \left(\frac{\mathrm{mes}(S(0))}{CKq}\right)^2, T_1 \right\}. \tag{12.18}$$

Hence, replacing T_1 in (12.17) with T satisfying (12.18) and maximizing the left-hand side of (12.17) over $[0, T]$, we obtain:

$$\|w_i^{(1)} - w_i^{(2)}\|_{C([0,T];R^2)} \le \frac{\sqrt{T}}{\sqrt{\mathrm{mes}(S(0))}} \|u^{(1)} - u^{(2)}\|_{(L^2(Q_T))^2}$$

$$+ \frac{CKq\sqrt{T}}{\mathrm{mes}(S(0))} \|w_i^{(1)} - w_i^{(2)}\|_{C([0,T];R^2)}.$$

In view of (12.18), if $w_i^{(1)}(t) \ne w_i^{(2)}(t)$ on $[0, T]$, then the above implies:

$$0 < \left(1 - \frac{CKq\sqrt{T}}{\mathrm{mes}(S(0))}\right) \|w_i^{(1)} - w_i^{(2)}\|_{C([0,T];R^2)} \le \frac{\sqrt{T}}{\sqrt{\mathrm{mes}(S(0))}} \|u^{(1)} - u^{(2)}\|_{(L^2(Q_T))^2}.$$

Thus, it follows that

$$\|w_i^{(1)} - w_i^{(2)}\|_{C([0,T];R^2)} \le \frac{\sqrt{T\mathrm{mes}(S(0))}}{\mathrm{mes}(S(0)) - CKq\sqrt{T}} \|u^{(1)} - u^{(2)}\|_{(L^2(Q_T))^2}. \tag{12.19}$$

Therefore, (12.18) and (12.19) imply that for every $u^{(1)}, u^{(2)} \in \mathscr{B}_q(0)$,

$$\|\mathbf{A}u^{(1)} - \mathbf{A}u^{(2)}\|_{[C([0,T];R^2)]^n} \le \frac{\sqrt{nT\mathrm{mes}(S(0))}}{\mathrm{mes}(S(0)) - CKq\sqrt{T}} \|u^{(1)} - u^{(2)}\|_{L^2(0,T;(H^2(\Omega))^2)}.$$

So, the operator $\mathbf{A}$ is continuous on $\mathscr{B}_q(0)$ for sufficiently small T as in (12.18).

Compactness. Furthermore, to show that $\mathbf{A}$ is compact, we will show that $\mathbf{A}$ maps any sequence in $\mathscr{B}_q(0)$ into a sequence in $[C([0,T];R^2)]^n$ which contains a convergent subsequence. To this end, consider any sequence $\left\{u^{(j)}\right\}_{j=1}^\infty$ in $\mathscr{B}_q(0)$. Using (12.13) with $w_i^{(j)}$ and $u^{(j)}$ in place of w_i and u, form the sequence $\left\{w_i^{(j)}\right\}_{j=1}^\infty$, $i = 1, \ldots, n$.

Let us now show that $\left\{w_i^{(j)}\right\}_{j=1}^{\infty}$ is uniformly bounded and equicontinuous. Indeed, applying (12.16) to an estimate like (12.9) and then maximizing over $[0,T]$ yields:

$$\|w_i^{(j)}\|_{C([0,T];R^2)} \leq \max_{i=1,\ldots,n} \{\|z_{i,0}\|_{R^2}\} + \frac{q\sqrt{T}}{\sqrt{\text{mes}(S(0))}}. \tag{12.20}$$

To show equicontinuity, consider any $t, t+h \in [0,T]$, e.g., when $h > 0$. Then for $i = 1,\ldots,n$:

$$\|w_i^{(j)}(t+h) - w_i^{(j)}(t)\|_{R^2} = \frac{1}{\text{mes}(S(0))} \left\| \int_t^{t+h} \int_{S(w_i^{(j)}(\tau))} u^{(j)}(x,\tau)dxd\tau \right\|_{R^2}$$

$$\leq \frac{\sqrt{h}}{\sqrt{\text{mes}(S(0))}} \|u^{(j)}\|_{(L^2(Q_T))^2}$$

$$\leq \frac{q\sqrt{h}}{\sqrt{\text{mes}(S(0))}} \quad j = 1,\ldots,$$

which implies the equicontinuity of $\left\{w_i^{(j)}\right\}_{j=1}^{\infty}$, $i = 1,\ldots,n$ (the case $h < 0$ is similar). Therefore, by Ascoli's Theorem, $\left\{Au^{(j)}\right\}_{j=1}^{\infty}$ contains a convergent subsequence in $[C([0,T];R^2)]^n$, i.e., A is compact.

Solution mapping for decoupled Stokes equations.

Now, consider the following decoupled Stokes initial boundary value problem:

$$\frac{\partial y_*}{\partial t} - v\Delta y_* + \nabla p_* = f(x,t) \quad \text{in } Q_T,$$

$$\text{div } y_* = 0 \quad \text{in } Q_T, \quad y_* = 0 \quad \text{in } \Sigma_T, \quad y_*|_{t=0} = y_0 \in H(\Omega). \tag{12.21}$$

For any $f \in (L^2(Q_T))^2$, it is known that the boundary value problem (12.21) possesses a unique solution y_* in $L^2(0,T;J_o(\Omega)) \cap L^2(0,T;(H^2(\Omega))^2)$ with the properties described in Theorem 12.1 (see, e.g. [97, 141]). Moreover, (see, e.g. (7) on p. 79 and (49) - (50) on p. 65 in [97]), there is a positive constant L such that:

$$\|y_*\|^2_{L^2(0,T;(H^2(\Omega))^2)} \leq L\|y_0\|^2_{H(\Omega)} + L\int_{Q_T} \|f(\cdot,\tau)\|^2_{R^2}d\tau. \tag{12.22}$$

Thus it follows that, given y_0, the operator

$$\mathbf{B} : (L^2(Q_T))^2 \longrightarrow J_0(Q_T) \cap L^2(0,T;(H^2(\Omega))^2), \quad \mathbf{B}f = y_*$$

is continuous.

The force term mapping $\mathbf{F}(z,v)$. Let $T > 0$ be as in (12.18). Given $u \in \mathscr{B}_q(0)$, let $F_*(w)$ denote the value of $F(z,v)$ in (12.3), with y_* from (12.21) and the w_i's from (12.13) in place of y and the z_i's, respectively. Consider the operator

$$\mathbf{F}: \ [C([0,T];R^2)]^n \longrightarrow (L^2(Q_T))^2, \mathbf{F}w = F_*(w).$$

We will show that $\mathbf{F}$ is continuous, but first we evaluate $\|F_*(w)\|_{(L^2(Q_T))^2}$.

Let $P(T)$ denote the upper bound in (12.20):

$$P(T) = \max_{i=1,\ldots,n} \{\|z_{i,0}\|_{R^2}\} + \frac{q\sqrt{T}}{\sqrt{\mathrm{mes}(S(0))}}. \tag{12.23}$$

Then, using (12.20) with w_i in place of $w_i^{(j)}$, we can evaluate the first term in square brackets in the first line of (12.3) as follows:

$$\left\| \xi_i(x,\tau)k_{i-1} \frac{(\|w_i(t) - w_{i-1}(t)\|_{R^2} - l_{i-1})}{\|w_i(t) - w_{i-1}(t)\|_{R^2}} (w_i(t) - w_{i-1}(t)) \right\|_{R^2}$$

$$\leq \max_{i=1,\ldots,n-1} \{k_i\} \left(2P(T) + \max_{i=1,\ldots n-1} \{l_i\} \right).$$

The second term in the same square brackets has a similar upper bound. Thus the first term of (12.3) is bounded above in the $(L^2(Q_T))^2$−norm by:

$$2(n-1)\sqrt{T\mathrm{mes}(\Omega)} \max_{i=1,\ldots,n-1} \{k_i\}$$

$$\left(2 \max_{i=1,\ldots,n} \{\|z_{i,0}\|_{R^2}\} + \frac{2q\sqrt{T}}{\sqrt{\mathrm{mes}(S(0))}} + \max_{i=1,\ldots n-1} \{l_i\} \right). \tag{12.24}$$

Using similar techniques, and making use of (12.15), we see that the term in square brackets of the second term of (12.3) satisfies the following estimate:

$$\| v_{i-1}(t)(\xi_{i-1}(x,t)A(w_{i-1}(t) - w_i(t))$$

$$-\xi_{i+1}(x,t) \frac{\|w_{i-1}(t) - w_i(t)\|_{R^2}^2}{\|w_{i+1}(t) - w_i(t)\|_{R^2}^2} A(w_{i+1}(t) - w_i(t))) \|_{R^2}$$

$$\leq |v_{i-1}(t)| \left(\|w_{i-1}(t) - w_i(t)\|_{R^2} + \frac{\|w_{i-1}(t) - w_i(t)\|_{R^2}^2}{\|w_{i+1}(t) - w_i(t)\|_{R^2}} \right)$$

$$< |v_{i-1}(t)| \left(\|w_{i-1}(t) - w_i(t)\|_{R^2} + \frac{\|w_{i-1}(t) - w_i(t)\|_{R^2}^2}{2r} \right)$$

$$\leq \|v_{i-1}\|_{L^\infty(0,T)} \left(2P(T) + \frac{2P(T)^2}{r} \right)$$

$$\leq 2 \max_{i=1,\dots,n-2} \{\|v_i\|_{L^\infty(0,T)}\} \left(P(T) + \frac{P(T)^2}{r} \right). \tag{12.25}$$

The third term of (12.3) (inside the sum) is also bounded above by (12.25). So it follows from (12.23) - (12.25) that (12.3) satisfies the estimate:

$$\|F_*(w)\|_{(L^2(Q_T))^2} < 2(n-1)\sqrt{T\mathrm{mes}(\Omega)} \max_{i=1,\dots,n} \{k_i\}$$

$$\times \left(2 \max_{i=1,\dots,n} \{\|z_{i,0}\|_{R^2}\} + \frac{2q\sqrt{T}}{\sqrt{\mathrm{mes}(S(0))}} + \max_{i=1,\dots,n-1} \{l_i\} \right)$$

$$+ 4(n-2)\sqrt{T\mathrm{mes}(\Omega)} \max_{i=1,\dots,n-2} \{\|v_i\|_{L^\infty(0,T)}\} [\max_{i=1,\dots,n} \{\|z_{i,0}\|_{R^2}\}$$

$$+ \frac{q\sqrt{T}}{\sqrt{\mathrm{mes}(S(0))}}$$

$$+ \frac{1}{r} \left(\max_{i=1,\dots,n} \{\|z_{i,0}\|_{R^2}\} + \frac{q\sqrt{T}}{\sqrt{\mathrm{mes}(S(0))}} \right)^2] \; \forall \, u \in \mathscr{B}_q(0). \tag{12.26}$$

We will now show that **F** is a continuous operator. Let $w^{(1)}, w^{(2)} \in [C([0,T];R^2)]^n$. Consider (12.3) with $w_i^{(j)}$ in place of z_i, $i = 1, \dots, n$, $j = 1, 2$.

We intend to evaluate $\|F_*(w^{(1)}) - F_*(w^{(2)})\|_{R^2}$. We will first write $F_*(w)$ in a form more amenable to the analysis we require. Notice that if we group the forces described in (12.3) by the characteristic function of the thick point $S(w_j(t))$ for $j = 1, \dots, n$, then $F_*(w)$ is the sum of the following terms:

$$\sum_{j=1}^{n-1} \xi_j(x,t) k_j \left[\left(1 - \frac{l_j}{\|w_{j+1}(t) - w_j(t)\|_{R^2}} \right) (w_{j+1}(t) - w_j(t)) \right], \tag{12.27}$$

$$\sum_{j=2}^{n} \xi_j(x,t) k_{j-1} \left[\left(1 - \frac{l_{j-1}}{\|w_{j-1}(t) - w_j(t)\|_{R^2}} \right) (w_{j-1}(t) - w_j(t)) \right], \tag{12.28}$$

$$\sum_{j=1}^{n-2} \xi_j(x,t) v_j(t) A(w_j(t) - w_{j+1}(t)) + \sum_{j=2}^{n-1} \xi_j(x,t) v_{j-1}(t) A(w_j(t) - w_{j-1}(t)), \tag{12.29}$$

$$\sum_{j=3}^{n} \xi_j(x,t) v_{j-2}(t) \frac{\|w_{j-2}(t) - w_{j-1}(t)\|^2}{\|w_j(t) - w_{j-1}(t)\|^2} A(w_{j-1}(t) - w_j(t)), \tag{12.30}$$

$$\sum_{j=2}^{n-1} \xi_j(x,t) v_{j-1}(t) \frac{\|w_{j-1}(t) - w_j(t)\|^2}{\|w_{j+1}(t) - w_j(t)\|^2} A(w_{j+1}(t) - w_j(t)). \tag{12.31}$$

Now, using the representation (12.27) – (12.31), we intend to evaluate the R^2-norm of $(F_*(w^{(1)}) - F(w^{(2)}))$. Making use of (12.15) and the inequality $\big|\,\|a\|_{R^2} - \|b\|_{R^2}\big| \leq \|a - b\|_{R^2}$, it follows that the term inside the square brackets in an expression similar to (12.27), but corresponding to the difference $(F_*(w^{(1)}) - F_*(w^{(2)}))$, satisfies the following inequality:

$$\| w^{(1)}_{j+1}(t) - w^{(2)}_{j+1}(t) + w^{(2)}_{j}(t) - w^{(1)}_{j}(t) + \frac{l_j(w^{(2)}_{j+1}(t) - w^{(2)}_{j}(t))}{\|w^{(2)}_{j+1}(t) - w^{(2)}_{j}(t)\|_{R^2}} - \frac{l_j(w^{(1)}_{j+1}(t) - w^{(1)}_{j}(t))}{\|w^{(1)}_{j+1}(t) - w^{(1)}_{j}(t)\|_{R^2}} \|_{R^2}$$

$$= \| w^{(1)}_{j+1}(t) - w^{(2)}_{j+1}(t) + w^{(2)}_{j}(t) - w^{(1)}_{j}(t) + l_j [\frac{(w^{(2)}_{j+1}(t) - w^{(1)}_{j+1}(t))}{\|w^{(2)}_{j+1}(t) - w^{(2)}_{j}(t)\|_{R^2}}$$

$$+ \frac{(w^{(1)}_{j}(t) - w^{(2)}_{j}(t))}{\|w^{(1)}_{j+1}(t) - w^{(1)}_{j}(t)\|_{R^2}} + w^{(1)}_{j+1} \frac{\|w^{(1)}_{j+1} - w^{(1)}_{j}(t)\|_{R^2} - \|w^{(2)}_{j+1}(t) - w^{(2)}_{j}(t)\|_{R^2}}{\|w^{(1)}_{j+1}(t) - w^{(1)}_{j}(t)\|_{R^2}\|w^{(2)}_{j+1}(t) - w^{(2)}_{j}(t)\|_{R^2}}$$

$$+ w^{(2)}_{j}(t) \frac{\|w^{(2)}_{j+1}(t) - w^{(2)}_{j}(t)\|_{R^2} - \|w^{(1)}_{j+1}(t) - w^{(1)}_{j}(t)\|_{R^2}}{\|w^{(1)}_{j+1}(t) - w^{(1)}_{j}(t)\|_{R^2}\|w^{(2)}_{j+1}(t) - w^{(2)}_{j}(t)\|_{R^2}}] \|_{R^2}$$

$$< \{\|w^{(1)}_{j+1}(t) - w^{(2)}_{j+1}(t)\|_{R^2} + \|w^{(1)}_{j}(t) - w^{(2)}_{j}(t)\|_{R^2} + l_j[\frac{\|w^{(1)}_{j+1}(t) - w^{(2)}_{j+1}(t)\|_{R^2}}{2r}$$

$$+ \frac{\|w^{(1)}_{j}(t) - w^{(2)}_{j}(t)\|_{R^2}}{2r} + \|w^{(1)}_{j+1}(t)\|_{R^2} \frac{\|w^{(1)}_{j+1}(t) - w^{(2)}_{j+1}(t) + w^{(2)}_{j}(t) - w^{(1)}_{j}(t)\|_{R^2}}{(2r)^2}$$

$$+ \|w^{(2)}_{j}(t)\|_{R^2} \frac{\|w^{(2)}_{j+1}(t) - w^{(1)}_{j+1}(t) + w^{(1)}_{j}(t) - w^{(2)}_{j}(t)\|_{R^2}}{(2r)^2}]\}$$

$$\leq \left[1 + \frac{l_j}{2r} + \frac{l_j P(T)}{2r^2}\right] \left(\|w^{(1)}_{j} - w^{(2)}_{j}(t)\|_{R^2} + \|w^{(1)}_{j+1}(t) - w^{(2)}_{j+1}(t)\|_{R^2}\right). \tag{12.32}$$

Similarly, notice that the term inside the square brackets in (12.28) corresponding to the difference $(F_*(w^{(1)}) - F_*(w^{(2)}))$ satisfies the following estimate:

$$\| w^{(1)}_{j-1}(t) - w^{(2)}_{j-1}(t) + w^{(2)}_{j}(t) - w^{(1)}_{j}(t) + \frac{l_{j-1}(w^{(2)}_{j-1}(t) - w^{(2)}_{j}(t))}{\|w^{(2)}_{j-1}(t) - w^{(2)}_{j}(t)\|_{R^2}} - \frac{l_{j-1}(w^{(1)}_{j-1}(t) - w^{(1)}_{j}(t))}{\|w^{(1)}_{j-1}(t) - w^{(1)}_{j}(t)\|_{R^2}} \|_{R^2}$$

$$< \left[1 + \frac{l_{j-1}}{2r} + \frac{l_{j-1} P(T)}{2r^2}\right]$$

$$\times \left(\|w^{(1)}_{j-1}(t) - w^{(2)}_{j-1}(t)\|_{R^2} + \|w^{(1)}_{j}(t) - w^{(2)}_{j}(t)\|_{R^2}\right). \tag{12.33}$$

In turn, the term inside the sum $(F_*(w^{(1)}) - F_*(w^{(2)}))$ corresponding to the first term of (12.29) satisfies the following estimate:

$$\left\| v_j(t) A(w^{(1)}_{j}(t) - w^{(1)}_{j+1}(t) - w^{(2)}_{j}(t) + w^{(2)}_{j+1}(t)) \right\|_{R^2}$$

$$\leq \max_{j=1,\ldots,n-2} \{\|v_j\|_{L^\infty(0,T)}\} (\|w^{(1)}_{j}(t) - w^{(2)}_{j}(t)\|_{R^2}$$

$$+ \|w^{(1)}_{j+1}(t) - w^{(2)}_{j+1}(t)\|_{R^2}). \tag{12.34}$$

The term in the sum $(F_*(w^{(1)}) - F_*(w^{(2)}))$ corresponding to the second term of (12.29) satisfies a similar estimate:

$$\left\| v_{j-1}(t)A(w_j^{(1)}(t) - w_{j-1}^{(1)}(t) - w_j^{(2)}(t) + w_{j-1}^{(2)}(t)) \right\|_{R^2}$$

$$\leq \max_{j=1,\ldots,n-2} \{\|v_j\|_{L^\infty(0,T)}\} (\|w_{j-1}^{(1)}(t) - w_{j-1}^{(2)}(t)\|_{R^2}$$

$$+ \|w_j^{(1)}(t) - w_j^{(2)}(t)\|_{R^2}). \tag{12.35}$$

Analogously, for the term in the sum $(F_*(w^{(1)}) - F_*(w^{(2)}))$ corresponding to (12.30), we obtain:

$$\| v_{j-2}(t)A[\frac{\|w_{j-2}^{(1)}(t) - w_{j-1}^{(1)}(t)\|_{R^2}^2}{\|w_j^{(1)}(t) - w_{j-1}^{(1)}(t)\|_{R^2}^2}(w_{j-1}^{(1)}(t) - w_j^{(1)}(t))$$

$$-\frac{\|w_{j-2}^{(2)}(t) - w_{j-1}^{(2)}(t)\|_{R^2}^2}{\|w_j^{(2)}(t) - w_{j-1}^{(2)}(t)\|_{R^2}^2}(w_{j-1}^{(2)}(t) - w_j^{(2)}(t))] \|_{R^2}$$

$$\leq |v_{j-2}(t)| \| \frac{\|w_{j-2}^{(1)}(t) - w_{j-1}^{(1)}(t)\|_{R^2}^2}{\|w_j^{(1)}(t) - w_{j-1}^{(1)}(t)\|_{R^2}^2}(w_{j-1}^{(1)}(t) - w_j^{(1)}(t))$$

$$-\frac{\|w_{j-2}^{(2)}(t) - w_{j-1}^{(2)}(t)\|_{R^2}^2}{\|w_j^{(2)}(t) - w_{j-1}^{(2)}(t)\|_{R^2}^2}(w_{j-1}^{(2)}(t) - w_j^{(2)}(t)) \|_{R^2}$$

$$= |v_{j-2}(t)| \| \frac{\|w_{j-2}^{(1)}(t) - w_{j-1}^{(1)}(t)\|_{R^2}^2 - \|w_{j-2}^{(2)}(t) - w_{j-1}^{(2)}(t)\|_{R^2}^2}{\|w_j^{(1)}(t) - w_{j-1}^{(1)}(t)\|_{R^2}^2}(w_{j-1}^{(1)}(t) - w_j^{(1)}(t))$$

$$+\|w_{j-2}^{(2)}(t) - w_{j-1}^{(2)}(t)\|_{R^2}^2 \left[\frac{w_{j-1}^{(1)}(t) - w_j^{(1)}(t)}{\|w_j^{(1)}(t) - w_{j-1}^{(1)}(t)\|_{R^2}^2} - \frac{w_{j-1}^{(2)}(t) - w_j^{(2)}(t)}{\|w_j^{(2)}(t) - w_{j-1}^{(2)}(t)\|_{R^2}^2} \right] \|_{R^2}$$

$$= |v_{j-2}(t)| \| \frac{\|w_{j-2}^{(1)}(t) - w_{j-1}^{(1)}(t)\|_{R^2}^2 - \|w_{j-2}^{(2)}(t) - w_{j-1}^{(2)}(t)\|_{R^2}^2}{\|w_j^{(1)}(t) - w_{j-1}^{(1)}(t)\|_{R^2}^2}(w_{j-1}^{(1)}(t) - w_j^{(1)}(t))$$

$$+\|w_{j-2}^{(2)}(t) - w_{j-1}^{(2)}(t)\|_{R^2}^2[\frac{w_{j-1}^{(1)}(t) - w_{j-1}^{(2)}(t)}{\|w_j^{(1)}(t) - w_{j-1}^{(1)}(t)\|_{R^2}^2} + \frac{w_j^{(2)}(t) - w_j^{(1)}(t)}{\|w_j^{(2)}(t) - w_{j-1}^{(2)}(t)\|_{R^2}^2}$$

$$+w_j^{(1)}(t)\frac{\|w_j^{(1)}(t) - w_{j-1}^{(1)}(t)\|_{R^2}^2 - \|w_j^{(2)}(t) - w_{j-1}^{(2)}(t)\|_{R^2}^2}{\|w_j^{(1)}(t) - w_{j-1}^{(1)}(t)\|_{R^2}^2 \|w_j^{(2)}(t) - w_{j-1}^{(2)}(t)\|_{R^2}^2}$$

$$+w_{j-1}^{(2)}(t)\frac{\|w_j^{(2)}(t) - w_{j-1}^{(2)}(t)\|_{R^2}^2 - \|w_j^{(1)}(t) - w_{j-1}^{(1)}(t)\|_{R^2}^2}{\|w_j^{(1)}(t) - w_{j-1}^{(1)}(t)\|_{R^2}^2 \|w_j^{(2)}(t) - w_{j-1}^{(2)}(t)\|_{R^2}^2}] \|_{R^2}. \tag{12.36}$$

Note that using the inequality $|\|a\|_{R^2} - \|b\|_{R^2}| \leq \|a - b\|_{R^2}$ again and (12.23), we obtain:

$$
\left| \|w_j^{(1)}(t) - w_{j-1}^{(1)}(t)\|_{R^2}^2 - \|w_j^{(2)}(t) - w_{j-1}^{(2)}(t)\|_{R^2}^2 \right|
$$

$$
= |(\|w_j^{(1)}(t) - w_{j-1}^{(1)}(t)\|_{R^2} + \|w_j^{(2)}(t) - w_{j-1}^{(2)}(t)\|_{R^2})
$$

$$
\times (\|w_j^{(1)}(t) - w_{j-1}^{(1)}(t)\|_{R^2} - \|w_j^{(2)}(t) + w_{j-1}^{(2)}(t)\|_{R^2})|
$$

$$
\leq 4P(T)(\|w_{j-1}^{(1)}(t) - w_{j-1}^{(2)}(t)\|_{R^2} + \|w_j^{(1)}(t) - w_j^{(2)}(t)\|_{R^2}). \quad (12.37)
$$

In view of (12.15), if we apply (12.37) to (12.36), we obtain:

$$
\| v_{j-2}(t)A[\frac{\|w_{j-2}^{(1)}(t) - w_{j-1}^{(1)}(t)\|_{R^2}^2}{\|w_j^{(1)}(t) - w_{j-1}^{(1)}(t)\|_{R^2}^2}(w_{j-1}^{(1)}(t) - w_j^{(1)}(t))
$$

$$
- \frac{\|w_{j-2}^{(2)}(t) - w_{j-1}^{(2)}(t)\|_{R^2}^2}{\|w_j^{(2)}(t) - w_{j-1}^{(2)}(t)\|_{R^2}^2}(w_{j-1}^{(2)}(t) - w_j^{(2)}(t))] \|_{R^2}
$$

$$
< \max_{j=1,\ldots,n-2}\{\|v_j\|_{L^\infty(0,T)}\}[\frac{2P(T)}{r}\|w_{j-2}^{(1)}(t) - w_{j-2}^{(2)}(t)\|_{R^2}
$$

$$
+ \left(\frac{2P(T)}{r} + \frac{(P(T))^2}{r^2} + \frac{2(P(T))^4}{r^4}\right)\|w_{j-1}^{(1)}(t) - w_{j-1}^{(2)}(t)\|_{R^2}
$$

$$
+ \left(\frac{(P(T))^2}{r^2} + \frac{2(P(T))^4}{r^4}\right)\|w_j^{(1)}(t) - w_j^{(2)}(t)\|_{R^2}]
$$

$$
\leq \max_{j=1,\ldots,n-2}\{\|v_j\|_{L^\infty(0,T)}\}\left(\frac{2P(T)}{r} + \frac{(P(T))^2}{r^2} + \frac{2(P(T))^4}{r^4}\right)
$$

$$
\times (\|w_{j-2}^{(1)}(t) - w_{j-2}^{(2)}(t)\|_{R^2}
$$

$$
+ \|w_{j-1}^{(1)}(t) - w_{j-1}^{(2)}(t)\|_{R^2} + \|w_j^{(1)}(t) - w_j^{(2)}(t)\|_{R^2}). \quad (12.38)
$$

Similarly, the term inside the sum $(F_*(w^{(1)}) - F_*(w^{(2)}))$ corresponding to (12.31) satisfies the following estimate:

$$
\| v_{j-1}(t)A[\frac{\|w_{j-1}^{(1)}(t) - w_j^{(1)}(t)\|_{R^2}^2}{\|w_{j+1}^{(1)}(t) - w_j^{(1)}(t)\|_{R^2}^2}(w_{j+1}^{(1)}(t) - w_j^{(1)}(t))
$$

$$
- \frac{\|w_{j-1}^{(2)}(t) - w_j^{(2)}(t)\|_{R^2}^2}{\|w_{j+1}^{(2)}(t) - w_j^{(2)}(t)\|_{R^2}^2}(w_{j+1}^{(2)}(t) - w_j^{(2)}(t))] \|_{R^2}
$$

$$
< \max_{j=1,\ldots,n-2}\{\|v_j\|_{L^\infty(0,T)}\}\left(\frac{2P(T)}{r} + \frac{(P(T))^2}{r^2} + \frac{2(P(T))^4}{r^4}\right)
$$

$$
\times (\|w_{j-1}^{(1)}(t) - w_{j-1}^{(2)}(t)\|_{R^2}
$$

$$
+ \|w_j^{(1)}(t) - w_j^{(2)}(t)\|_{R^2} + \|w_{j+1}^{(1)}(t) - w_{j+1}^{(2)}(t)\|_{R^2}). \quad (12.39)
$$

Finally, denote

$$M(T) = 4 \max_{j=1,\ldots,n-1} \{k_j\} \left[1 + \max_{j=1,\ldots,n-1} \{l_j\} \left(\frac{1}{2r} + \frac{P(T)}{2r^2} \right) \right]$$
$$+ 6 \max_{j=1,\ldots,n-2} \{ \|v_j\|_{L^\infty(0,T)} \} \left(1 + \frac{2P(T)}{r} + \frac{(P(T))^2}{r^2} + \frac{2(P(T))^4}{r^4} \right).$$

Maximizing the right-hand sides of (12.32) - (12.34)–(12.35), (12.38), and (12.39) over $[0, T]$, and summing them respectively yields:

$$\|F_*(w^{(1)}) - F_*(w^{(2)})\|_{R^2} \le M(T) \sum_{j=1}^{n} \|w_j^{(1)} - w_j^{(2)}\|_{C([0,T];R^2)}. \tag{12.40}$$

Note from (12.23) that $P(T)$ is nonincreasing as $T \to 0^+$ (where T satisfies (12.18)); hence $M(T)$ is positive and nonincreasing at $T \to 0^+$. Thus,

$$\|F_*(w^{(1)}) - F_*(w^{(2)})\|_{(L^2(Q_T))^2} \le nM(T)\sqrt{T\,\mathrm{mes}(\Omega)}\|w^{(1)} - w^{(2)}\|_{[C([0,T];R^2)]^n},$$

hence $\mathbf{F}$ is a continuous operator for T satisfying (12.18).

12.5 Proof of the Main Result

Before proceeding to the proof of Theorem 12.1, we summarize the main results of the previous section. In subsection 12.4, we proved that for sufficiently small $T > 0$ satisfying (12.18), the operators

$$\mathbf{A} : \mathscr{B}_q(0) \longrightarrow [C([0,T];R^2)]^n, \quad \mathbf{A}u = w = (w_1, \ldots, w_n),$$
$$\mathbf{F} : [C([0,T];R^2)]^n \longrightarrow (L^2(Q_T))^2, \quad \mathbf{F}w = F_*(w),$$

and

$$\mathbf{B} : (L^2(Q_T))^2 \longrightarrow L^2(0,T;J_o(\Omega)) \bigcap L^2(0,T;(H^2(\Omega))^2), \quad \mathbf{B}f = y_*$$

are all continuous, and additionally, that the mapping A is also compact. As a result,

$$\mathbf{BFA} : B_q(0) \longrightarrow L^2(0,T;J_o(\Omega)) \bigcap L^2(0,T;(H^2(\Omega))^2), \quad \mathbf{BFA}u = y_*$$

is continuous and compact.

12.5.1 Existence: Fixed Point Argument

Select the value of q to be any positive number larger than $\sqrt{L}\|y_0\|_{H(\Omega)}$, and choose $T > 0$ as in (12.18) so that Lemma 12.1 holds. In view of (12.22) (with F_* in place of f), we select $T^* \in (0, \min\{T, 1\})$ small enough so that the continuous and compact operator **BFA** maps the closed ball $B_q(0)$ into itself.

To show this, observe that if $T \le 1$, then (12.26) can be rewritten as follows:

$$\|F_*(w)\|_{(L^2(Q_T))^2} < C_1 \sqrt{T},$$

where the positive constant C_1 is independent of T. Select T^* so that:

$$0 < T^* < \min\left\{ \frac{q^2 - L\|y_0\|_{H(\Omega)}^2}{LC_1^2}, T, 1 \right\}. \tag{12.41}$$

Since T^* satisfies (12.18), if we replace T with T^* in (12.22), (12.24)–(12.26), we obtain:

$$\|\textbf{BFA}u\|_{L^2(0,T^*;(H^2(\Omega))^2)}^2 \le L\|y_0\|_{H(\Omega)}^2 + L\|F_*(w)\|_{(L^2(Q_{T^*}))^2}^2$$
$$< L\|y_0\|_{H(\Omega)}^2 + LC_1^2 T^* < L\|y_0\|_{H(\Omega)}^2 + q^2 - L\|y_0\|_{H(\Omega)}^2 = q^2.$$

Hence, **BFA** maps $\mathscr{B}_q(0)$ into itself if (12.41) is satisfied.

Thus, by Schauder's Fixed Point Theorem, **BFA** has a fixed point y which is a solution of the system (12.1) - (12.3), and which satisfies all of the requirements of Theorem 12.1. As usual, we may select ∇p in $L^2(0, T^*; G(\Omega))$ to complement the solution $y \in L^2(0, T^*; J_o(\Omega))$ in Theorem 12.1. This completes the proof of existence for Theorem 12.1.

12.5.2 Uniqueness

To prove that the solution found in Section 12.5.1 is unique, we will argue by contradiction. Namely, suppose, e.g., that there are two different solutions

$$\left\{ z^{(1)} = (z_1^{(1)}, \dots, z_n^{(1)}), y^{(1)}, p^{(1)} \right\}$$

and

$$\left\{ z^{(2)} = (z_1^{(2)}, \dots, z_n^{(2)}), y^{(2)}, p^{(2)} \right\}$$

to (12.1) - (12.3), satisfying the properties described in Theorem 12.1 on some time interval $[0, T]$, where T satisfies the inequality (12.41). Without loss of generality, we assume the two solutions are different right from $t = 0$.

By (12.19), with $z_i^{(j)}$ in place of $w_i^{(j)}$, $i = 1,\dots,n$, and $y^{(j)}$ in place of $u^{(j)}$, both for $j = 1,2$, we see that for any $T_0 \in (0,T]$, and each $i = 1,\dots,n$:

$$\|z_i^{(1)} - z_i^{(2)}\|_{C([0,T_0];R^2)} \leq \frac{\sqrt{T_0 \mathrm{mes}(S(0))}}{\mathrm{mes}(S(0)) - CKq\sqrt{T_0}} \|y^{(1)} - y^{(2)}\|_{(L^2(Q_{T_0}))^2}. \quad (12.42)$$

Let us now evaluate $\|y^{(1)} - y^{(2)}\|_{(L^2(Q_{T_0}))^2}$. Note that $(y^{(1)} - y^{(2)})$ satisfies the following Stokes initial-value problem:

$$\frac{\partial(y^{(1)} - y^{(2)})}{\partial t} = v\Delta(y^{(1)} - y^{(2)}) + (F(z^{(1)}, v) - F(z^{(2)}, v)) - \nabla(p^{(1)} - p^{(2)}) \text{ in } Q_{T_0},$$

$$\mathrm{div}\,(y^{(1)} - y^{(2)}) = 0 \text{ in } Q_{T_0}, \quad (y^{(1)} - y^{(2)}) = 0 \text{ in } \Sigma_{T_0}, \quad (y^{(1)} - y^{(2)})|_{t=0} = 0.$$

According to (12.22) we have:

$$\|y^{(1)} - y^{(2)}\|_{(L^2(Q_{T_0}))^2}^2 \leq L \int_0^{T_0} \int_\Omega \|F(z^{(1)}, v) - F(z^{(2)}, v)\|_{R^2}^2 dx dt. \quad (12.43)$$

In turn, replace $w^{(j)}$ with $z^{(j)}$, $j = 1,2$, and $F_*(w)$ with $F(z,v)$ in (12.27)–(12.40). Then, maximizing the right-hand sides of resulting estimates like (12.32)–(12.34), (12.35, (12.38), and (12.39) over $[0, T_0]$, and summing them respectively yields:

$$\|F(z^{(1)}, v) - F(z^{(2)}, v)\|_{R^2} \leq M(T) \sum_{j=1}^n \|z_j^{(1)} - z_j^{(2)}\|_{C([0,T_0];R^2)}. \quad (12.44)$$

Recall that $M(T)$ is nonincreasing at $T \to 0^+$. Hence, combining (12.42) - (12.44) yields:

$$\|y^{(1)} - y^{(2)}\|_{(L^2(Q_{T_0}))^2}^2$$

$$\leq \frac{nM(T)T_0\sqrt{L\mathrm{mes}(S(0))\mathrm{mes}(\Omega)}}{\mathrm{mes}(S(0)) - CKq\sqrt{T_0}} \|y^{(1)} - y^{(2)}\|_{(L^2(Q_{T_0}))^2}. \quad (12.45)$$

Now, select T_0 as follows:

$$0 < T_0 < \min\left\{ \frac{\mathrm{mes}(S(0))}{4nM(T)\sqrt{L\mathrm{mes}(S(0))\mathrm{mes}(\Omega)}}, \left(\frac{\mathrm{mes}(S(0))}{2CKq}\right)^2, T \right\}. \quad (12.46)$$

This choice of T_0 implies that the following inequality holds:

$$nM(T)T_0\sqrt{L\mathrm{mes}(S(0))\mathrm{mes}(\Omega)} + \frac{1}{2}Ckq\sqrt{T_0} < \frac{1}{2}\mathrm{mes}(S(0)).$$

So, it follows from (12.45), (12.46) that:

$$\|y^{(1)} - y^{(2)}\|_{(L^2(Q_{T_0}))^2} < \frac{1}{2}\|y^{(1)} - y^{(2)}\|_{(L^2(Q_{T_0}))^2}.$$

Therefore $y^{(1)} \equiv y^{(2)}$ on $[0, T_0]$, and thus by (12.42), $z_i^{(1)} \equiv z_i^{(2)}$ for $i = 1, \ldots, n$ on $[0, T_0]$. Contradiction. This ends the proof of Theorem 12.1. □

12.6 Further Comments on Modeling Philosophy

Our modeling approach to swimming phenomenon (introduced in [76] and represented in this section by equations (12.1)–(12.3)) can be viewed as an attempt to describe it by a hybrid system of equations, consisting of:

(a) a pde for a fluid, excited in a bounded domain by actions of swimmer's *internal forces*, which is coupled with
(b) an (integro-)ordinary differential equation allowing one to track the motion of the swimmer in space.

The latter is the principal element of our modeling technique because it allows us to attack the associated swimming controllability problem.

We will now discuss a 2-D swimming model from the interesting recent paper [134] which, in the above sense, can be viewed as an implementation of a similar modeling approach to the case when the swimmer is a one-piece solid body and the fluid is governed by the Navier-Stokes equation. (We refer the reader to [134] for the complete account of all details).

A model of a fish-like solid swimmer in the fluid governed by the 2-D Navier-Stokes equation. Let Ω be an open bounded set in R^2. The authors of [134] assumed that the solid swimmer (e.g., a fish) occupies a domain $\mathscr{S}_0 \in R^2$ in a reference configuration (say, initially when $t = 0$) and that at time $t > 0$ it occupies the domain $\mathscr{S}(\xi(t), \theta(t), t)$, which can be described as follows:

$$\mathscr{S}(\xi(t), \theta(t), t) = P_{\theta(t)}\mathscr{S}^*(t) + \xi(t), \tag{12.47}$$

where $\xi(t)$ and $\theta(t)$ are respectively the position of swimmer's center of mass and the angle giving its orientation at time $t > 0$. P_θ stands for the matrix associated with the rotation of angle θ. The parameters $\theta(t)$ and $\xi(t)$ are unknown (to be determined from the model equations (12.49)–(12.53) below). Furthermore, it is assumed that

$$\mathscr{S}^*(t) = X^*(\mathscr{S}_0, t), \tag{12.48}$$

and $y \to X^*(y, t)$ is C^∞-diffeomorphism from $\mathscr{S}_0$ onto $\mathscr{S}^*(t)$, while for every $y \in \mathscr{S}_0$, the mapping $t \to X^*(y, t)$ is of class C^∞ and $X^*(y, 0) = y$. It is assumed that the mapping X^* represents the *known* undulatory deformation of the swimmer.

Assumptions (12.47)–(12.48) mean that the *swimming motion has no affect on the shape of swimmer's body, except for its uniform rotation* described by the angle parameter θ (see the discussion in the end of this subsection).

For every $t \geq 0$ denote by Y^* the inverse of X^*, that is, the diffeomorphism satisfying

$$X^*(Y^*(x,t),t) = x, \quad Y^*(X^*(y,t),y) = y$$

for every $x \in \mathscr{S}^*(t), t \geq 0$ and $y \in \mathscr{S}_0$.

Denote by w^* the undulatory velocity of swimmer, written as a vector field on $\mathscr{S}^*(t)$:

$$w^*(x,t) = \frac{\partial X^*(y,t)}{\partial t}\Big|_{y=Y^*(x,t)}, \quad x \in \mathscr{S}^*(t), \, t \geq 0.$$

It is also assumed that the volume and mass of swimmer are preserved:

$$\int_{\mathscr{S}^*(t)} dx = \int_{\mathscr{S}_0} dx, \quad t \geq 0,$$

$$\int_{\mathscr{S}^*(t)} \rho^*(x,t)dx = \int_{\mathscr{S}_0} \rho^*(x,0)dx, \quad t \geq 0,$$

where $\rho^*(x,t)$ is the density field of $\mathscr{S}^*(t)$.

To ensure that the motion of swimmer is generated by its internal forces, one also needs to assume that its linear and angular momenta are conserved:

$$\int_{\mathscr{S}^*(t)} \rho^*(x,t)w^*(x,t)dx = 0, \quad t \geq 0,$$

$$\int_{\mathscr{S}^*(t)} \rho^*(x,t)x^\perp \cdot w^*(x,t)dx = 0, \quad t \geq 0,$$

where $x^\perp$ stands to the vector in R^2 perpendicular to x.

Let $\mathscr{F}(\xi(t), \theta(t), t) = \Omega \backslash \mathrm{cl}\{\mathscr{S}(\xi(t), \theta(t), t)\}$, where "cl" stands for closure. Assuming that the density of fluid is equal to one, the following system was proposed in [134] to describe the self-propelling motion of the above-described swimmer:

$$\frac{\partial u}{\partial t} + u \cdot \nabla u + \nabla p - \nu \Delta u = 0, \quad \mathrm{div}\, u = 0 \quad \text{in } \mathscr{F}(\xi, \theta, t), \ t \in (0, T), \quad (12.49)$$

$$u = 0 \quad \text{in } \partial \Omega, \quad t \in (0, T),$$

$$u = \xi' + \theta'(x - \xi)^\perp + w \quad \text{on } \partial \mathscr{S}(\xi, \theta, t), \ t \in (0, T), \quad (12.50)$$

$$m\xi'' = - \int_{\partial \mathscr{S}(\xi, \theta, t)} \sigma(u, p) n\, d\Gamma \quad \text{in } (0, T), \quad (12.51)$$

$$(I\theta')' = - \int_{\partial \mathscr{S}(\xi,\theta,t)} (x-\xi)^{\perp} \cdot \sigma(u,p)n\,d\Gamma \quad \text{in } (0,T), \tag{12.52}$$

$$\xi(0) = \xi_0, \quad \theta(0) = \theta_0, \quad \xi'(0) - g_0, \quad \theta'(0) = \hat{\theta}_0,$$
$$u = u_0 \quad \text{in } \mathscr{F}(\xi(0),\theta(0),0). \tag{12.53}$$

In the above the prime stands for the differentiation operator with respect to t, m is the mass of the swimmer, and $I(t)$ is its moment of inertia with respect to the axis orthogonal to the plane of the motion and passing through $\xi(t)$:

$$I(t) = \int_{\mathscr{S}(\xi(t),\theta(t),t)} \rho(x,t) \, |\, x - \xi(t)\,|^2 \, dx,$$

where

$$\rho(x,t) = \rho^*(P_{-\theta(t)}(x-\xi(t)),t), \quad t \ge 0, x \in \mathscr{S}(\xi(t),\theta(t),t).$$

The symbol $\sigma(u,p)$ is the stress tensor field and $n(x,t)$ denotes the unit normal to $\partial \mathscr{S}(\xi(t),\theta(t),t)$ oriented towards the solid, and

$$w = P_{\theta(t)} w^*(P_{-\theta(t)}(x-\xi(t)),t).$$

In other words, the formula (12.50) means that the Eulerian velocity field of swimmer can be decomposed into a part due to the rigid motion (the first two terms) and w due to the known undulatory deformation.

Let us state now the main wellposedness result from [134]. In what follows let $\mathscr{F}(\xi(t),\theta(t),t) = \mathscr{F}(t), \mathscr{S}(\xi(t),\theta(t),t) = \mathscr{S}(t)$. Assume that $T > 0$. A quadruplet (u,p,ξ,θ) with

$$u \in L^2(0,T;[H^2(\mathscr{F}(t))]^2) \bigcap H^1(0,T;[L^2(\mathscr{F}(t))]^2) \bigcap C([0,T];[H^1(\mathscr{F}(t))]^2),$$
$$p \in L^2(0,T;H^1(\mathscr{F}(t))), \quad \xi \in [H^2(0,T)]^2, \quad \theta \in H^2(0,T)$$

such that the distance from $\mathscr{S}(t)$ to $\partial \Omega$ is positive for all $t \in [0,T]$ and which satisfies (12.47)–(12.53) is called a strong solution to (12.47)–(12.53) on $[0,T]$.

Theorem 12.2. *Under the above assumptions, assume that $u_0 \in [H^1(\mathscr{F}(t))]^2$ and that*

$$\operatorname{div} u_0 = 0 \quad \text{in } \mathscr{F}(0), \quad u_0 = 0 \text{ on } \partial \Omega,$$
$$u_0 = g_0 + \hat{\theta}_0 x^{\perp} + w^*(x,0) \text{ on } \partial \mathscr{S}(0), \quad \operatorname{dist}\{\mathscr{S}(0),\partial\Omega\} > 0.$$

Then there exists $T_{\max} > 0$ such that for every $T \in (0,T_{\max})$ the system (12.47)–(12.53) admits a unique strong solution on $[0,T]$. One of the following alternatives holds: either $T_{\max} = \infty$ or

$$\lim_{t \to T_{\max}} \operatorname{dist}\{\mathscr{S}(t),\partial\Omega\} = 0.$$

Comparison of two models. Let us try to compare the models (12.47)–(12.53) and (12.1)–(12.3).

First of all, the role of equation (12.1) in the latter model is obviously played by (12.49) in the former one. Relation (12.50) describes the fluid velocity field on the surface $\partial \mathscr{S}(\xi, \theta, t)$ of the swimmer as equal to swimmer's velocity at the same points, which can be viewed as an analogue of system (12.2) describing the velocity of points $z_i(t)$ in model (12.1)–(12.3) as the averaged fluid velocity at these points. Next, equations (12.51)–(12.52) in model (12.47)–(12.53) ensure that the linear and angular momenta in this model are conserved, while equation (12.3) for the force term carries the same information for model (12.1)–(12.3). Equation (12.51) is also the equation for the center of swimmer's mass $\xi(t)$. The respective motion of the center of mass of swimmer in model (12.1)–(12.3) is given by the sum $(z_1(t) + \ldots + z_n(t))/n$, whose value is determined from system (12.2).

In terms of differences, it should be noted that the allowed prescribed deformation of the one-piece swimmer in (12.47)–(12.53) is given through the diffeomorphism X^*. In comparison, the motion of swimmer in model (12.1)–(12.3) (or (11.1)–(11.3) from [76] in Chapter 11) is linked to its internal forces, described explicitly in (12.3). (We use this force representation to study controllability.)

The proofs of respective wellposedness results for models (12.47)–(12.53) (see [134] for details) and (12.1)–(12.3) are quite different from the technical point of view, having in mind that the latter deals with the nonstationary Stokes equation, while the former is formulated as the mixed problem for the Navier-Stokes equation with moving boundary. Nonetheless, the wellposedness result established in [134] methodologically is similar to Theorem 12.1 and Remark 12.1, namely, the respective solution to (12.47)–(12.53) exists and unique up to the contact with the boundary $\partial \Omega$.

The model (12.47)–(12.53) assumes that the swimming motion has no other affect on the shape of swimmer's body than its uniform rotation described by angle θ in (12.47). In particular, this excludes the case when swimmer's body parts may come to a contact with each other, which in turn may affect the duration of the time-interval for the wellposedness result (T_{max} in Theorem 12.2).

The situation is different for model (12.1)–(12.3). Equation (12.2) allows independent distortion of positions of all points $z_i(t), i = 1, \ldots, n$ due to the motion of the fluid. This may "unpredictably" change the distances between them and the angles between segments $z_1 z_2, z_2 z_3, \ldots, z_{n-1} z_n$. In particular, it may result in folding the swimmer's body up to the point when some of its parts will "collide". This possibility was addressed in Theorem 12.1 and Remark 12.1 of this section and also in Remark 11.1 in Chapter 11. Furthermore, we are particularly concerned below with the question of how the swimmer's internal forces, applied to one part of its body, may distort its "planned" shape elsewhere due to "unwanted" interaction between swimmer's body parts as transmitted through the surrounding fluid, see Theorems 13.1 and 13.2 and Lemma 13.2 in Chapter 13.

Chapter 13
Geometric Aspects of Controllability for a Swimming Phenomenon

Abstract We investigate how the geometric shape of the swimming object affects the forces acting upon it in a fluid, particularly, in the case when the object at hand consists of either small rectangles or discs.

13.1 Introduction

We intend now to analyze the swimming capabilities of the "basic" model (11.1)–(11.3) from Chapter 11. *We will assume now that the "thick points" forming the swimming object are of the same shape as $S(0)$ but can be different from each other in their space orientation.* (Note that this does not change the wellposedness results of Chapter 12.) Respectively, below (and in Chapter 15), we replace the notation $S(z_i)$ with $S_i(z_i), i = 1, \ldots, n$, where the subscript i is to indicate that the "thick points" forming the object at hand may differ from each other in the space orientation.

The main goal of this chapter is to describe explicitly the forces acting in the fluid on each of the supports $S_i(z_i(t)), 1 = 1, \ldots, n$ of the swimming object at hand at every moment of time when $S_i(z_i(t))$'s are either "small" rectangles or "small"discs. As the reader will see it below, these forces determine the actual trajectory of the object moving in the fluid and thus its controllability properties.

We emphasize again that without a fluid the center of mass of a swimming-by-itself object will not move, because the sum of all internal forces generated by it is equal to zero. However, when a swimming object interacts with the fluid, the sum of the aforementioned internal forces *projected onto the fluid velocity space* may not be zero, which will then result in its motion. If so, one should think that the object at hand will have the *best swimming capabilities* if at any moment of time the span of the sum of all its possible averaged (over the corresponding supports $S_i(z_i(t))$'s) internal forces projected on the fluid velocity space covers the whole space R^2 in which it lies. We will regard this property below as the *force controllability*.

Our discussion from now on will deal only with the (more interesting, as the reader will see later) case when $S_i(z_i), i = 1, \ldots, n$ are underline{rectangles}. The respective results for the case when they are discs will easily follow at no extra cost.

A.Y. Khapalov, *Controllability of Partial Differential Equations Governed by Multiplicative Controls*, Lecture Notes in Mathematics 1995, DOI 10.1007/978-3-642-12413-6_13, © Springer-Verlag Berlin Heidelberg 2010

13.2 Main Results

From Corollary 12.1 in Chapter 12 it follows that the unique solution to (11.1)–(11.3), lies in the space $J_0(Q_T)$ at all times, while admitting the following implicit Fourier series representation:

$$y(x,t) = \sum_{k=1}^{\infty} e^{-\lambda_k t} \left(\int_{\Omega} y_0^T \omega_k dx \right) \omega_k(x)$$

$$+ \sum_{k=1}^{\infty} \int_0^t e^{-\lambda_k(t-\tau)} \left(\int_{\Omega} (F(z,v))^T \omega_k dx d\tau \right) \omega_k(x). \qquad (13.1)$$

Here the 2-D vector functions $\omega_k, k = 1,\dots$ and the real numbers $-\lambda_k, k = 1,\dots$ denote respectively the orthonormalized in $(L^2(\Omega))^2$ eigenfunctions and eigenvalues of the spectral problem associated with (11.1):

$$v \Delta \omega_k - \nabla p_k = -\lambda_k \omega_k \quad \text{in } \Omega, \quad \text{div } \omega_k = 0 \text{ in } \Omega, \quad \omega_k = 0 \text{ in } \partial\Omega.$$

The series in (13.1) and the series obtained from it by differentiation once with respect to t and twice with respect to the spatial variables converge in $(L^2(\Omega))^2$ uniformly for $t \geq 0$ (e.g., [96, 141]). The functions $\{\omega_k\}_{k=1}^{\infty}$ also form a basis in $J_0(\Omega) \cap H(\Omega)$.

If we now denote the orthogonal projection operator from the space $(L^2(\Omega))^2$ onto $J_0(\Omega)$ by P, we can rewrite (13.1) also as:

$$y(x,t) = \sum_{k=1}^{\infty} e^{-\lambda_k t} \left(\int_{\Omega} y_0^T \omega_k dx \right) \omega_k(x)$$

$$+ \sum_{k=1}^{\infty} \int_0^t e^{-\lambda_k(t-\tau)} \left(\int_{\Omega} (PF(z,v))^T \omega_k dx d\tau \right) \omega_k(x). \qquad (13.2)$$

Before we will give the formal definition of force controllability, we would also like to state the following result, which highlights its importance.

Lemma 13.1. *Let on some time-interval $[t_0, T], t_0 \geq 0$ the sets $S_i(z_i(t)), i = 1,\dots,n$ do not change their orientation and controls $v_i, i = 1,\dots,n-1$ remain constant. Then we have the following formulas for solutions to the system (11.1)–(11.3) from Chapter 11 under the Assumption P from Chapter 12 as $t \to t_0+$:*

$$\frac{dz_i(t)}{dt} = \frac{1}{\text{mes}\{S(0)\}} \int_{S_i(z_i(t_0))} y(x,t_0)dx + \frac{t-t_0}{\text{mes}\{S(0)\}} \int_{S_i(z_i(t_0))} (PF(y,z,v))(x,t_0)dx$$

$$+ u(t-t_0) + o(t-t_0)$$

$$\times \left(\max_{i=1,\dots,n-1} |v_i| + \max_{i=1,\dots,n-1} k_i \right), \quad i = 1,\dots,n, \qquad (13.3)$$

where

$$\| u(t-t_0) \|_{R^2} \le \frac{\text{mes}^{1/2}\{S(z_i(t_0))\cap S_i(z_i(t))\}}{\text{mes}\{S(0)\}}\mu(t,y(\cdot,t_0))$$

$$+\frac{2\max^{1/2}\{\text{mes}\{S_i(z_i(t_0))\backslash S_i(z_i(t)), \text{mes}\, S_i(z_i(t))\backslash S_i(z_i(t_0))\}}{\text{mes}\{S(0)\}}$$

$$\times \| y(\cdot,t_0) \|_{(L^2(\Omega))^2}, \qquad (13.4)$$

$$\mu(t,y(\cdot,t_0)),\ o(t-t_0)/(t-t_0) \to 0 \text{ as } t \to t_0+,$$

$$0 \le \mu(t,y(\cdot,t_0)) \le C\| y(\cdot,t_0) \|_{(L^2(\Omega))^2},$$

and $C > 0$ denotes a (generic) constant.

Thus, if "sufficiently large" control forces $v_1,\ldots,v_{n-1}$ are applied, *the direction in which each of the points $z_i(t), i = 1,\ldots,n$ and, hence, its center of mass $z_c(t) = (1/n)\sum_{i=1}^{n} z_i(t)$ will move from its current position is primarily determined by the projections of the swimming object's forces on the fluid velocity space <u>at this moment</u>, averaged over its corresponding parts $S_i(z_i(t)), i = 1,\ldots,n$.* This observation gives rise to the following definition.

Denote by F_v the part of F in (11.3) which includes controls $v_1,\ldots,n-1$:

$$F_v(z,v) = \sum_{i=2}^{n+1} \xi_{i-1}(x,t)(v_{i-1}(t)A(z_i(t)-z_{i-1}(t)) + v_{i-2}(t)A(z_{i-2}(t)-z_{i-1}(t))).$$

Definition 13.1 (Force Controllability). Given the moment $t > 0$ and the state $\{y(t),\ z_i(t),\ i = 1,\ldots,n\}$, we will say that the system (11.1)–(11.3) is *force-controllable with respect to its point $z_i(t)$ at time t* if

$$\left\{ \int_{S_i(z_i(t))} PF_v(z,v)dx \mid v_i \in R, i = 1,\ldots,n-1 \right\} = R^2. \qquad (13.5)$$

If

$$\left\{ \sum_{i=1}^{n} \int_{S_i(z_i(t))} PF_v(z,v)dx \mid v_i \in R, i = 1,\ldots,n-1 \right\} = R^2. \qquad (13.6)$$

we will say that the system (11.1)–(11.3) is <u>force-controllable</u> or force-controllable with respect to its center of mass $z_c(t)$ at time t.

Hypothetically, if system (11.1)–(11.3) is force controllable at every moment on some interval $[0,T]$, then due to (13.3)–(13.4), one can steer the center of the swimming object any way one wishes. Of course, the rigorous verification of this property is a separate (global controllability) problem. However, if one considers "concrete" motions of the swimming object at hand such as, e.g., "forward" or "turning" motions, this verification can be significantly simplified – we refer the reader to Chapter 15 below, dedicated to such motions for a particular swimming model of type (11.1)–(11.3).

Theorem 13.1 below, provides qualitative formulas for the terms appearing in (13.3)–(13.6) in the case when

$$S(0) = S_0 = \{(x = (x_1, x_2) \mid -p < x_1 < p, \quad q < x_2 < q\}, \tag{13.7}$$

where p and q are "small" positive numbers.

Theorem 13.1. *Let $b = (b_1, b_2)$ be a given 2-D vector and $S(0) = S_0$ as in (13.7) lie in Ω. Let $q, p, q^{1-a}/p \to 0+$ for some $a \in (0,1)$. Then*

$$\frac{1}{\text{mes}\{S(0)\}} \int_{S(0)} (Pb\xi)(x)dx = (b_1, 0) + \left(O(q^a) + O(q^{1-a}/p) + O(p)\right) \parallel b \parallel_{R^2}$$

$$\tag{13.8}$$

as $q, p, q^{1-a}/p \to 0+$, where $\xi(x)$ is the characteristic function of $S(0)$.

In the above and below, the notation $O(s)$ means that $\parallel O(s) \parallel_{R^2} \le Cr$ as $s \to 0+$ for some positive constant C.

The result of Theorem 13.1 is illustrated bellow by Figs. A and B and can be interpreted as follows: If a constant force (i.e., b) acts upon a small narrow rectangle in a fluid, it will try to move this rectangle in the direction of the least resistance – parallel to its longer side.

Figure A shows a rectangle S_0 which is under the action of a constant force b when there is *no fluid around*. Figure B on the right shows the average force

$$b_* = \frac{1}{rq} \int_{S_0} (Pb\xi)(x)dx,$$

which will act on S_0 as a result if it is placed *in the fluid*.

In fact, Theorem 13.1 and Lemma 13.2 below expose what we can call *a "dual" effect of swimmer's internal forces on its motion*, which has to be addressed in an incompressible fluid. Indeed, we can view rectangle S_0 as a part of swimmer's body. Then Theorem 13.1 will describe the *direct* effect of force b applied to S_0. In addition to that, Lemma 13.2 below evaluates how the effect of this force will propagate to the rest of the fluid. In particular, this can help to answer the question on *how an internal force applied to one part of swimmer's body will "distort" the motion of its other parts, namely, as a distortion transmitted through the surrounding fluid.*

Figure A   Figure B

(Note that this question is principally impossible to address if a swimming model is finite-dimensional, i.e., described by ODE's, see the discussion in Chapter 10 in the above.)

The next result will be established as an immediate consequence of the proof of Theorem 13.1.

Theorem 13.2. *Let $b = (b_1, b_2)$ be a given 2-D vector and $S(0) = S_0$ be a disk of radius r lying in Ω. Then*

$$\frac{1}{\text{mes}\,\{S(0)\}} \int\limits_{S(0)} (Pb\xi)(x)dx = \frac{1}{2}(b_1, b_2) + O(r) \, \| \, b \, \|_{R^2} \qquad (13.9)$$

as $r \to 0+$, where $\xi(x)$ is the characteristic function of $S(0)$.

We would also like to point out here that, from the mathematical viewpoint, the simplest possible shape for a swimmer is a collection of discs (balls in 3-D). In this case, due to the symmetry of a disc (ball), it is easier to evaluate the result of action of a force applied to it in a fluid. However, this symmetry also means that such a swimmer is the "least shape efficient" in the sense that one cannot "re-orient" a disc/ball.

Remark 13.1 (Interrelation between the force controllability and local and global controllability). In the next Chapter 14 we will introduce the definition of local controllability for system (11.1)–(11.3) and will show that conditions (13.5) and (13.6) are sufficient to ensure the respective *local controllability* property for any $T \in [0, T_*]$ for some $T_* > 0$. These conditions are easily verifiable when the object at hand consists of either small rectangles or small discs, thus providing a foundation for the further study of the *global controllability* of (11.1)–(11.3) in Chapter 15.

We also have the following result for the local "point" controllability.

Theorem 13.3. *Let $S(0)$ in model (11.1)–(11.3) (see Chapter 11) be of the form $S(0) = rS_*(0), r \in (0, 1)$, where $S_*(0)$ is given. Suppose that for some $i \in \{2, \ldots, n - 1\}$ the three adjacent points $z_{i-1}(0), z_i(0)$, and $z_{i+1}(0)$ in the swimming object's skeleton do not lie on a straight line and that $S_k(z_k(t)), k = i - 1, i, i + 1$ have the same orientation, say, as $S_*(0)$ on some time-interval $[0, T^*]$. Then the system (11.1)–(11.3) is force controllable with respect to $z_i(0)$ at time $t = 0$ for sufficiently small r.*

The remainder of Chapter 13 is organized as follows. In the Section 13.3 we discuss some auxiliary results, which are further used in the proofs of Theorems 13.1 and 13.2, given in the Section 13.4. In the Section 13.5 we prove Lemma 13.1 and Theorem 13.3 is proven in the Section 13.6.

13.3 Some Estimates for Forces Acting upon Small Sets in a Fluid

In this section we discuss two auxiliary results needed to prove Theorems 13.1 and 13.2. They deal with the general sets $S(0)$ (introduced in Chapter 11).

Lemma 13.2. *Let $b = (b_1, b_2)$ be a given 2-D vector. Let $S(0) \subset \Omega$ be a nonzero measure set as in (11.1)–(11.3) of Chapter 11 which is strictly separated from $\partial\Omega$ and lies in an r-neighborhood $(r > 0)$ of the origin. Then for any subset Q of Ω of positive measure of diameter 2r (that is, it fits some disc of radius r) which lies strictly outside of $S(0)$ and is strictly separated from $\partial\Omega$ we have:*

$$\frac{1}{\operatorname{mes}\{Q\}} \int_Q (Pb\xi)(x)dx = O(r), \quad \| O(r) \|_{R^2} \leq C \frac{\| b \|_{R^2}}{d_*^4} \operatorname{mes}^{1/2}\{S_0\} \quad (13.10)$$

as $r \to 0+$, where $\xi(x)$ is the characteristic function of $S(0)$, $C > 0$ is a (generic) constant and d_* is the smallest out of the distances from Q to S_0 and to $\partial\Omega$.

We can interpret Lemma 13.1 as that the effect of the force b, supported on S_0, on similarly sized sets that are *strictly separated from* $S(0)$ is "negligible" as the measure of $S(0)$ decreases and the distance between these sets and S_0 increases.

Proof of Lemma 13.2.
Step 1. Recall first that (e.g., [141]):

$$J_o(\Omega) = \{u \in (L^2(\Omega))^2, \ \operatorname{div} u = 0, \ \gamma_\nu u |_{\partial\Omega} = 0\}, \quad (13.11)$$
$$G(\Omega) = \{u \in (L^2(\Omega))^2, \ u = \nabla p, \ p \in H^1(\Omega)\}, \quad (13.12)$$

where ν is the unit vector normal to the boundary $\partial\Omega$ (pointing outward) and $\gamma_\nu u |_{\partial\Omega}$ is the restriction of $u \cdot \nu$ to $\partial\Omega$.

Hence, we can decompose $b\xi(x)$ as follows:

$$(Pb\xi)(x) = b\xi(x) - \nabla w(x), \quad (13.13)$$

where w solves the following generalized Neumann problem:

$$\Delta w = \operatorname{div} b\xi(x) \quad \text{in } \Omega, \quad \frac{\partial w}{\partial \nu}\Big|_{\partial\Omega} = 0. \quad (13.14)$$

Note that, since, ξ vanishes in Q, $Pb\xi = -\nabla w$ in Q, and

$$\frac{1}{\operatorname{mes}\{Q\}} \int_Q (Pb\xi)(x)dx = -\frac{1}{\operatorname{mes}\{Q\}} \int_Q \nabla w dx. \quad (13.15)$$

Hence, to prove Lemma 13.2, it suffices to show that

$$\frac{1}{\operatorname{mes}\{Q\}} \int_Q \nabla w \, dx = O(r) \quad \text{as } r \to 0+ . \tag{13.16}$$

Step 2. Green's formula. To derive (13.16) we intend to use the generalized version of the classical Green's formula representing solutions of the boundary problems (13.14), namely:

$$
2\pi w(x) = -\int_{\partial\Omega} w(\eta) \frac{\partial}{\partial v} \left(\ln \frac{1}{\sqrt{(x_1 - \eta_1)^2 + (x_2 - \eta_2)^2}} \right) d\eta
$$

$$
- \int_\Omega \Delta w(y) \left(\ln \frac{1}{\sqrt{(x_1 - y_1)^2 + (x_2 - y_2)^2}} \right) dy
$$

$$
= - \int_{\partial\Omega} w(\eta) \frac{\partial}{\partial v} \left(\ln \frac{1}{\sqrt{(x_1 - \eta_1)^2 + (x_2 - \eta_2)^2)}} \right) d\eta
$$

$$
+ \int_{S(0)} b^T \nabla \left(\ln \frac{1}{\sqrt{(x_1 - y_1)^2 + (x_2 - y_2)^2}} \right) dy, \tag{13.17}
$$

where $y = (y_1, y_2)$ and the term in the 2-nd line is understood only formally. Here and below, when we write ∇ within some integral we mean that the corresponding differentiation is conducted with respect to the integration variables.

Indeed, to do that we can, making use of the integration by parts, first establish (13.17) for a sequence of solutions to (13.14) generated by a sequence of continuously differentiable functions $g_n(x)$ on the right, which converge to $\xi(x)$ in $L^2(\Omega)$, and then pass to the limit as $n \to \infty$. Note that the 2-nd integral on the right in (13.17) is well defined near the "bad point" (x_1, x_2), which can be shown by switching to the polar coordinates near it. Here and below we interpret the improper integral over the given domain E for a function with a discontinuity at x as the limit of the integrals over $E \backslash B_s(x)$ as $s \to 0+$, where $B_s(x)$ is a ball of radius s with center at x.

To show (13.16), we intend to evaluate the gradients of the terms in (13.17) and their integrals over the set Q.

Step 3. Evaluation of the integral of the gradient of the 1-st terms on the right in (13.17) over Q. Recall first that

$$\| w \|_{L^2(\partial\Omega)} \le L_0 \| w \|_{H^1(\Omega)},$$

where L_0 depends on the $\partial\Omega$.

Then, if $\{\alpha_k\}_{k=1}^{\infty}$ ($\alpha_k \to -\infty$ as $k \to \infty$) and $\{p_k\}_{k=1}^{\infty}$ are the negative eigenvalues and orthonormalized in $L^2(\Omega)$ eigenfunctions associated with the spectral problem

$$\Delta p = \alpha p, \quad \frac{\partial p}{\partial v}\Big|_{\partial\Omega} = 0,$$

(13.14) yields:

$$w(x) = \sum_{k=1}^{\infty} \frac{1}{\sqrt{-\alpha_k}}\left(\int_{\Omega} \xi(x)b^T \frac{\nabla p_k}{\sqrt{-\alpha_k}} dx\right) p_k(x) + K,$$

where without loss of generality we can set $K = 0$ (since in (13.16) we only deal with ∇w, we are interested only in the 1-st term here). Hence, noticing that $\{\frac{\nabla p_k}{\sqrt{-\alpha_k}}\}_{k=1}^{\infty}$ is an orthonormal sequence in $(L^2(\Omega))^2$, we derive from the Bessel inequality that

$$\| w_1 \|_{H^1(\Omega)} \le C \| b \|_{R^2} \ \mathrm{mes}^{1/2}\{S(0)\}, \qquad (13.18)$$

where C denotes a (generic) positive constant.

Furthermore, for $i, j = 1, 2, i \neq j$ and $x \neq y$:

$$\frac{\partial}{\partial y_i}\left(\ln\frac{1}{\sqrt{(x_1-y_1)^2+(x_2-y_2)^2}}\right) = \frac{x_i-y_i}{(x_1-y_1)^2+(x_2-y_2)^2}, \qquad (13.19)$$

$$\frac{\partial^2}{\partial y_i\partial x_i}\left(\ln\frac{1}{\sqrt{(x_1-y_1)^2+(x_2-y_2)^2}}\right) = \frac{-(x_i-y_i)^2+(x_j-y_j)^2}{((x_1-y_1)^2+(x_2-y_2)^2)^2}, \quad (13.20)$$

$$\frac{\partial^2}{\partial y_i\partial x_j}\left(\ln\frac{1}{\sqrt{(x_1-y_1)^2+(x_2-y_2)^2}}\right) = \frac{-2(x_i-y_i)(x_j-y_j)}{((x_1-y_1)^2+(x_2-y_2)^2)^2}. \quad (13.21)$$

Denote next by d_0 the distance between the set Q and $\partial\Omega$:

$$d_0 = \inf_{x\in Q, y\in\partial\Omega} \sqrt{(x_1-y_1)^2+(x_2-y_2)^2}. \qquad (13.22)$$

Then, combining (13.18)–(13.21), we obtain:

$$\left\| \int_Q \nabla \int_{\partial\Omega} w(\eta)\frac{\partial}{\partial v}\left(\ln\frac{1}{\sqrt{(x_1-\eta_1)^2+(x_2-\eta_2)^2}}\right) d\eta dx \right\|_{R^2}$$

$$\le \frac{L_0 C}{d_0^4} \| b \|_{R^2} \ \mathrm{mes}\{Q\}\mathrm{mes}^{1/2}\{S(0)\}\mathrm{mes}^{1/2}\{\partial\Omega\}, \qquad (13.23)$$

where, again, C is a (generic) positive constant.

Remark 13.2. Analogously, we can show that

$$\left\| \int_{S(0)} \nabla \int_{\partial\Omega} w(\eta) \frac{\partial}{\partial v} \left(\ln \frac{1}{\sqrt{(x_1 - \eta_1)^2 + (x_2 - \eta_2)^2}} \right) d\eta dx \right\|_{R^2}$$

$$\leq \frac{L_0 C}{d_*^4} \| b \|_{R^2} \, \mathrm{mes}^{3/2} \{S(0)\} \mathrm{mes}^{1/2} \{\partial\Omega\}, \tag{13.24}$$

where d_* denotes the distance between the set $S(0)$ and $\partial\Omega$.

Step 4. Evaluation of the integral of the gradient of the 2-nd term in (13.17) over Q.
Denote next by d_1 the distance between the set $S(0)$ and Q:

$$d_1 = \inf_{x \in S(0), y \in Q} \sqrt{(x_1 - y_1)^2 + (x_2 - y_2)^2}.$$

Then for $d_1 > 0$ we have, similar to (13.23), from (13.17) it follows that

$$\left\| \int_Q \nabla \int_{S(0)} b^T \nabla \left(\ln \frac{1}{\sqrt{(x_1 - y_1)^2 + (x_2 - y_2)^2}} \right) dydx \right\|_{R^2}$$

$$\leq \frac{C}{d_1^4} \| b \|_{R^2} \, \mathrm{mes} \{S(0)\} \, \mathrm{mes} \{Q\}, \tag{13.25}$$

where C is a (generic) positive constant. Combining (13.23) and (13.25) yields
(13.16), which provides the result of Lemma 13.2. $\qquad\Box$

Lemma 13.3. *Let $S(0)$ be open and lie in Ω. Assume that it is strictly separated
from $\partial\Omega$ and lies in an r-neighborhood $(r > 0)$ of the origin as well. Let $b = (b_1, b_2)$
be a given 2-D vector. Then*

$$\frac{1}{\mathrm{mes}\{S(0)\}} \int_{S(0)} (Pb\xi)(x)dx = b - \frac{1}{\mathrm{mes}\{S(0)\}} \int_{S(0)} g(x)dx + O(r), \tag{13.26}$$

as $r \to 0+$, where $\xi(x)$ is the characteristic function of $S(0)$, $g = (g_1, g_2)$ and

$$g_1(x) = \frac{1}{2\pi} b_1 \left(\int_{S(0) \setminus B_h(x)} \frac{-(x_1 - y_1)^2 + (x_2 - y_2)^2}{((x_1 - y_1)^2 + (x_2 - y_2)^2)^2} dy + \pi \right)$$

$$- \frac{1}{\pi} b_2 \int_{S(0) \setminus B_h(x)} \frac{(x_1 - y_1)(x_2 - y_2)}{((x_1 - y_1)^2 + (x_2 - y_2)^2)^2} dy, \tag{13.27}$$

$$g_2(x) = \frac{1}{2\pi} b_2 \left(\int_{S(0)\setminus B_h(x)} \frac{-(x_2-y_2)^2 + (x_1-y_1)^2}{((x_1-y_1)^2 + (x_2-y_2)^2)^2} dy + \pi \right)$$

$$-\frac{1}{\pi} b_1 \int_{S(0)\setminus B_h(x)} \frac{(x_1-y_1)(x_2-y_2)}{((x_1-y_1)^2 + (x_2-y_2)^2)^2} dy, \qquad (13.28)$$

and $B_h(x) = \{y \mid \|x-y\|_{\mathbb{R}^2} < h\}$ is any ball of some radius $h = h(x) > 0$ with center at x that lies in $S(0)$.

Proof of Lemma 13.3.
Step 1. Due to (13.13) and similar to (13.15),

$$\frac{1}{\text{mes}\{S(0)\}} \int_{S(0)} (Pb\xi)(x)dx = b - \frac{1}{\text{mes}\{S(0)\}} \int_{S(0)} \nabla w dx. \qquad (13.29)$$

Furthermore, in view of Remark 13.2 (see (13.24)), to evaluate the 2-nd term on the right in (13.29), it is sufficient to evaluate the integral of the gradient of the 2-nd term in (13.17) over $S(0)$.

Step 2. Consider any point $x = (x_1, x_2) \in S(0)$. Since we assumed that $S(0)$ is open, for any $x \in S(0)$ there exists an $h = h(x) > 0$ such that $B_{h(x)}(x) \subset S(0)$ and $B_{h(x)}(x)$ is strictly separated from $\partial S(0)$.
Then, from (13.17):

$$\int_{S(0)} b^T \nabla \left(\ln \frac{1}{\sqrt{(x_1-y_1)^2 + (x_2-y_2)^2}} \right) dy$$

$$= \int_{S(0)\setminus B_{h(x)}(x)} b^T \nabla \left(\ln \frac{1}{\sqrt{(x_1-y_1)^2 + (x_2-y_2)^2}} \right) dy$$

$$+ \int_{B_{h(x)}(x)} b^T \nabla \left(\ln \frac{1}{\sqrt{(x_1-y_1)^2 + (x_2-y_2)^2}} \right) dy.$$

Now note that, in view of (13.19):

$$\int_{B_{h(x)}(x)} \left(\ln \frac{1}{\sqrt{(x_1-y_1)^2 + (x_2-y_2)^2}} \right)_{y_1} dy$$

$$= \lim_{s \to 0+} \int_{B_{h(x)}(x)\setminus B_s(x)} \frac{x_1-y_1}{(x_1-y_1)^2 + (x_2-y_2)^2} dy$$

$$= -\lim_{s \to 0+} \int_0^{2\pi} \int_s^{h(x)} \cos\theta d\rho d\theta = 0. \qquad (13.30)$$

These and similar calculations for the integration with respect to y_2 within the circle $B_h(x)$, yield that

$$\int_{B_{h(x)}(x)} b^T \nabla \left(\ln \frac{1}{\sqrt{(x_1-y_1)^2+(x_2-y_2)^2}} \right) dx = 0.$$

Thus,

$$\int_{S(0)} b^T \nabla \left(\ln \frac{1}{\sqrt{(x_1-y_1)^2+(x_2-y_2)^2}} \right) dy$$

$$= \int_{S(0)\backslash B_{h(x)}(x)} \left(b_1 \frac{x_1-y_1}{(x_1-y_1)^2+(x_2-y_2)^2} + b_2 \frac{x_2-y_2}{(x_1-y_1)^2+(x_2-y_2)^2} \right) dy. \quad (13.31)$$

Step 3. We intend now to calculate the gradient of the 2-nd term in (13.17) represented as in (13.31).

Fix any $x = (x_1, x_2) \in S(0)$. Due to our selection of h for the given x, for "small" Δx_1:

$$B_{h((x_1+\Delta x_1, x_2))}((x_1+\Delta x_1, x_2)) \subset S(0).$$

Furthermore, notice that as in the second equality in (13.30):

$$\int_{B_{h((x_1+\Delta x_1, x_2))}((x_1+\Delta x_1, x_2))\backslash B_{h(x)}((x_1+\Delta x_1, x_2))} \left(\ln \frac{1}{\sqrt{(x_1+\Delta x_1-y_1)^2+(x_2-y_2)^2}} \right)_{y_1} dy$$
$$= 0.$$

Taking this into account, we obtain from (13.31):

$$\frac{\partial}{\partial x_1} \int_{S(0)} b^T \nabla \left(\ln \frac{1}{\sqrt{(x_1-y_1)^2+(x_2-y_2)^2}} \right) dy$$

$$= \lim_{\Delta x_1 \to 0} \frac{1}{\Delta x_1}$$

$$\times \left(\int_{S(0)\backslash B_{h((x_1+\Delta x_1, x_2))}((x_1+\Delta x_1, x_2))} b^T \nabla \left(\ln \frac{1}{\sqrt{(x_1+\Delta x_1-y_1)^2+(x_2-y_2)^2}} \right) dy \right.$$

$$\left. - \int_{S(0)\backslash B_{h(x)}(x)} b^T \nabla \left(\ln \frac{1}{\sqrt{(x_1-y_1)^2+(x_2-y_2)^2}} \right) dy \right)$$

$$= \lim_{\Delta x_1 \to 0} \frac{1}{\Delta x_1} \left(\int_{S(0)\backslash B_{h(x)}((x_1+\Delta x_1, x_2))} b^T \nabla \left(\ln \frac{1}{\sqrt{(x_1+\Delta x_1 - y_1)^2 + (x_2 - y_2)^2}} \right) dy \right.$$

$$\left. - \int_{S(0)\backslash B_{h(x)}(x)} b^T \nabla \left(\ln \frac{1}{\sqrt{(x_1 - y_1)^2 + (x_2 - y_2)^2}} \right) dy \right)$$

$$= b_1 \frac{\partial}{\partial x_1} \left(\int_{S(0)\backslash B_h(x)} \frac{x_1 - y_1}{(x_1 - y_1)^2 + (x_2 - y_2)^2} dy \right)$$

$$+ b_2 \frac{\partial}{\partial x_1} \left(\int_{S(0)\backslash B_h(x)} \frac{x_2 - y_2}{(x_1 - y_1)^2 + (x_2 - y_2)^2} dy \right),$$

where $h = h(x)$ in the last line is now treated as independent of x when calculating the derivatives.

Step 4. Let us calculate the derivative in the 1-st term in the last expression. To simplify notations, we will further write h instead of $h(x)$.

$$\frac{\partial}{\partial x_1} \int_{S(0)\backslash B_h(x)} \frac{x_1 - y_1}{(x_1 - y_1)^2 + (x_2 - y_2)^2} dy$$

$$= \frac{\partial}{\partial x_1} \left\{ \int_{-r}^{x_1 - h\beta(y_1)} \int_{\alpha(y_1)} \frac{x_1 - y_1}{(x_1 - y_1)^2 + (x_2 - y_2)^2} dy_2 dy_1 \right.$$

$$+ \int_{x_1 + h\alpha(y_1)}^{r} \int^{\beta(y_1)} \frac{x_1 - y_1}{(x_1 - y_1)^2 + (x_2 - y_2)^2} dy_2 dy_1$$

$$+ \int_{x_1 - h}^{x_1 + h} \left[\int_{\alpha(y_1)}^{x_2 - \sqrt{h^2 - (x_1 - y_1)^2}} \frac{x_1 - y_1}{(x_1 - y_1)^2 + (x_2 - y_2)^2} dy_2 \right.$$

$$\left. + \int_{x_2 + \sqrt{h^2 - (x_1 - y_1)^2}}^{\beta(y_1)} \frac{x_1 - y_1}{(x_1 - y_1)^2 + (x_2 - y_2)^2} dy_2 \right] dy_1 \right\}$$

$$= \int_{S(0)\backslash B_h(x)} \frac{-(x_1 - y_1)^2 + (x_2 - y_2)^2}{((x_1 - y_1)^2 + (x_2 - y_2)^2)^2} dy + 2 \int_{x_1 - h}^{x_1 + h} \frac{(x_1 - y_1)^2}{h^2 \sqrt{h^2 - (x_1 - y_1)^2}} dy_1$$

$$= \int_{S(0)\backslash B_h(x)} \frac{-(x_1 - y_1)^2 + (x_2 - y_2)^2}{((x_1 - y_1)^2 + (x_2 - y_2)^2)^2} dy + \pi. \tag{13.32}$$

Similar calculations also yield:

$$\frac{\partial}{\partial x_2} \int\limits_{S(0)\backslash B_h(x)} \frac{x_2 - y_2}{(x_1 - y_1)^2 + (x_2 - y_2)^2} dy$$

$$= \int\limits_{S(0)\backslash B_h(x)} \frac{-(x_2 - y_2)^2 + (x_1 - y_1)^2}{((x_1 - y_1)^2 + (x_2 - y_2)^2)^2} dy + \pi, \qquad (13.33)$$

$$\frac{\partial}{\partial x_j} \int\limits_{S(0)\backslash B_h(x)} \frac{x_i - y_i}{(x_1 - y_1)^2 + (x_2 - y_2)^2} dy$$

$$= - \int\limits_{S(0)\backslash B_h(x)} 2\frac{(x_i - y_i)(x_j - y_j)}{((x_1 - y_1)^2 + (x_2 - y_2)^2)^2} dy, \quad i \neq j, i, j = 1, 2. \quad (13.34)$$

Formulas (13.32)–(13.34) provide the formula for g in (13.27)–(13.28) (see Step 1 of the proof). This completes the proof of Lemma 13.3. □

13.4 Proofs of Theorems 13.1 and 13.2

13.4.1 Proof of Theorem 13.1: Forces Acting upon Small Narrow Rectangles in a Fluid

Without loss of generality, we can assume that $q < p$. In view of (13.26), to establish (13.8) we just need to evaluate the integrals in (13.27)–(13.28) for the case when $S(0) = S_0$ as in (13.7). In this case we select $B_h(x)$ in Lemma 13.3 as

$$B_{h(x)}(x) = \{y \mid \|x - y\|_{R^2} < \frac{1}{2}\min\{\|x\|_{R^2}, q - \|x\|_{R^2}\} = h(x)\} \subset S(0).$$

Step 1. We begin with the 1-st term on the right in (13.27):

$$\int\limits_{S_0\backslash B_h(x)} \frac{-(x_1 - y_1)^2 + (x_2 - y_2)^2}{((x_1 - y_1)^2 + (x_2 - y_2)^2)^2} dy$$

$$= \int\limits_{-p}^{x_1 - h} \int\limits_{-q}^{q} \frac{-(x_1 - y_1)^2 + (x_2 - y_2)^2}{((x_1 - y_1)^2 + (x_2 - y_2)^2)^2} dy$$

$$+ \int\limits_{x_1+h-q}^{p} \int\limits_{}^{q} \frac{-(x_1-y_1)^2 + (x_2-y_2)^2}{((x_1-y_1)^2 + (x_2-y_2)^2)^2} dy$$

$$+ \int\limits_{x_1-h}^{x_1+hx_2-\sqrt{h^2-(x_1-y_1)^2}} \int\limits_{-q}^{} \frac{-(x_1-y_1)^2 + (x_2-y_2)^2}{((x_1-y_1)^2 + (x_2-y_2)^2)^2} dy$$

$$+ \int\limits_{x_1-h}^{x_1+h} \int\limits_{x_2+\sqrt{h^2-(x_1-y_1)^2}}^{q} \frac{-(x_1-y_1)^2 + (x_2-y_2)^2}{((x_1-y_1)^2 + (x_2-y_2)^2)^2} dy. \qquad (13.35)$$

Step 2. Making use of (13.19)–(13.21), we further obtain from (13.35):

$$\int\limits_{-p}^{x_1-h} \int\limits_{-q}^{q} \frac{-(x_1-y_1)^2 + (x_2-y_2)^2}{((x_1-y_1)^2 + (x_2-y_2)^2)^2} dy + \int\limits_{x_1+h-q}^{p} \int\limits_{}^{q} \frac{-(x_1-y_1)^2 + (x_2-y_2)^2}{((x_1-y_1)^2 + (x_2-y_2)^2)^2} dy$$

$$= \int\limits_{-q}^{q} \frac{y_1 - x_1}{(x_1-y_1)^2 + (x_2-y_2)^2} \Big|_{y_1=-p}^{y_1=x_1-h} dy_2$$

$$+ \int\limits_{-q}^{q} \frac{y_1 - x_1}{(x_1-y_1)^2 + (x_2-y_2)^2} \Big|_{y_1=x_1+h}^{y_1=p} dy_2$$

$$= \int\limits_{-q}^{q} \frac{-h}{h^2 + (y_2-x_2)^2} dy_2 + \int\limits_{-q}^{q} \frac{p+x_1}{(p+x_1)^2 + (y_2-x_2)^2} dy_2$$

$$+ \int\limits_{-q}^{q} \frac{p-x_1}{(p-x_1)^2 + (y_2-x_2)^2} dy_2 - \int\limits_{-q}^{q} \frac{h}{h^2 + (y_2-x_2)^2} dy_2$$

$$= -\tan^{-1} \frac{y_2 - x_2}{h} \Big|_{y_2=-q}^{y_2=q} + \tan^{-1} \frac{y_2 - x_2}{p+x_1} \Big|_{y_2=-q}^{y_2=q}$$

$$+ \tan^{-1} \frac{y_2 - x_2}{p-x_1} \Big|_{y_2=-q}^{y_2=q} - \tan^{-1} \frac{y_2 - x_2}{h} \Big|_{y_2=-q}^{y_2=q}$$

$$= -\tan^{-1} \frac{q - x_2}{h} + \tan^{-1} \frac{-q - x_2}{h} + \tan^{-1} \frac{q - x_2}{p+x_1} - \tan^{-1} \frac{-q - x_2}{p+x_1}$$

$$+ \tan^{-1} \frac{q - x_2}{p-x_1} - \tan^{-1} \frac{-q - x_2}{p-x_1} - \tan^{-1} \frac{q - x_2}{h} + \tan^{-1} \frac{-q - x_2}{h}. \qquad (13.36)$$

Step 3. Similarly to the derivation of (13.36), for the remaining two terms in (13.35) we have:

$$\int_{x_1-h}^{x_1+hx_2-\sqrt{h^2-(x_1-y_1)^2}} \int_{-q} \frac{-(x_1-y_1)^2+(x_2-y_2)^2}{((x_1-y_1)^2+(x_2-y_2)^2)^2} dy$$

$$+ \int_{x_1-h}^{x_1+h} \int_{x_2+\sqrt{h^2-(x_1-y_1)^2}}^{q} \frac{-(x_1-y_1)^2+(x_2-y_2)^2}{((x_1-y_1)^2+(x_2-y_2)^2)^2} dy$$

$$= - \int_{x_1-h}^{x_1+h} \frac{y_2-x_2}{(x_1-y_1)^2+(x_2-y_2)^2} \Big|_{-q}^{x_2-\sqrt{h^2-(x_1-y_1)^2}} dy_1$$

$$- \int_{x_1-h}^{x_1+h} \frac{y_2-x_2}{(x_1-y_1)^2+(x_2-y_2)^2} \Big|_{x_2+\sqrt{h^2-(x_1-y_1)^2}}^{q} dy_1$$

$$= \int_{x_1-h}^{x_1+h} \frac{\sqrt{h^2-(x_1-y_1)^2}}{h^2} dy_1 + \int_{x_1-h}^{x_1+h} \frac{-q-x_2}{(x_1-y_1)^2+(x_2+q)^2} dy_1$$

$$- \int_{x_1-h}^{x_1+h} \frac{q-x_2}{(x_1-y_1)^2+(q-x_2)^2} dy_1 + \int_{x_1-h}^{x_1+h} \frac{\sqrt{h^2-(x_1-y_1)^2}}{h^2} dy_1$$

$$= \pi - \tan^{-1}\frac{h}{q+x_2} + \tan^{-1}\frac{-h}{q+x_2} - \tan^{-1}\frac{h}{q-x_2} + \tan^{-1}\frac{-h}{q-x_2}, \quad (13.37)$$

where we also used the formula:

$$\int_{x_1-h}^{x_1+h} \frac{\sqrt{h^2-(x_1-y_1)^2}}{h^2} dy_1 = \frac{\pi}{2}.$$

Step 4. Making use of the equality

$$\tan^{-1}s + \tan^{-1}\frac{1}{s} = \frac{\pi}{2} \quad \forall s > 0 \ (\tan^{-1}s \in (0,\pi/2)),$$

(13.36) and (13.37) yield for the expression in (13.35):

$$\int_{S_0\backslash B_h(x)} \frac{-(x_1-y_1)^2+(x_2-y_2)^2}{((x_1-y_1)^2+(x_2-y_2)^2)^2} dy$$

$$= -\pi + \tan^{-1}\frac{q-x_2}{p+x_1} + \tan^{-1}\frac{q+x_2}{p+x_1}$$

$$+ \tan^{-1}\frac{q-x_2}{p-x_1} + \tan^{-1}\frac{q+x_2}{p-x_1}. \quad (13.38)$$

To proceed further with the proof of Theorem 13.1, we will need to evaluate (see (13.26) and the 1-st term on the right in (13.27)) the term

$$
\frac{1}{\text{mes}\,\{S(0)\}} \times \int_{S_0} \left(-\pi + \tan^{-1}\frac{q-x_2}{p+x_1} + \tan^{-1}\frac{q+x_2}{p+x_1} \right.
$$

$$
\left. + \tan^{-1}\frac{q-x_2}{p-x_1} + \tan^{-1}\frac{q+x_2}{p-x_1} \right) dx \qquad (13.39)
$$

Step 5. We start with the 2-nd term in (13.39) (under the sign of integral).
Recall that

$$
\tan^{-1}(s) \in \left(-\frac{\pi}{2}, \frac{\pi}{2} \right) \quad \forall s. \qquad (13.40)
$$

Denote

$$
A(p,q) = \{(x=(x_1,x_2)\mid -q<x_2<q,\ -p+q^{1-a}<x_1<p-q^{1-a}\}.
$$

Under the assumption of Theorem 13.1 that $a \in (0,1)$, $q,p,q^{1-a}/p \to 0+$, without loss of generality, we can assume that $A(p,q) \subset S(0)$ with

$$
\text{mes}\,A(p,q) = 4q(p-q^{1-a}) = 4pq(1-q^{1-a}/p). \qquad (13.41)
$$

Furthermore, for $x \in A(p,q)$

$$
0 < \frac{q-x_2}{p+x_1} < q^a, \qquad (13.42)
$$

and hence

$$
\tan^{-1}\frac{q-x_2}{p+x_1} \le \tan^{-1}q^a, \quad x \in A(p,q). \qquad (13.43)
$$

Then, making use of (13.40)–(13.43), we obtain:

$$
\left| \frac{1}{\text{mes}\,\{S(0)\}} \int_{S_0} \tan^{-1}\frac{q-x_2}{p+x_1}dx \right| = \frac{1}{4pq} \int_{-p}^{p} \int_{-q}^{q} \tan^{-1}\frac{q-x_2}{p+x_1}dx_2dx_1
$$

$$
= \frac{1}{4pq} \int_{A(p,q)} \tan^{-1}\frac{q-x_2}{p+x_1}dx + \int_{S(0)\backslash A(p,q)} \tan^{-1}\frac{q-x_2}{p+x_1}dx
$$

$$
\le (1-q^{1-a}/p)\tan^{-1}q^a + \frac{4q^{2-a}}{4pq}\frac{\pi}{2}
$$

$$
= O(\tan^{-1}q^a) + O(q^{1-a}/p) \qquad (13.44)
$$

as $p,q^{1-a}/p \to 0+$.

Step 6. In a similar way we can evaluate the remaining terms in (13.39), which will result in:

$$\frac{1}{\text{mes}\{S(0)\}} \int_{S_0} \left(\int_{S_0 \backslash B_h(x)} \frac{-(x_1-y_1)^2+(x_2-y_2)^2}{((x_1-y_1)^2+(x_2-y_2)^2)^2} dy \right) dx = \frac{1}{\text{mes}\{S(0)\}}$$

$$\times \int_{S_0} \left(-\pi + \tan^{-1}\frac{q-x_2}{p+x_1} + \tan^{-1}\frac{q+x_2}{p+x_1} + \tan^{-1}\frac{q-x_2}{p-x_1} + \tan^{-1}\frac{q+x_2}{p-x_1} \right) dx$$

$$= -\pi + O(q^a) + O(q^{1-a}/p) \tag{13.45}$$

as $q, p, q^{1-a}/p \to 0+$.

Step 7. Furthermore, since we assumed that $S(0)$ is a rectangle, due to the antisymmetry of the linear functions and our "symmetric" choice of $h(x)$ in the beginning of this proof (i.e., $h((x_1,x_2)) = h((-x_1,x_2)) = h((x_1,-x_2)) = h((-x_1,-x_2))$):

$$\int_{S(0)} \int_{S(0)\backslash B_h(x)} \frac{(x_i-y_i)(x_j-y_j)}{((x_1-y_1)^2+(x_2-y_2)^2)^2} dydx = 0, \quad i,j = 1,2, \ i \neq j. \tag{13.46}$$

Step 8. We now need to evaluate the remaining terms in (13.27)–(13.28). To this end, similar to (13.38)–(13.39), for the 1-st term on the right in (13.28) we have from (13.45):

$$\frac{1}{\text{mes}\{S(0)\}} \int_{S_0} \left(\int_{S_0 \backslash B_h(x)} \frac{-(x_2-y_2)^2+(x_1-y_1)^2}{((x_1-y_1)^2+(x_2-y_2)^2)^2} dy \right) dx$$

$$= \pi + O(q^a) + O(q^{1-a}/p) \tag{13.47}$$

as $q, p, q^{1-a}/p \to 0+$.

Step 9. Combining (13.45), (13.46) and (13.47) yield that in (13.27)–(13.28) for our $S(0)$ we have:

$$\frac{1}{\text{mes}\{S(0)\}} \int_{S_0} g_1 dx = O(q^a) + O(q^{1-a}/p), \tag{13.48}$$

$$\frac{1}{\text{mes}\{S(0)\}} \int_{S_0} g_2 dx = b_2 + O(q^a) + O(q^{1-a}/p), \tag{13.49}$$

Combining (13.48)–(13.49) with (13.26) yields (13.8), which completes the proof of Theorem 13.1. □

13.4.2 Proof of Theorem 13.2: Forces Acting upon Small Discs in a Fluid

It follows immediately from Lemma 13.3, because, due to the symmetry of a disc, all the integrals in (13.27) are equal to zero.   □

13.5 Proof of Lemma 13.1: General Formula for Micromotions

Without loss of generality we can assume that $t_0 = 0$ so that $y(x, t_0) = y_0(x)$.
Step 1. We will show first that the first term in (13.1)–(13.2) generates the first term in (13.3) and u in (13.4), namely, that,

$$\frac{1}{\text{mes}\{S(0)\}} \int_{S_i(z_i(t))} \left(\sum_{k=1}^{\infty} e^{-\lambda_k t} \left(\int_\Omega y_0^T \omega_k ds \right) \omega_k(x) \right) dx$$

$$= \frac{1}{\text{mes}\{S(0)\}} \int_{S_i(z_i(0))} y_0 dx + u(t). \qquad (13.50)$$

Indeed, making use of (13.1)–(13.2),

$$\left\| \frac{1}{\text{mes}\{S(0)\}} \int_{S_i(z_i(t))} \left(\sum_{k=1}^{\infty} e^{-\lambda_k t} \left(\int_\Omega y_0^T \omega_k ds \right) \omega_k(x) \right) dx - \frac{1}{\text{mes}\{S(0)\}} \int_{S_i(z_i(0))} y_0 dx \right\|_{R^2}$$

$$\leq \frac{1}{\text{mes}\{S(0)\}} \left\| \int_{S_i(z_i(0)) \cap S_i(z_i(t))} \sum_{k=1}^{\infty} \left(e^{-\lambda_k t} - 1 \right) \int_\Omega y_0^T \omega_k ds \omega_k(x) \, dx \right\|_{R^2}$$

$$+ \frac{1}{\text{mes}\{S(0)\}} \left\| \int_{S_i(z_i(0)) \backslash S_i(z_i(t))} \sum_{k=1}^{\infty} \int_\Omega y_0^T \omega_k ds \omega_k(x) \, dx \right\|_{R^2}$$

$$+ \frac{1}{\text{mes}\{S(0)\}} \left\| \int_{S_i(z_i(t)) \backslash S_i(z_i(0))} \sum_{k=1}^{\infty} e^{-\lambda_k t} \int_\Omega y_0^T \omega_k ds \omega_k(x) \, dx \right\|_{R^2}$$

$$\leq \frac{\mu(t, y_0)}{\text{mes}\{S(0)\}} \text{mes}^{1/2} \{S_i(z_i(0)) \cap S_i(z_i(t))\}$$

$$+ \frac{2 \max^{1/2}\{\text{mes}\{S_i(z_i(0)) \backslash S_i(z_i(t)), \text{ mes } S_i(z_i(t)) \backslash S_i(z_i(0))\}}{\text{mes}\{S(0)\}}$$

$$\times \| y_0 \|_{(L^2(\Omega))^2}, \qquad (13.51)$$

where $\mu(t, y_0) \to 0$ as $t \to 0+$, $0 \leq \mu(t, y_0) \leq C \| y_0 \|_{(L^2(\Omega))^2}$, and $C > 0$ denotes a (generic) constant.

Step 2. Note that all the terms in the solution formula (13.1)–(13.2) associated with the forcing term admit the following representation:

$$\sum_{k=1}^{\infty} \int_0^t e^{-\lambda_k(t-\tau)} \left(\int_{\Omega} \xi_j(s,\tau) w^T(\tau) \omega_k ds d\tau \right) \omega_k(x) \tag{13.52}$$

for $j \in \{1,\ldots,n\}$ and some $2-$D function $w(t)$ as given in (11.3). Making use of (13.52), to show that (13.3), (13.4) holds, we need to evaluate, e.g., the following expression:

$$\left\| \frac{1}{\text{mes}\{S(0)\}} \int_{S_i(z_i(t))} \left(\sum_{k=1}^{\infty} \int_0^t e^{-\lambda_k(t-\tau)} \left(\int_{\Omega} \xi_i(s,\tau) w^T(\tau) \omega_k ds d\tau \right) \omega_k(x) \right) dx \right.$$

$$\left. - \frac{t}{\text{mes}\{S(0)\}} \int_{S_i(z_i(0))} \sum_{k=1}^{\infty} \left(\int_{\Omega} \xi_i(s,0) w^T(0) \omega_k ds \right) \omega_k(x) dx \right\|_{R^2}$$

$$\leq \frac{1}{\text{mes}\{S(0)\}} \left\| \int_{S_i(z_i(0)) \cap S_i(z_i(t))} \left(\sum_{k=1}^{\infty} \int_0^t e^{-\lambda_k(t-\tau)} \left(\int_{\Omega} [\xi_i(s,\tau) w(\tau) \right. \right. \right.$$

$$\left. \left. \left. - \xi_i(s,0) w(0)]^T \omega_k ds d\tau \right) \omega_k(x) \right) dx \right\|_{R^2}$$

$$+ \frac{1}{\text{mes}\{S(0)\}} \left\| \int_{S_i(z_i(0)) \cap S_i(z_i(t))} \left(\sum_{k=1}^{\infty} \int_0^t [e^{-\lambda_k(t-\tau)} - 1] \right. \right.$$

$$\left. \left. \times \left(\int_{\Omega} \xi_i(s,0) w^T(0) \omega_k ds d\tau \right) \omega_k(x) \right) dx \right\|_{R^2} + \frac{1}{\text{mes}\{S(0)\}}$$

$$\times \left\| \int_{S_i(z_i(t)) \backslash S_i(z_i(0))} \left(\sum_{k=1}^{\infty} \int_0^t e^{-\lambda_k(t-\tau)} \left(\int_{\Omega} \xi_i(s,\tau) w^T(\tau) \omega_k ds d\tau \right) \omega_k(x) \right) dx \right\|_{R^2}$$

$$+ \frac{t}{\text{mes}\{S(0)\}} \left\| \int_{S_i(z_i(0)) \backslash S_i(z_i(t))} \left(\sum_{k=1}^{\infty} \left(\int_{\Omega} \xi_i(s,\tau) w^T(0) \omega_k ds \right) \omega_k(x) \right) dx \right\|_{R^2}$$

$$\leq \frac{t^{1/2} \text{mes}^{1/2}\{S_i(z_i(0)) \cap S_i(z_i(t))\}}{\text{mes}\{S(0)\}} \| \xi_i(\cdot,\cdot) w(\cdot) - \xi_i(\cdot,0) w(0) \|_{(L^2(Q_t))^2}$$

$$+ \frac{1}{\text{mes}^{1/2}\{S(0)\}} \left(\sum_{k=1}^{\infty} t^2 \left[\frac{1-e^{-\lambda_k t}}{\lambda_k t} - 1 \right]^2 \left(\int_{\Omega} \xi_i(s,0) w^T(0) \omega_k ds \right)^2 \right)^{1/2}$$

$$+ \frac{t^{1/2} \text{mes}^{1/2}\{S_i(z_i(t)) \backslash S_i(z_i(0))\}}{\text{mes}\{S(0)\}} \| \xi_i(\cdot,\cdot) w(\cdot) \|_{(L^2(Q_t))^2}$$

$$+t\,\frac{\mathrm{mes}^{1/2}\left\{S_i(z_i(0))\backslash S_i(z_i(t))\right\}}{\mathrm{mes}\left\{S(0)\right\}}\,\|\,\xi_i(\cdot,0)w(0)\,\|_{(L^2(\Omega))^2}$$

$$\leq t\,\frac{\mathrm{mes}\left\{S_i(z_i(0))\cap S_i(z_i(t))\right\}}{\mathrm{mes}\left\{S(0)\right\}}\left(\|\,w(\cdot)-w(0)\,\|_{(C[0,t])^2}\right)+t\frac{1}{\mathrm{mes}^{1/2}\left\{S(0)\right\}}\gamma(t)$$

$$+t\,\frac{\mathrm{mes}^{1/2}\left\{S_i(z_i(t))\backslash S_i(z_i(0))\right\}}{\mathrm{mes}\left\{S(0)\right\}}\mathrm{mes}^{1/2}\left\{S_i(z_i(0))\cap S_i(z_i(t))\right\}\|\,w(\cdot)\|_{(C[0,t])^2}$$

$$+t\,\frac{\mathrm{mes}^{1/2}\left\{S_i(z_i(0))\backslash S_i(z_i(t))\right\}}{\mathrm{mes}\left\{S(0)\right\}}\mathrm{mes}^{1/2}\left\{S_i(z_i(0))\cap S_i(z_i(t))\right\}\|\,w(0)\,\|_{R^2}$$

$$+t\,\frac{\mathrm{mes}^{1/2}\left\{S_i(z_i(t))\backslash S_i(z_i(0))\right\}}{\mathrm{mes}^{1/2}\left\{S(0)\right\}}\,\|\,w(\cdot)\|_{(C[0,t])^2}$$

$$+t\,\frac{\mathrm{mes}^{1/2}\left\{S_i(z_i(0))\backslash S_i(z_i(t))\right\}}{\mathrm{mes}^{1/2}\left\{S(0)\right\}}\,\|\,w(0)\,\|_{R^2},\tag{13.53}$$

where we used the estimate

$$\left|\,\frac{1-e^{-s}}{s}-1\,\right|<1,\quad s>0$$

and also $\gamma(t)\to 0+$ as $t\to 0+$.

The estimates (13.51) and (13.53) yield (13.3)–(13.4). This ends the proof of Lemma 13.1.   □

13.6 Proof of Theorem 13.3: Sufficient Conditions for the Force Controllability

Select any point $i\in\{2,\dots,n-1\}$. Without loss of generality, we can *assume that* $z_i(0)=0$, *i.e., it is the origin.* Recall that in this theorem $S_k(z_k(t))=rS_*(0)$, $k=i-1,i,i+1$ on some $[0,T^*]$.

Step 1. As the reader will see in Chapter 14, the sufficient condition for the force controllability (and for the local controllability) of system (11.1)–(11.3) with respect to its point $z_i^*(T)$ at a "small" time $T>0$ by means of two controls v_{i-1} and v_i only (that is, $v_j=0$, $j\neq i-1,i$), is that the matrix

$$\Big(\int_{rS_*(z_i(0))}P\left[\xi_{i-1}(\cdot,0)A(z_i(0)-z_{i-1}(0))\right]dx+$$

$$\int_{rS_*(z_i(0))}P\left[\xi_i(\cdot,0)A(z_{i-1}(0)-z_i(0))\right]dx,$$

$$\int\limits_{rS_*(z_i(0))} P\left[\xi_i(\cdot,0)A(z_{i+1}(0)-z_i(0))\right]dx$$

$$+ \int\limits_{rS_*(z_i(0))} P\left[\xi_{i+1}(\cdot,0)A(z_i(0)-z_{i+1}(0))\right]dx) \tag{13.54}$$

is not degenerate (see also (11.3) and (13.5) for its equivalent form).

Due to Lemma 13.2, we can rewrite (13.54) as

$$\frac{1}{\text{mes}\{rS_*(0)\}}\left(\int\limits_{rS_*(0)} F_{1r}dx, \int\limits_{rS_*(0)} F_{2r}dx\right)+O(r), \tag{13.55}$$

where the expression $O(r)$ has the same meaning as in the above but in the matrix space, and where, to simplify further notations, we denoted :

$$b_{(1)} = A(z_{i-1}(0)-z_i(0)), \quad b_{(2)} = A(z_{i+1}(0)-z_i(0)),$$
$$F_{1r} = P\left[\xi_i(\cdot,0)b_{(1)}\right], \quad F_{2r} = P\left[\xi_i(\cdot,0)b_{(2)}\right].$$

To prove Theorem 13.3, we intend to show next that under its assumptions, that is, the points $z_k(0), k = i-1, i, i+1$ do not line on the same straight line, the determinant of the first matrix in (13.55) is bounded from below by a positive number as $r \to 0+$.

Step 2. From (13.13) we have:

$$F_{kr} = b_{(k)}\xi_i(\cdot,0) - \nabla w_{kr}, \quad k = 1,2, \tag{13.56}$$

where w_{kr}'s satisfy the boundary problems like (13.14).

We will now analyze the generic equation as in (13.13) with $S(0) = S_*(0)$. Denote $F = Pb\xi$. In this case (13.13) will look like:

$$F = b\xi(x) - \nabla w(x). \tag{13.57}$$

Multiplication of (13.57) by F^T and further integration over Ω yields:

$$\| F \|^2_{(L^2(\Omega))^2} = b^T \int\limits_{S_*(0)} F dx. \tag{13.58}$$

We claim that for any $b \in R^2$ (and hence the associated F in (13.57)):

$$\frac{1}{\text{mes}\{S_*(0)\}} \| F \|^2_{(L^2(\Omega))^2} = \frac{1}{\text{mes}\{S_*(0)\}} b^T \int\limits_{S_*(0)} F dx \geq m_0 \| b \|^2_{R^2} \tag{13.59}$$

for some constant $m_0 > 0$.

Indeed, if not, then there is a sequence $b_j, j = 1, \ldots, \| b_j \|_{R^2} = 1$ such that for corresponding $F = F_{(j)}, j = 1, \ldots$ (calculated as in (13.57)):

$$\| F_{(j)} \|_{(L^2(\Omega))^2} \to 0, \quad b_j \to b_0, \quad \| b_0 \|_{R^2} = 1 \text{ in } R^2 \text{ as } j \to \infty.$$

If so, passing to the limit as $j \to \infty$ in the spaces $J_0(\Omega)$ and $G(\Omega)$ in the equation as in (13.57) with $b = b_j$ yields:

$$b_0 \xi(x) = \nabla w_0$$

for some $\nabla w_0 \in G(\Omega)$, $w_0 \in H^1(\Omega)$, see (13.12). Contradiction.

Step 3. For any given b, we will now evaluate what happens to (13.59) when $S_*(0)$ is replaced with $rS_*(0)$ and $r \to 0+$. To clarify further notations (when passing to the limit), we will be using the notations F_r and w_r in place of F in this case:

$$\frac{1}{\text{mes}\{rS_*(0)\}} b^T \int_{rS_*(0)} F_r(x)dx = \frac{1}{\text{mes}\{rS_*(0)\}} b^T \int_{S_*(0)} F_r(ru)r^2 du$$

$$= \frac{1}{\text{mes}\{S_*(0)\}} b^T \int_{S_*(0)} \hat{F}_r(u)du, \qquad (13.60)$$

where we made a substitution $u = (1/r)x$ and set $\hat{F}_r(u) = F_r(ru)$.

In a similar way, introduce

$$\nabla \hat{w}_r(u) = \nabla w_r(ru)$$

and denote by $\xi_{rS_*(0)}$ and by $\xi_{S_*(0)}$ the characteristic functions of respective sets.

Noticing now that

$$r\nabla w_r(x) = \nabla \hat{w}_r(u), \quad x = ur,$$

where the differentiation on the left is with respect to $x = (x_1, x_2)$ and it is with respect to $u = (u_1, u_2)$ on the right, we derive from the equation like (13.57) in x for the set $rS_*(0)$, namely,

$$F_r(x) = b\xi_{rS_*(0)}(x) - \nabla w_r(x), \qquad (13.61)$$

the following equation in u:

$$\hat{F}(u) = b\xi_{S_*(0)}(u) - \nabla\left(\frac{1}{r}\hat{w}(u)\right). \qquad (13.62)$$

Combining (13.60) with the estimate (13.59) applied to the equation (13.62), we derive that for solutions to (13.61):

$$\frac{1}{\mathrm{mes}\{rS_*(0)\}}\, \| F_r \|^2_{(L^2(\Omega))^2} = \frac{1}{\mathrm{mes}\{rS_*(0)\}} b^T \int_{rS_*(0)} F_r dx$$

$$= \frac{1}{\mathrm{mes}\{S_*(0)\}} b^T \int_{S_*(0)} \hat{F}_r(u) du$$

$$\geq m_0 \| b \|^2_{R^2} \quad \forall b \in R^2 \quad \text{as } r \to 0+. \quad (13.63)$$

Step 4. Denote now (see (13.55)):

$$\eta_{1r} = \frac{1}{\mathrm{mes}\{rS_*(0)\}} \int_{rS_*(z_i(0))} F_{1r} dx, \quad \eta_{2r} = \int_{rS_*(z_i(0))} F_{2r} dx,$$

and introduce the matrix $\mathbf{A}_r = (\eta_{1r}, \eta_{2r})$.

Then for any real numbers α, β, $\alpha^2 + \beta^2 = 1$, we can write the following equation similar to (13.61) (recall that we assumed that $z_i(0) = 0$):

$$(\alpha F_{1r} - \beta F_{2r}) = (\alpha b_{(1)} - \beta b_{(2)})\xi_{rS_*(0)} - (\alpha \nabla w_{1r} - \beta \nabla w_{2r}).$$

Estimate (13.63), applied for this equation, gives:

$$\| \alpha b_{(1)} - \beta b_{(2)} \|_{R^2} \| \alpha \eta_{1r} + \beta \eta_{2r} \|_{R^2} \geq (\alpha b_{(1)} - \beta b_{(2)})^T (\alpha \eta_{1r} + \beta \eta_{2r})$$

$$\geq m_0 \| \alpha b_{(1)} - \beta b_{(2)} \|^2_{R^2},$$

from which, recalling that, by the assumptions of Theorem 13.3, $b_{(1)}$ and $b_{(2)}$ are not co-linear, we conclude that

$$\| \mathbf{A}_r(\alpha, \beta)^T \|_{R^2} = \| \alpha \eta_{1r} + \beta \eta_{2r} \|_{R^2} \geq m_0 \| \alpha b_{(1)} - \beta b_{(2)} \|_{R^2} \geq m_1 > 0 \quad (13.64)$$

for some constant $m_1 > 0$ and any $\alpha^2 + \beta^2 = 1$.

In turn, (13.64) implies that the smallest eigenvalue of the matrix $\mathbf{A}_r^T \mathbf{A}_r$ and hence the determinant of matrix $\mathbf{A}_r$ is bounded from below by a positive number as $r \to 0+$ as we planned to show in Step 1. This ends the proof of Theorem 13.3. $\square$

Chapter 14
Local Controllability for a Swimming Model

Abstract We study the local controllability of the "basic" mathematical model (11.1)–(11.3) from Chapter 11 under an extra assumption on the geometric regularity of the sets $S_r(z_i)$'s forming the swimming object at hand.

14.1 The Model and Its Well-Posedness

In this chapter we intend to study the local controllability of the swimming model (11.1)–(11.3) from Chapter 11, namely (we rewrite it here for the reader's convenience):

$$\frac{\partial y}{\partial t} = v \Delta y + F(z,v) - \nabla p \quad \text{in} \quad Q_T = \Omega \times (0,T), \tag{14.1}$$

$$\text{div}\, y = 0 \text{ in } Q_T, \ y = 0 \text{ in } \Sigma_T = \partial \Omega \times (0,T), \ y\mid_{t=0} = y_0 \in H(\Omega) \bigcap (H^2(\Omega))^2,$$

$$\frac{dz_i}{dt} = \frac{1}{\text{mes}\,\{S(0)\}} \int\limits_{S(z_i(t))} y(x,t)dx, \ z_i(0) = z_{i0}, \ i = 1,\ldots,n, \ n > 2, \tag{14.2}$$

$$(F(z,v))(x,t) = \sum_{i=2}^{n+1} \xi_{i-1}(x,t)[k_{i-1} \frac{(\parallel z_i(t) - z_{i-1}(t) \parallel_{R^2} - l_{i-1})}{\parallel z_i(t) - z_{i-1}(t) \parallel_{R^2}} (z_i(t) - z_{i-1}(t))$$

$$+ k_{i-2} \frac{(\parallel z_{i-2}(t) - z_{i-1}(t) \parallel_{R^2} - l_{i-2})}{\parallel z_{i-2}(t) - z_{i-1}(t) \parallel_{R^2}} (z_{i-2}(t) - z_{i-1}(t))]$$

$$+ \sum_{i=2}^{n+1} \xi_{i-1}(x,t)(v_{i-1}(t)A(z_i(t) - z_{i-1}(t))$$

$$+ v_{i-2}(t)A(z_{i-2}(t) - z_{i-1}(t))). \tag{14.3}$$

A.Y. Khapalov, *Controllability of Partial Differential Equations Governed by Multiplicative Controls*, Lecture Notes in Mathematics 1995, DOI 10.1007/978-3-642-12413-6_14, © Springer-Verlag Berlin Heidelberg 2010

Assumption 1. *We assume throughout Chapter 14 that the set $S(0)$ introduced in Chapter 11 can also be described as follows:*

$$S(0) = \{x \mid -r < x_1 < r, \ \alpha(x_1) < x_2 < \beta(x_1)\}, \tag{14.4}$$

where α and β are the given continuously differentiable functions.

Alternatively, $S(0)$ may consist of finitely many non-overlapping sets as in Assumption 1.

14.2 A Concept of Controllability for a Swimming Model and the Main Results

For the given initial datum

$$\{y_0, \ z_i(0), \ i = 1, \ldots, n\}, \tag{14.5}$$

denote by

$$\{y^*(x,t), \ z_i^*(t), \ i = 1, \ldots, n\} \tag{14.6}$$

the solution pair to (14.1)–(14.3) generated by the zero controls $v_i = 0$, $i = 1$, $\ldots, n-1$.

We also distinguish the following equilibrium initial state for system (14.1)–(14.3):

$$\{y_0 = 0, \ z_i(0), \ i = 1, \ldots, n \text{ such that } l_{i-1} = \| z_i(0) - z_{i-1}(0) \|_{R^2}, \ i = 2, \ldots, n\}, \tag{14.7}$$

in which case the fluid "rests" and the apparatus does not move for any $t > 0$, that is,

$$\{y^*(x,t) \equiv 0, \ z_i^*(t) \equiv z_i(0), \ i = 1, \ldots, n\}, \ t \geq 0. \tag{14.8}$$

We intend to approach the general issue of local controllability for (14.1)–(14.3) by asking first a "rather simple" question:

Given the equilibrium initial datum (14.7) in (14.1)–(14.3), can we move at least one point, say, z_i, in the skeleton of our swimmer anywhere within some neighborhood of its initial equilibrium position $z_i(0)$ at some pre-assigned moment $T > 0$?

We will call this problem the *local controllability problem with respect to z_i near equilibrium* at time T. However, the question of main interest, associated with the actual motion of any object, is:

Given the equilibrium initial datum (14.7) in (14.1)–(14.3), can we move the "center of mass" of our swimmer, namely, the point

$$z_c(t) = \frac{1}{n} \sum_{i=1}^{n} z_i(t),$$

anywhere within some neighborhood of its initial equilibrium position

$$z_c(0) = \frac{1}{n} \sum_{i=1}^{n} z_i(0)$$

at some pre-assigned moment $T > 0$?

We will call this problem the *local controllability problem with respect to $z_c(0)$ near equilibrium at time T.*

In the case of the general (not necessarily equilibrium) initial conditions (14.5), the motion of the swimmer, associated with the zero-controls v_i's is, what we can call, a "drifting" (uncontrolled) motion (14.6), generated, on the one hand, by the given initial fluid condition y_0, and, on the other hand, by the elastic forces in the first two lines on the right in (14.3) trying to return our swimmer to its natural equilibrium position, that is, when the distances between the adjacent points z_i's are exactly l_i's. In this case, our goal is to try investigate the *local controllability of system (14.1)–(14.3) near the "drifting" trajectory $z_i^*(t)$, $i = 1, \ldots, n$:*

Given the initial datum in (14.1)–(14.3), can we move at least one point, say, z_i of the swimmer or its center of mass z_c anywhere within some neighborhood respectively of $z_i^*(T)$ or of

$$z_c^*(T) = \frac{1}{n} \sum_{i=1}^{n} z_i^*(T),$$

along its uncontrolled drifting trajectory (14.5)–(14.6) for some $T > 0$?

We will call these two problems the *local controllability problems near the drifting positions respectively of $z_i^*(T)$ and of $z_c^*(T)$.*

Our strategy in this section is centered around the following propositions.

Proposition 14.1. *Assume that in (14.1)–(14.3) only two controls are active, say, v_j and v_l, where $j \neq l$ and $j, l \in \{1, \ldots, n-1\}$, while $v_k = 0$ for $k = 1, \ldots, n-1, k \neq j, l$. Assume further that v_j and v_l are independent of time. Then, if for some $i \in \{1, \ldots, n\}$ there exists a $T > 0$ such that the matrix*

$$\left(\frac{dz_i(T)}{dv_j} \Big|_{v'_m s = 0}, \frac{dz_i(T)}{dv_l} \Big|_{v'_m s = 0} \right), \tag{14.9}$$

is non-degenerate, then the system (14.1)–(14.3) is locally controllable near its drifting position $z_i^*(T)$ in (14.6). Namely, there is an $\varepsilon > 0$ such that

$$B_\varepsilon(z_i^*(T)) \subset \{z_i(T) \mid v_j, v_l \in R, v_k = 0 \text{ for } k = 1, \ldots, n-1, k \neq j, l\}. \tag{14.10}$$

In particular, for the initial equilibrium position (14.7) condition (14.9) implies the local controllability with respect to z_i near equilibrium at time T.

In other words, (14.10) means that the set of all possible positions of $z_i(T)$ when controls v_i's run over R will include some ε-neighborhood of $z_i^*(T)$.

In (14.9) and anywhere below the subscript $v'_m s = 0$ indicates that the corresponding expressions are calculated for $v_m = 0, m = 1, \ldots, n-1$.

Proof of Proposition 14.1. This is an immediate consequence of the Inverse Function Theorem, which, in view of (14.9), implies that the mapping

$$R^2 \ni (v_j, v_l) \rightarrow z_i(T) \in R^2,$$

defined on some (open) neighborhood of the origin, has the inverse mapping, defined on some (open) neighborhood of $z_i^*(T)$, that is, (14.10) holds.

The same argument implies a similar result for the motion of the center of mass $z_c(t)$.

Proposition 14.2. *Assume that in (14.1)–(14.3) only two controls are active, say, v_j and v_l, where $j \neq l$ and $j, l \in \{1, \ldots, n-1\}$, while $v_k = 0$ for $k = 1, \ldots, n-1, k \neq j, l$. Assume further that v_j and v_l are independent of time. Then, if there exists a $T > 0$ such that the matrix*

$$\left(\frac{dz_c(T)}{dv_j} \Big|_{v'_m s = 0}, \frac{dz_c(T)}{dv_l} \Big|_{v'_m s = 0} \right), \tag{14.11}$$

is non-degenerate, then the system (14.1)–(14.3) is locally controllable near its drifting position of the center of mass $z_c^*(T)$. Namely, there is an $\varepsilon > 0$ such that

$$B_\varepsilon(z_c^*(T)) \subset \{z_c(T) \mid v_j, v_l \in R, v_k = 0 \text{ for } k = 1, \ldots, n-1, k \neq j, l\}.$$

In particular, for the initial equilibrium position (14.7) we have the local controllability with respect to z_c near equilibrium at time T.

Our *main results* below deal with the conditions under which the matrices in (14.9) and (14.11) in Propositions 14.1 and 14.2 are non-degenerate (the general scheme of our proof is described in the beginning of Section 14.3). To formulate them, we will need to introduce some notations first.

Recall that the unique solution to (14.1)–(14.3) admits the following *implicit* representation (see (13.1)–(13.2) in Chapter 13):

$$y(x,t) = \sum_{k=1}^{\infty} e^{-\lambda_k t} \left(\int_\Omega y_0^T \omega_k dq \right) \omega_k(x)$$

$$+ \sum_{k=1}^{\infty} \int_0^t e^{-\lambda_k(t-\tau)} \left(\int_\Omega F^T(y, z, v) \omega_k dq d\tau \right) \omega_k(x). \tag{14.12}$$

Here and below, where appropriate, we use $q = (q_1, q_2)$ to denote the space variable in the internal integration.

Denote the projection of the sum of two rotation forces in the last line of (14.3), generated at the initial moment $t = 0$ by the unit control input $v_j = 1$, on the *divergence-free* space $J_0(\Omega)$ by

$$F_j(x) = F_{j,1}(x) + F_{j,2}(x), \quad j = 1, \ldots, n-1, \tag{14.13}$$

where

$$F_{j,1}(x) = \sum_{k=1}^{\infty} \left[\int_{\Omega} \left(\xi_j(q,0) A(z_{j+1}(0) - z_j(0)) \right)^T \omega_k(q) dq \right] \omega_k(x)$$

$$= \sum_{k=1}^{\infty} \left[\left(A(z_{j+1}(0) - z_j(0)) \right)^T \int_{S_r(z_j(0))} \omega_k(q) dq \right] \omega_k(x), \tag{14.14}$$

$$F_{j,2}(x) = \sum_{k=1}^{\infty} \left[\int_{\Omega} \left(-\xi_{j+1}(q,0) A(z_{j+1}(0) - z_j(0)) \right)^T \omega_k(q) dq \right] \omega_k(x)$$

$$= -\sum_{k=1}^{\infty} \left[\left(A(z_{j+1}(0) - z_j(0)) \right)^T \int_{S_r(z_{j+1}(0))} \omega_k(q) dq \right] \omega_k(x), \tag{14.15}$$

Here we used the fact that $\{\omega_k\}_{k=1}^{\infty}$ form an orthonormalized basic in $J_0(\Omega) \subset (L^2(\Omega))^2$.

Assumption 2. *Let the* $[2 \times 2]$*-matrix*

$$\left(\int_{S(z_i(0))} F_j(x)dx, \quad \int_{S(z_i(0))} F_l(x)dx \right) \tag{14.16}$$

be nondegenerate for some $i \in \{1, \ldots, n\}$ *and* $l, j \in \{1, \ldots, n-1\}$.

Theorem 14.1. *Let* $i \in \{1, \ldots, n\}$, $l, j \in \{1, \ldots, n-1\}$ *and Assumption 2 hold. Then there exists a* $T^* > 0$ *such that the matrix (14.9) is non-degenerate for any* $T \in (0, T^*]$ *and Proposition 14.1 holds. Namely, we have the local controllability of system (14.1)–(14.4) near its drifting position* $z_i^*(T)$. *In particular, for the equilibrium position (14.7)–(14.8) condition (14.16) implies the local controllability with respect to* z_i *near equilibrium at time* T.

The argument of Theorem 14.1 establishes that

$$\frac{dz_i(t)}{dv_j} \Big|_{v_m' s=0} = \frac{t^2}{2mes\{S(0)\}} \int_{S(z_i(0))} F_j(x)dx + t^2 O(t), \quad j = 1, \ldots, n-1, \tag{14.17}$$

which allows us to apply (14.16) to ensure that (14.9) in Proposition 14.1 is non-degenerate.

Theorem 14.2. *As we noted in Chapter 13, Assumption 2 holds for any point $z_i, i = 2, \ldots, n-1$ in the original position of the swimming object with controls acting in the adjacent points z_{i-1}, z_i and z_{i+1} only, provided that* (a) *the vectors*

$$z_{i+1}(0) - z_i(0) \text{ and } z_i(0) - z_{i-1}(0) \tag{14.18}$$

are nonparallel and (b) *that the sets $S(z_j(0)), j = i-1, i, i+1$ are identical and sufficiently small. The above does not require Assumption 1 to hold.*

Due to (14.17), at no extra cost, Theorem 14.1 implies the respective statement for the center of mass $z_c(t)$.

Theorem 14.3. *Let $l, j \in \{1, \ldots, n-1\}$ and the matrix*

$$\sum_{i=1}^{n} \left(\int\limits_{S(z_i(0))} F_j(x)dx, \quad \int\limits_{S(z_i(0))} F_l(x)dx \right)$$

be non-degenerate. Then there exists a $T^* > 0$ such that for any $T \in (0, T^*]$ Proposition 14.2 holds. Namely, we have the local controllability of system (14.1)–(14.4) with respect to the position of center of mass $z_c(T)$ near its drifting position $z_c^*(T)$. In particular, for the equilibrium position (14.7)–(14.8) we have the local controllability with respect to z_c near equilibrium at time T.

Discussion of Theorems 14.1 and 14.3. Note that the two columns in (14.16) multiplied by $\text{mes}^{-1}\{S(0)\}$ describe the "average" forces induced respectively by the forces $F_j(x)$ and $F_l(x)$ over the region $S(z_i(0))$. Thus, the sufficient conditions for the local controllability near the drifting position $z_i^*(T)$ in Theorem 14.1 require that these average forces are not co-linear. Respectively, for the local controllability of the center of mass z_c Theorem 14.3 also requires that the sums of such average forces over all "thick points" in the skeleton of the swimmer generated respectively by the unit controls $v_l = 1$ and $v_j = 1$ are not co-linear.

Remark 14.1 (Supports of $F_j(x)$'s). In spite of the fact that the rotation forces in (14.3) have only local supports this does not have to be so for their projections (as in (14.13)–(14.14)) on the solenoidal part $J_0(\Omega)$ of $(L^2(\Omega))^2$, associated with incompressible fluids.

Recall now that the space $(L^2(\Omega))^2$ is the direct sum of the spaces $J_0(\Omega)$ and $G(\Omega)$. In (14.13)–(14.15) we denoted the projections of the functions

$$\xi_j(x,0)\left(A(z_{j+1}(0) - z_j(0))\right) \text{ and } -\xi_{j+1}(x,0)\left(A(z_{j+1}(0) - z_j(0))\right), \, j = 1, \ldots, n-1$$

on $J_o(\Omega)$ by $F_{j,1}(x)$ and $F_{j,2}(x)$. Denote now the projections of the aforementioned functions on the space $G(\Omega)$ respectively by $F_{j,1}^{\perp}(x)$ and $F_{j,2}^{\perp}(x)$. Since (e.g., [96], p. 28; [141], p. 15):

$$J_o(\Omega) = \{u \in (L^2(\Omega))^2, \ \mathrm{div}\,u = 0, \ \gamma_v u \,|_{\partial\Omega} = 0\}, \tag{14.19}$$

$$G(\Omega) = \{u \in (L^2(\Omega))^2, \ u = \nabla p, \ p \in H^1(\Omega)\}, \tag{14.20}$$

where v is the unit vector normal to the boundary $\partial\Omega$ (pointing outward) and $\gamma_v u \,|_{\partial\Omega}$ is the restriction of $u \cdot v$ to $\partial\Omega$, we have

$$F_{j,1}^{\perp}(x) = \nabla w_{j,1}(x), \quad F_{j,2}^{\perp}(x) = \nabla w_{j,2}(x), \quad j = 1,\dots,n-1$$

for some functions $w_{j,1}, w_{j,2} \in H^1(\Omega), j = 1,\dots,n-1$. Thus,

$$F_{j,1}(x) = \xi_j(x,0)\left(A(z_{j+1}(0) - z_j(0))\right) - \nabla w_{j,1}(x), \tag{14.21}$$

$$F_{j,2}(x) = -\xi_{j+1}(x,0)\left(A(z_{j+1}(0) - z_j(0))\right) - \nabla w_{j,2}(x). \tag{14.22}$$

Furthermore, $w_{j,1}$ and $w_{j,2}$ solve the following two generalized Neumann problems:

$$\Delta w_{j,1} = \mathrm{div}\,\xi_j(x,0)\left(A(z_{j+1}(0) - z_j(0))\right) \text{ in } \Omega, \ \frac{\partial w_{j,1}}{\partial v}\,|_{\partial\Omega} = 0, \tag{14.23}$$

$$\Delta w_{j,2} = -\mathrm{div}\,\xi_{j+1}(x,0)\left(A(z_{j+1}(0) - z_j(0))\right) \text{ in } \Omega, \ \frac{\partial w_{j,2}}{\partial v}\,|_{\partial\Omega} = 0. \tag{14.24}$$

Indeed, the equation (14.23), e.g., can be obtained by applying divergence to (14.21) and recalling that $F_{j,1} \in J_o(\Omega)$, which in particular implies that $\mathrm{div}F_{j,1} = 0$. In turn, the boundary condition in (14.23) follows from (14.21) by recalling that, due to (14.19), $\gamma_v F_{j,1}\,|_{\partial\Omega} = 0$ and that $\xi_j(x,0)$ vanishes outside of $S_r(z_j(0))$, which provides $\gamma_v \nabla w_{j,1}\,|_{\partial\Omega} = \frac{\partial w_{j,1}}{\partial v}\,|_{\partial\Omega} = 0$.

14.3 Preliminary Results

Our plan to prove Theorem 14.1 is as follows:

1. We intend to use Propositions 14.1 and 14.2 involving derivatives $\frac{dz_i(T)}{dv_j}, i = 1,\dots,n, j = 1,\dots,n-1$. In order to evaluate them, in Section 14.3 we differentiate the implicit solution formula (14.12) with respect to v_j's.
2. In Section 14.4 the results of Section 14.3 are presented as a vector Volterra equation for the aforementioned $\frac{dz_i(T)}{dv_j}$'s and suitable asymptotic analysis is used to qualitatively evaluate them for "small" T's.
3. Making use of all of the above, to obtain the qualitative estimates for the terms in (14.16), we complete the proof of Theorem 14.1 in Section 14.5.

In this section we intend to derive a number of auxiliary formulas. (We remind the reader that everywhere below, for simplicity of calculations we assume that $S(0)$ has the form as in (14.4).)

14.3.1 Solution Formula

Let us rewrite (14.12) as follows:

$$y(x,t) = \sum_{k=1}^{\infty} e^{-\lambda_k t} \left(\int_{\Omega} y_0^T \omega_k dx \right) \omega_k(x)$$

$$+ \sum_{i=2}^{n+1} (P_i(x,t) + v_{i-1}(t)Q_i(x,t) + v_{i-2}(t)R_i(x,t)), \qquad (14.25)$$

where *here and below we always assume* that (besides all other imposed assumptions, if there are such) T is selected "sufficiently small" as in Theorem 12.1 to ensure the wellposedness of system (14.1)–(14.4) at hand on $[0,T]$, and

$$P_i(x,t) = \sum_{k=1}^{\infty} (\int_0^t e^{-\lambda_k(t-\tau)} [k_{i-1} \frac{(\| z_i(\tau) - z_{i-1}(\tau) \|_{R^2} - l_{i-1})}{\| z_i(\tau) - z_{i-1}(\tau) \|_{R^2}} (z_i(\tau) - z_{i-1}(\tau))^T$$

$$+ k_{i-2} \frac{(\| z_{i-2}(\tau) - z_{i-1}(\tau) \|_{R^2} - l_{i-2})}{\| z_{i-2}(\tau) - z_{i-1}(\tau) \|_{R^2}} (z_{i-2}(\tau) - z_{i-1}(\tau))^T]$$

$$\times \left(\int_{\Omega} \xi_{i-1}(q,\tau)\omega_k dq \right) d\tau) \omega_k(x), \qquad (14.26)$$

$$Q_i(x,t) = \sum_{k=1}^{\infty} \left(\int_0^t e^{-\lambda_k(t-\tau)} (A(z_i(\tau) - z_{i-1}(\tau)))^T \left(\int_{\Omega} \xi_{i-1}(q,\tau)\omega_k dq \right) d\tau \right)$$

$$\times \omega_k(x), \qquad (14.27)$$

$$R_i(x,t) = \sum_{k=1}^{\infty} \left(\int_0^t (e^{\lambda_k(t-\tau)} (A(z_{i-2}(\tau) - z_{i-1}(\tau)))^T \left(\int_{\Omega} \xi_{i-1}(q,\tau)\omega_k dq \right) d\tau \right)$$

$$\times \omega_k(x). \qquad (14.28)$$

14.3.2 Differentiation with Respect to v_j's

We assume from now on that the functions $v_j, j = 1, \ldots, n-1$ are constant in time. Below we formally differentiate various expressions with respect to $v_j, j = 1, \ldots, n-1$. The validity of these calculations will be discussed in the next section.

Derivatives of z_i's. In view of (14.2) and (14.4), for any $t \in [0, T]$:

$$z_i(t) = z_{i0} + \frac{1}{\text{mes}\,\{S(0)\}} \int_0^t \int\limits_{S(z_i(\tau))} y(x, \tau)dxd\tau.$$

$$= z_{i0} + \frac{1}{\text{mes}\,\{S(0)\}} \int_0^t \int_{z_{i,1}(\tau)-r}^{z_{i,1}(\tau)+r}$$

$$\times \int_{z_{i,2}(\tau)+\alpha(x_1-z_{i,1}(\tau))}^{z_{i,2}(\tau)+\beta(x_1-z_{i,1}(\tau))} y(x, \tau)dx_2dx_1d\tau. \qquad (14.29)$$

Denote $z_i(t) = (z_{i,1}(t), z_{i,2}(t))$. Then differentiating (14.29) with respect to v_j, we obtain:

$$\frac{dz_i(t)}{dv_j} = \frac{1}{\text{mes}\,\{S(0)\}} \int_0^t \int_{z_{i,1}(\tau)-r}^{z_{i,1}(\tau)+r} (y(x_1, z_{i,2}(\tau) + \beta(x_1 - z_{i,1}(\tau)), \tau)$$

$$\times \left(\frac{dz_{i,2}(\tau)}{dv_j} - \beta'(x_1 - z_{i,1}(\tau))\frac{dz_{i,1}(\tau)}{dv_j} \right)$$

$$- y(x_1, z_{i,2}(\tau) + \alpha(x_1 - z_{i,1}(\tau)), \tau) \left(\frac{dz_{i,2}(\tau)}{dv_j} - \alpha'(x_1 - z_{i,1}(\tau))\frac{dz_{i,1}(\tau)}{dv_j} \right))dx_1d\tau$$

$$+ \frac{1}{\text{mes}\,\{S(0)\}} \int_0^t \frac{dz_{i,1}(\tau)}{dv_j} \int_{z_{i,2}(\tau)+\alpha(r)}^{z_{i,2}(\tau)+\beta(r)} y(z_{i,1}(\tau) + r, x_2, \tau)dx_2d\tau$$

$$- \frac{1}{\text{mes}\,\{S(0)\}} \int_0^t \frac{dz_{i,1}(\tau)}{dv_j} \int_{z_{i,2}(\tau)+\alpha(-r)}^{z_{i,2}(\tau)+\beta(-r)} y(z_{i,1}(\tau) - r, x_2, \tau)dx_2d\tau$$

$$+ \frac{1}{\text{mes}\,\{S(0)\}} \int_0^t \int\limits_{S(z_i(\tau))} \frac{dy(x, \tau)}{dv_j}dxd\tau. \qquad (14.30)$$

Derivatives of y. Making use of (14.25), we obtain:

$$\frac{dy}{dv_j}\Big|_{v'_{ms}=0} = \frac{d}{dv_j} \left(\sum_{i=2}^{n+1} (P_i + v_{i-1}Q_i + v_{i-2}R_i) \right) \Big|_{v'_{ms}=0}$$

$$= Q_{j+1}\Big|_{v'_{ms}=0} + R_{j+2}\Big|_{v'_{ms}=0} + \sum_{i=2}^{n+1} \left(\frac{d}{dv_j}P_i + v_{i-1}\frac{d}{dv_j}Q_i + v_{i-2}\frac{d}{dv_j}R_i \right)\Big|_{v'_{ms}=0}$$

$$= Q_{j+1}\Big|_{v'_{ms}=0} + R_{j+2}\Big|_{v'_{ms}=0} + \sum_{i=2}^{n+1} \left(\frac{d}{dv_j}P_i \right)\Big|_{v'_{ms}=0}. \qquad (14.31)$$

Derivatives of P_i's. Making use of (14.26), while noticing that

$$\int_\Omega \xi_{i-1}(x,\tau)\omega_k dx = \int_{z_{i-1,1}(\tau)-r}^{z_{i-1,1}(\tau)+r} \int_{z_{i-1,2}(\tau)+\alpha(x_1-z_{i-1,1}(\tau))}^{z_{i-1,2}(\tau)+\beta(x_1-z_{i-1,1}(\tau))} \omega_k(x)dx_2 dx_1,$$

we obtain:

$$\frac{d}{dv_j}P_i(x,t)\,|_{v'_m s=0} = \sum_{k=1}^{\infty}[\int_0^t e^{-\lambda_k(t-\tau)} \times \{k_{i-1}\Theta_{1,i,j}(\tau)+k_{i-2}\Theta_{2,i,j}(\tau)\}$$

$$\times \left(\int_\Omega \xi_{i-1}(q,\tau)\omega_k dq\right) d\tau]\omega_k(x)\,|_{v'_m s=0}$$

$$+\sum_{k=1}^{\infty}\left(\int_0^t e^{-\lambda_k(t-\tau)}\Theta_{3,i}(\tau)\,\Theta_{4,i,j}(\tau)d\tau\right)$$

$$\times \omega_k(x)\,|_{v'_m s=0}, \tag{14.32}$$

where

$$\Theta_{1,i,j}(\tau) = \left(\frac{dz_i(\tau)}{dv_j}-\frac{dz_{i-1}(\tau)}{dv_j}\right)^T\left(1-\frac{l_{i-1}}{\|z_i(\tau)-z_{i-1}(\tau)\|_{R^2}}\right)$$

$$+l_{i-1}(z_i(\tau)-z_{i-1}(\tau))^T$$

$$\times \frac{1}{\|z_i(\tau)-z_{i-1}(\tau)\|_{R^2}^3}<\frac{dz_i(\tau)}{dv_j}-\frac{dz_{i-1}(\tau)}{dv_j},z_i(\tau)-z_{i-1}(\tau)>_{R^2},$$

$$\Theta_{2,i,j}(\tau) = \left(\frac{dz_{i-2}(\tau)}{dv_j}-\frac{dz_{i-1}(\tau)}{dv_j}\right)^T\left(1-\frac{l_{i-2}}{\|z_{i-2}(\tau)-z_{i-1}(\tau)\|_{R^2}}\right)$$

$$+l_{i-2}(z_{i-2}(\tau)-z_{i-1}(\tau))^T$$

$$\times \frac{1}{\|z_{i-2}(\tau)-z_{i-1}(\tau)\|_{R^2}^3}<\frac{dz_{i-2}(\tau)}{dv_j}-\frac{dz_{i-1}(\tau)}{dv_j},z_{i-2}(\tau)-z_{i-1}(\tau)>_{R^2},$$

$$\Theta_{3,i}(\tau) = k_{i-1}\left(\frac{(\|z_i(\tau)-z_{i-1}(\tau)\|_{R^2}-l_{i-1})}{\|z_i(\tau)-z_{i-1}(\tau)\|_{R^2}}(z_i(\tau)-z_{i-1}(\tau))^T\right)$$

$$+k_{i-2}\left(\frac{(\|z_{i-2}(\tau)-z_{i-1}(\tau)\|_{R^2}-l_{i-2})}{\|z_{i-2}(\tau)-z_{i-1}(\tau)\|_{R^2}}(z_{i-2}(\tau)-z_{i-1}(\tau))^T\right),$$

$$\Theta_{4,i,j}(\tau) = \int_{z_{i-1,1}(\tau)-r}^{z_{i-1,1}(\tau)+r}\omega_k(x_1,z_{i-1,2}(\tau)+\beta(x_1-z_{i-1,1}(\tau)))$$

$$\times\left(\frac{dz_{i-1,2}(\tau)}{dv_j}-\beta'(x_1-z_{i-1,1}(\tau))\frac{dz_{i-1,1}(\tau)}{dv_j}\right)dx_1$$

$$-\int_{z_{i-1,1}(\tau)-r}^{z_{i-1,1}(\tau)+r}\omega_k(x_1,z_{i-1,2}(\tau)+\alpha(x_1-z_{i-1,1}(\tau)))$$

$$\times \left(\frac{dz_{i-1,2}(\tau)}{dv_j} - \alpha'(x_1 - z_{i-1,1}(\tau)) \frac{dz_{i-1,1}(\tau)}{dv_j} \right) dx_1$$

$$+ \frac{dz_{i-1,1}(\tau)}{dv_j} \int_{z_{i-1,2}(\tau)+\alpha(r)}^{z_{i-1,2}(\tau)+\beta(r)} \omega_k(z_{i-1,1}(\tau) + r, x_2) dx_2$$

$$- \frac{dz_{i-1,1}(\tau)}{dv_j} \int_{z_{i-1,2}(\tau)+\alpha(-r)}^{z_{i-1,2}(\tau)+\beta(-r)} \omega_k(z_{i-1,1}(\tau) - r, x_2) dx_2.$$

To better understand the terms in the 2-nd sum in (14.32), denote

$$V_i(\tau) = \Theta_{3,i}(\tau) \big|_{v'_m s = 0} .$$

Then we can rewrite, e.g., the term in the 2-nd sum in (14.32) associated with the factor $\frac{dz_{i-1,2}(\tau)}{dv_j}$ in the 1-st line in the expression for $\Theta_{4,i,j}(\tau)$ as follows:

$$\int_0^t \frac{dz_{i-1,2}(\tau)}{dv_j} \left[\sum_{k=1}^{\infty} e^{-\lambda_k(t-\tau)} \{(\Phi_1(\tau))(V_i(\tau)\omega_k)\} \right] d\tau \omega_k(x) \big|_{v'_m s = 0}, \qquad (14.33)$$

where we denoted

$$(\Phi_1(t))(\psi) = \int_{z_{i-1,1}(t)-r}^{z_{i-1,1}(t)+r} \psi(x_1, z_{i-1,2}(t) + \beta(x_1 - z_{i-1,1}(t))) dx_1.$$

Note that $\Phi_1(t) \in H^{-1}(\Omega)$ for any $t \in [0,T]$, where the space $H^{-1}(\Omega)$ is dual of $H_0^1(\Omega)$, namely, it is the space of all linear bounded functionals on $H_0^1(\Omega)$. (As usual, we endow the latter space with the norm $\| \psi \|_{H_0^1(\Omega)} = \{ \int_\Omega (\psi_{x_1}^2 + \psi_{x_2}^2) dx \}^{1/2}$.)

Indeed, regardless of t, for any $\psi \in H_0^1(\Omega)$, due to the continuous embedding of $H_0^1(a,b)$ into $C[a,b]$ for any finite interval $[a,b]$, for any $t \in [0,T]$ we have:

$$| (\Phi_1(t))(\psi) | \leq \int_{z_{i-1,1}(t)-r}^{z_{i-1,1}(t)+r} | \psi(x_1, z_{i-1,2}(t) + \beta(x_1 - z_{i-1,1}(t))) | dx_1$$

$$\leq C_1 \int_{z_{i-1,1}(t)-r}^{z_{i-1,1}(t)+r} \| \psi(x_1, \cdot) \|_{H_0^1(\{\xi | (x_1, \xi) \in \Omega\}} dx_1$$

$$\leq \sqrt{2r} C_1 \| \psi \|_{H_0^1(\Omega)},$$

where C_1 is a positive constant. Thus, for any $t \in [0, T]$:

$$\| \Phi_1(t) \|_{H^{-1}(\Omega)} = \sup_{\| \psi \|_{H_0^1(\Omega)} = 1, \psi \in H_0^1(\Omega)} | (\Phi_1(t))(\psi) | < \sqrt{2r} C_1. \qquad (14.34)$$

In the next subsection we will need the following observation.

Remark 14.2. Consider any vector $\kappa \in R^2$. Then, similar to the derivation of (14.34), we can show that for any $t \in [0, T]$ the expression

$$(\Phi_1(t)\kappa)(\phi) = (\Phi_1(t))(\kappa^T \phi), \quad \text{where } \phi \in (H_0^1(\Omega))^2,$$

defines a linear bounded functional on $(H_0^1(\Omega))^2$ and

$$\sup_{\| \phi \|_{(H_0^1(\Omega))^2} = 1, \phi \in (H_0^1(\Omega))^2} | (\Phi_1(t)\kappa)(\phi) | \leq \| \kappa \|_{R^2} \| \Phi_1(t) \|_{H^{-1}(\Omega)}$$

$$\leq \sqrt{2r} C_1 \| \kappa \|_{R^2}.$$

On the other hand, if $\phi \in H(\Omega) \subset (H_0^1(\Omega))^2$, then it admits the following representation:

$$\phi = \sum_{k=1}^{\infty} a_k \omega_k, \quad \| \phi \|_{H(\Omega)} = \| \phi \|_{(H_0^1(\Omega))^2} = \left(\sum_{k=1}^{\infty} \frac{\lambda_k}{\nu} a_k^2 \right)^{1/2}.$$

In this case, we can introduce the space $H'(\Omega)$ of linear bounded functionals $\mathscr{P}$ on $H(\Omega)$ making use of the duality product

$$\mathscr{P}(\phi) = \sum_{k=1}^{\infty} a_k b_k, \quad \text{where } b_k = \mathscr{P}(\omega_k), \quad k = 1, \ldots$$

with the norm

$$\| \mathscr{P} \|_{H'(\Omega)} = \left(\sum_{k=1}^{\infty} \frac{\nu}{\lambda_k} b_k^2 \right)^{1/2},$$

equivalent to the regular one on this space (i.e., analogous to that in (14.34)). Thus, in particular, for any $\kappa \in R^2$ we have for any $t \in [0, T]$:

$$\sum_{k=1}^{\infty} \frac{\nu}{\lambda_k} \{ (\Phi_1(t)\kappa)(\omega_k) \}^2 \leq C_* \left(\sup_{\| \phi \|_{H(\Omega)} = 1, \phi \in H(\Omega)} | (\Phi_1(t)\kappa)(\phi) | \right)^2$$

$$\leq C_* \left(\sup_{\| \phi \|_{(H_0^1(\Omega))^2} = 1, \phi \in (H_0^1(\Omega))^2} | (\Phi_1(t)\kappa)(\phi) | \right)^2$$

$$\leq C_* \left(\sqrt{2r} C_1 \| \kappa \|_{R^2} \right)^2, \qquad (14.35)$$

where $C_* > 0$ is some (generic) constant.

14.3.3 Kernels

Let us now consider in detail the contribution of (14.33) to the expression in (14.30), multiplied by mes $\{S(0)\}$, when $v'_m s = 0$, namely:

$$\int_0^t \int_0^\tau \frac{dz_{i-1,2}(s)}{dv_j} \int_{S(z_i(\tau))} \left[\sum_{k=1}^\infty e^{-\lambda_k(\tau-s)} \{ (\Phi_1(s)V_i(s))(\omega_k) \} \omega_k(x) \right] dx\, ds\, d\tau$$

$$= \int_0^t \frac{dz_{i-1,2}(s)}{dv_j} \left\{ \int_s^t \int_{S(z_i(\tau))} \left[\sum_{k=1}^\infty e^{-\lambda_k(\tau-s)} \{ (\Phi_1(s)V_i(s))(\omega_k) \} \omega_k(x) \right] dx\, d\tau \right\} ds.$$

In the above, the factor at $\frac{dz_{i-1,2}(s)}{dv_j}$ can be regarded as a 2-D kernel $K(t,s)$, $(t,s) \in (0,T) \times (0,T)$, vanishing for $s > t$.

Lemma 14.1. *The kernel K is an element of $(L^\infty((0,T) \times (0,T)))^2$.*

Proof of Lemma 14.1. Indeed, for almost all $(s,\tau) \in (0,T) \times (0,T)$:

$$\| K(t,s) \|_{R^2}^2 = \| \int_s^t \int_{S(z_i(\tau))} \left[\sum_{k=1}^\infty e^{-\lambda_k(\tau-s)} \{ (\Phi_1(s)V_i(s))(\omega_k) \} \omega_k(x) \right] dx\, d\tau \|_{R^2}^2$$

$$\leq T \operatorname{mes}\{S_r(0)\} \sum_{k=1}^\infty \left(\int_s^t e^{-2\lambda_k(\tau-s)} d\tau \right) \{ (\Phi_1(s)V_i(s))(\omega_k) \}^2$$

$$\leq CT \sum_{k=1}^\infty \frac{1}{\lambda_k} \{ (\Phi_1(s)V_i(s))(\omega_k) \}^2$$

$$\leq \hat{C}T \left(\sqrt{2r}C_1 \| V_i \|_{C([0,T];R^2)} \right)^2 \tag{14.36}$$

for some (generic) positive constants C and $\hat{C}$, while C_1 is from (14.35) in Remark 14.2. In (14.36) we also used the following type of estimates, employing Bessel's inequality:

$$\| \int_\omega \left(\sum_{k=1}^\infty a_k \omega_k(x) \right) dx \|_{R^2}^2 = \| \int_\Omega \left(\sum_{k=1}^\infty a_k \omega_k(x) \right) \xi_\omega(x) dx \|_{R^2}^2$$

$$\leq \| \sum_{k=1}^\infty a_k \omega_k \|_{(L^2(\Omega))^2}^2 \| \xi_\omega \|_{L^2(\Omega)}^2$$

$$= \left(\sum_{k=1}^\infty a_k^2 \right) \operatorname{mes}\{\omega\},$$

where $\xi_\omega(x)$ is the characteristic function of a set $\omega \subset \Omega$. This ends the proof of Lemma 14.1. $\square$

The assertion of Lemma 14.1 can be established for all other kernels associated with $\frac{dz_i(s)}{dv_j}$'s in (14.32) and (14.30).

14.4   Volterra Equations for $\frac{dz_i(\tau)}{dv_j}$'s

Below we will deal only with the terms $\frac{dz_i(t)}{dv_j}|_{v'_m s=0}$. Therefore, to simplify further notations, we will *omit the subscript* $|_{v'_m s=0}$ *from now on.*

Volterra equations. (14.30) can be rewritten as the following vector Volterra equation:

$$\begin{pmatrix} \frac{dz_1(t)}{dv_j} \\ \vdots \\ \frac{dz_n(t)}{dv_j} \end{pmatrix} + \int_0^t \mathbf{B}_j(t,s) \begin{pmatrix} \frac{dz_1(s)}{dv_j} \\ \vdots \\ \frac{dz_n(s)}{dv_j} \end{pmatrix} ds = \begin{pmatrix} \Xi_{1j}(t) \\ \vdots \\ \Xi_{nj}(t) \end{pmatrix}, \quad j = 1,\ldots,n-1, \quad t \in [0,T],$$

(14.37)

where,

$$\Xi_{ij}(t) = \frac{1}{\text{mes}\,\{S(0)\}} \int_0^t \int_{S(z_i(\tau))} (Q_{j+1} + R_{j+2})\, dx d\tau \,|_{v'_m s=0}$$

and $\mathbf{B}_j(t,s), j = 1,\ldots,n-1$ are respectively the vector- and matrix-functions, defined by (14.30) along (14.31)–(14.36). We will need the following asymptotic result.

Lemma 14.2.

$$\Xi_{ij}(t) = \frac{t^2}{2\text{mes}\,\{S(0)\}} \left[\sum_{k=1}^\infty \left(\int_{S(z_i(0))} \omega_k dx \right) \left(\int_{S(z_j(0))} \omega_k^T(x) dx \right) \right] A(z_{j+1}(0) - z_j(0))$$

$$+ \frac{t^2}{2\text{mes}\,\{S_r(0)\}} \left[\sum_{k=1}^\infty \left(\int_{S(z_i(0))} \omega_k dx \right) \left(\int_{S(z_{j+1}(0))} \omega_k^T(x) dx \right) \right] A(z_j(0) - z_{j+1}(0))$$

$$+ t^2 O(t)$$

(14.38)

as $t \to 0+$, *where* $i = 1,\ldots,n$, $j = 1,\ldots,n-1$ *and* $O(t)$ *stands for a vector-function whose* R^2-*norm tends to zero as* $t \to 0+$.

Proof of Lemma 14.2.
Step 1. In view of (14.27) and (14.28), we have:

$$
\text{mes}\,\{S(0)\}\Xi_{ij}(t) = \int\limits_0^t\int\limits_0^\tau\int\limits_\Omega\left[\sum_{k=1}^\infty e^{-\lambda_k(\tau-s)}\left(\int\limits_\Omega (A(z_{j+1}(s)-z_j(s)))^T\,\xi_j(q,s)\omega_k dq\right)\omega_k(x)\right]
$$

$$
\times\,\xi_i(x,\tau)dxds d\tau
$$

$$
+\int\limits_0^t\int\limits_0^\tau\int\limits_\Omega\left[\sum_{k=1}^\infty e^{\lambda_k(\tau-s)}\left(\int\limits_\Omega (A(z_j(s)-z_{j+1}(s)))^T\,\xi_{j+1}(q,s)\omega_k dq\right)\omega_k(x)\right]
$$

$$
\times\,\xi_i(x,\tau)dxds d\tau. \tag{14.39}
$$

Step 2. We will further deal with the first term in (14.39). Let us show that the following representation holds for the expression in the square brackets in this term:

$$
\mathbf{W}(s,\tau) = \|\sum_{k=1}^\infty e^{-\lambda_k(\tau-s)}\left(\int\limits_\Omega (A(z_{j+1}(s)-z_j(s)))^T\,\xi_j(q,s)\omega_k dq\right)\omega_k
$$

$$
-\sum_{k=1}^\infty\left(\int\limits_\Omega (A(z_{j+1}(0)-z_j(0)))^T\,\xi_j(q,0)\omega_k dq\right)\omega_k \|^2_{(L^2(\Omega))^2}
$$

$$
\leq O(t) \tag{14.40}
$$

as $t\to 0+$ uniformly over $0\leq s\leq\tau\leq t$.
  Indeed,

$$
\mathbf{W}(s,\tau)\leq 2\sum_{k=1}^\infty e^{-2\lambda_k(\tau-s)}
$$

$$
\times\left[\int\limits_\Omega ((A(z_{j+1}(s)-z_j(s)))^T\xi_j(q,s)-(A(z_{j+1}(0)-z_j(0)))^T\xi_j(q,0))\,\omega_k dq\right]^2
$$

$$
+2\sum_{k=1}^N (e^{-\lambda_k(\tau-s)}-1)^2\left(\int\limits_\Omega (A(z_{j+1}(0)-z_j(0)))^T\,\xi_j(q,0)\omega_k dq\right)^2
$$

$$
+2\sum_{k=N+1}^\infty (e^{-2\lambda_k(\tau-s)}-1)^2\left(\int\limits_\Omega (A(z_{j+1}(0)-z_j(0)))^T\,\xi_j(q,0)\omega_k dq\right)^2.
$$

$$
\tag{14.41}
$$

Step 3. Due to Bessel's inequality,

$$\left\| \sum_{k=1}^{\infty} \left(\int_{\Omega} \left(A(z_{j+1}(0) - z_j(0)) \right)^T \xi_j(q,0) \omega_k dq \right) \omega_k \right\|_{(L^2(\Omega))^2}^2$$

$$= \sum_{k=1}^{\infty} \left(\int_{\Omega} \left(A(z_{j+1}(0) - z_j(0)) \right)^T \xi_j(q,0) \omega_k dq \right)^2$$

$$\leq \| \xi_j(\cdot,0) \left(A(z_{j+1}(0) - z_j(0)) \right) \|_{(L^2(\Omega))^2}^2 .$$

Therefore for every $\varepsilon > 0$ there is an $N = N(\varepsilon)$ such that the last sum on the right in (14.41) can be made smaller than $\varepsilon/3$ regardless of $0 \leq s \leq \tau \leq t$.

In turn, for this $N(\varepsilon)$, determined by ε, there is a $t_* = t_*(\varepsilon) > 0$ such that, by continuity of the exponential function, the second sum on the right in (14.41) can be made smaller than $\varepsilon/3$ as well for any $0 \leq s \leq \tau \leq t \leq t_*$.

Step 4. Now recall that continuity of solutions in time in Theorem 12.1 yields that

$$\| A(z_{j+1}(s) - z_j(s)) - A(z_{j+1}(0) - z_j(0)) \|_{R^2} = O(s)$$

as $s \to 0+$. In turn, (14.4), (12.5) implies that for any $k = 1,\ldots,n$:

$$\| \xi_k(\cdot,s) - \xi_k(\cdot,0) \|_{L^2(\Omega)}^2$$

$$= \int_{(S_r(z_k(s)) \cup S_r(z_k(0))) \setminus (S_r(z_k(s)) \cap S_r(z_k(0)))} dx$$

$$= O(\| z_k(s) - z_k(0) \|_{R^2}) = O(s) \qquad (14.42)$$

as $s \to 0+$. Therefore, for any selected-above ε there is a $t_{**} = t_{**}(\varepsilon) > 0$ such that the first term on the right in (14.41) can be made smaller than $\varepsilon/3$ for any $0 \leq s \leq \tau \leq t \leq t_{**}$. Indeed,

$$\sum_{k=1}^{\infty} e^{-2\lambda_k(\tau-s)} \times \left[\int_{\Omega} \left((A(z_{j+1}(s)-z_j(s)))^T \xi_j(q,s) - (A(z_{j+1}(0)-z_j(0)))^T \xi_j(q,0) \right) \omega_k dq \right]^2$$

$$\leq \sum_{k=1}^{\infty} \left[\int_{\Omega} \left((A(z_{j+1}(s) - z_j(s)))^T \xi_j(q,s) - (A(z_{j+1}(0) - z_j(0)))^T \xi_j(q,0) \right) \omega_k dq \right]^2$$

$$\leq \| \xi_j(\cdot,s)(A(z_{j+1}(s) - z_j(s))) - \xi_j(\cdot,0)(A(z_{j+1}(0) - z_j(0))) \|_{(L^2(\Omega))^2}^2 = O(s)$$

as $s \to 0+$ uniformly over $0 \leq s \leq \tau \leq t$.

Combining all the above, we obtain that $W(s,\tau)$ in (14.40), (14.41) can be made smaller than $\varepsilon > 0$ for any $t \in [0, \min\{t_*, t_{**}\}]$, which yields (14.40).

Step 5. Applying (14.42) to $\xi_i(x, \tau)$ in the first term in (14.39) and making use of (14.40) yields the assertion of Lemma 14.2 for the first term in ((14.38)). The proof for the second term is similar. This ends the proof of Lemma 14.2.   □

Auxiliary estimates. It is well known, e.g., [92], that (as a special form of Fredholm equation) (14.37) admits a unique solution in $(L^2(0,T))^{2n}$. This will allow us to prove, making use of the classical methods, that dz_i/dv_j's indeed exist in $(L^2(0,T))^2$, and all the above calculations leading to (14.37) are valid. (Namely, one needs, based on (14.29), to write the Volterra equations for the expressions $\Delta z_i/\Delta v_j$'s and then pass to the limit as Δv_j tend to zero to obtain (14.37).)

Furthermore, it is easy to see that there exists a ("small") $T_1 > 0$ such that for any $T \in (0, T_1]$:

$$\| \left(\frac{dz_1}{dv_j}, \ldots, \frac{dz_n}{dv_j}\right) \|_{(L^2(0,T))^{2n}} \leq C \| (\Xi_{1j}, \ldots, \Xi_{nj}) \|_{(L^2(0,T))^{2n}}, \tag{14.43}$$

where $C > 0$ is a (generic) positive constant independent of $T \in [0, T_1]$.

Moreover, since in (14.37) the integral terms with $(L^2(0,T))^2$-derivatives dz_i/dv_j's in them and the right-hand sides are actually continuous functions (which can be shown, in particular, making use of Lemma 14.1), we have:

$$\left(\frac{dz_1}{dv_j}, \ldots, \frac{dz_n}{dv_j}\right) \in C([0,T]; R^{2n})$$

and, similar to (14.43), there exists a ("small") $T_2 \in [0, T_1]$ such that for any $T \in (0, T_2]$:

$$\| \left(\frac{dz_1}{dv_j}, \ldots, \frac{dz_n}{dv_j}\right) \|_{C([0,T];R^{2n})} \leq C \| (\Xi_1, \ldots, \Xi_n) \|_{C([0,T];R^{2n})}, \tag{14.44}$$

where, to simplify notations, we again used the generic notation C for the constant.

Applying ((14.38)) and (14.44) to (14.37), we obtain that

$$\| \left(\frac{dz_1}{dv_j}, \ldots, \frac{dz_n}{dv_j}\right) \|_{C([0,T];R^{2n})} \leq Lt^2 \tag{14.45}$$

for some constant $L > 0$ as $t \to 0+$.

14.5 Proof of Theorem 14.1: Local Controllability

From (14.37), making use of ((14.38)), (14.45), and of (14.13)–(14.14), we derive that

$$\frac{dz_i(t)}{dv_j} = \frac{t^2}{2\text{mes}\{S(0)\}} \int_{S(z_i(0))} F_j(x)dx + t^2 O(t) \tag{14.46}$$

for $j = 1, \ldots, n-1, t \in [0, T]$.

Thus, the matrix

$$\left(\frac{dz_i(t)}{dv_j}, \frac{dz_i(t)}{dv_l} \right)$$

is not degenerate under Assumption 2 of Theorem 14.1 for sufficiently small t. We can now select any such "small" number t as a $T > 0$ in Proposition 14.1 and obtain the statement of Theorem 14.1 from this proposition. This ends the proof of Theorem 14.1.   □

Chapter 15
Global Controllability for a "Rowing" Swimming Model

Abstract We consider a swimmer whose body consists of three small narrow rectangles connected by flexible links. Our goal is to study its swimming capabilities when it applies a rowing motion in a fluid governed in a bounded domain by the nonstationary Stokes equation. Our approach explores an idea that a body in a fluid will move in the direction of least resistance determined by its geometric shape. Respectively, we assume that the means by which we can affect the motion of swimmer are the change of the spatial orientation of the aforementioned rectangles and the direction and strength of rowing motion. The main results are derived in the framework of mathematical controllability theory for pde's and are based on a constructive technique allowing one to calculate an incremental motion of swimmer.

15.1 Model Equations and Controllability Question

In this chapter we intend to discuss the swimming capabilities of a particular version of model (11.1)–(11.3) from Chapter 11. Namely, we consider a swimmer (which can be viewed as a possible engineering design for a robot), which consists of three *"small" narrow rectangles*, connected by flexible links as shown on Fig. 1.

<u>Controllability question:</u> *We would like to investigate whether our swimmer can actually swim anywhere in the given domain Ω (where, of course, its size permits) when we vary two sets of parameters:*

- *Bilinear or multiplicative controls – the coefficients $v_i, i = 1, 2, 3, 4$ in equations (15.1)–(15.3) below, defining the magnitudes and directions of the forces shown on Fig. 1 and*
- <u>*Geometric controls*</u> *– the orientation of rectangles on Fig. 1 (see also Figs. 2–7).*

The first type of the aforementioned parameters – a set of functions – lies in a typical "medium" for controls in control theory. On the other hand, they enter the system equations in a *non-standard highly nonlinear way – as coefficients*

A.Y. Khapalov, *Controllability of Partial Differential Equations Governed by Multiplicative Controls*, Lecture Notes in Mathematics 1995, DOI 10.1007/978-3-642-12413-6_15, © Springer-Verlag Berlin Heidelberg 2010

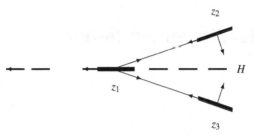

Figure 1 : All forces active

(as opposed to the classical additive controls). The 2-nd type of control parameters is motivated by the nature of swimming phenomenon, namely, by the fact that a swimmer can "choose"/control its shape.

The swimmer on Fig. 1 can also be viewed as a "modified scallop" with four hinges – one at point z_1, to allow the (folding/unfolding) leg motion, and three others at points z_1, z_2, and z_3, to allow the rotation of rectangles about these points as their respective centers (compare it to the abstract scallop in Chapter 10).

The governing model equations for the swimmer on Fig. 1 are as follows:

$$\frac{\partial y}{\partial t} = \nu \Delta y + F(z,v) - \nabla p \quad \text{in } Q_T = \Omega \times (0,T), \tag{15.1}$$

$$\text{div} y = 0 \text{ in } Q_T, \quad y = 0 \quad \text{in } \Sigma_T = \partial\Omega \times (0,T), \quad y \mid_{t=0} = y_0 \text{ in } \Omega,$$

$$\frac{dz_i}{dt} = \frac{1}{rq} \int\limits_{S_i(z_i(t))} y(x,t)dx, \quad z_i(0) = z_{i0}, \quad i = 1,2,3, \tag{15.2}$$

where $z(t) = (z_1(t), z_2(t), z_3(t)) \in (R^2)^3$, $v(t) = (v_1(t), v_2(t), v_3(t), v_4(t)) \in R^4$, and

$$(F(z,v))(x,t) = \xi_1(x,t)v_3(z_2(t) - z_1(t)) + \xi_2(x,t)v_3(z_1(t) - z_2(t))$$

$$+ \xi_1(x,t)v_4(z_3(t) - z_1(t)) + \xi_3(x,t)v_4(z_1(t) - z_3(t))$$

$$- \xi_1(x,t)v_1(t) \begin{pmatrix} 0 & 1 \\ -1 & 0 \end{pmatrix} (z_2(t) - z_1(t))$$

$$+ \xi_2(x,t)v_1(t) \begin{pmatrix} 0 & 1 \\ -1 & 0 \end{pmatrix} (z_2(t) - z_1(t))$$

$$- \xi_1(x,t)v_2(t) \begin{pmatrix} 0 & 1 \\ -1 & 0 \end{pmatrix} (z_3(t) - z_1(t))$$

$$+ \xi_3(x,t)v_2(t) \begin{pmatrix} 0 & 1 \\ -1 & 0 \end{pmatrix} (z_3(t) - z_1(t)). \tag{15.3}$$

We assume that rectangles on Fig. 1 have identical dimensions $r \times q, q \ll r$. The sets representing their positions at time t are denoted by $S_i(z_i(t)), i = 1, 2, 3$, while their characteristic functions by $\xi_i(x, t), i = 1, 2, 3$:

$$\xi_i(x, t) = \begin{cases} 1, & \text{if } x \in S_i(z_i(t)), \\ 0, & \text{if } x \in \Omega \setminus S_i(z_i(t)), \end{cases} \quad i = 1, 2, 3.$$

Symbols $z_i(t)$'s denote the centers of $S_i(z_i)$'s. We assume that we can change the orientations of $S_i(z_i(t))$'s, which are regarded as the aforementioned *geometric controls*. Throughout this chapter we will only consider the situations when that rectangles $S_i(z_i(t)), i = 1, 2, 3$ are "small" and stay strictly away from each other and from $\partial\Omega$. This can be assured by proper "control actions" which we intend to apply.

We assume that in its "*natural shape*" the distances from points z_2 and z_3 to z_1 are l-units long. The first two lines in (15.3) describe the forces (we further call them – "structural") acting along the straight lines connecting $z_2(t)$ and $z_3(t)$ to $z_1(t)$. Their goal is to preserve the original shape of the swimming object, that is, to control the distances between rectangles $S_i(z_i)$'s (as shown on Figs. 1 and 4). These forces can also be viewed as elastic Hooke's forces with variable rigidity coefficients. The last two lines in (15.3) represent the rotation forces, acting perpendicularly to the lines z_1z_2 and z_1z_3, see Figs. 1 and 2. The magnitudes and directions of the rotation and structural forces are determined by the coefficients $v_i, i = 1, 2, 3, 4$, which we regard as *multiplicative controls*.

As we discussed it in Chapter 11, all the forces in (15.1)–(15.3) are internal relative to the swimmer. The internal linear momentum of our swimmer remains constant, but the rotation forces in (15.1)–(15.3), in general, can generate a non-zero torque. To explicitly preserve the internal angular momentum of swimmer in (15.1)–(15.3), one should additionally assume that (see Chapter 12):

$$v_2 = -v_1 \frac{\| z_2(t) - z_1(t) \|_{R^2}^2}{\| z_3(t) - z_1(t) \|_{R^2}^2}. \tag{15.4}$$

Wellposedness of models (15.1)–(15.3)/(15.4). Throughout this section we assume, as usual, that $y_0 \in H(\Omega)$. The wellposedness of systems (15.1)–(15.3)/(15.4) was established in Chapter 12.

We now slightly reformulate Assumption 2 of Chapter 12 for the specific purposes of this section, which we, as usual, assume below.

Assumption 1. *The sets* $S_i(z_i(0)), i = 1, 2, 3$ *are strictly separated from each other and from* $\partial\Omega$ *and the rectangles* $S_i(z_i(t)), i = 1, 2, 3$ *can change their orientation only finitely many times in any given time-interval* $[0, T]$.

Provided that $q < r$, the rectangular shape of sets $S_i(z_i(t))$'s implies the following Lipschitz type condition:

$$\int_{(S_i(z_i(t))\cup S_i(z_i(\hat{t})))\backslash(S_i(z_i(\hat{t}))\cap S_i(z_i(t)))} dx \leq 4r \parallel z_i(t) - z_i(\hat{t}) \parallel_{R^2} \qquad (15.5)$$

for any $\hat{t} > t$ in any time-interval in $(0, T)$, if rectangles $S_i(z_i(t))$ and $S_i(z_i(\hat{t}))$ are of the same orientation.

The remainder of Chapter 15 is organized as follows. In Section 15.2 we formulate and discuss the main global controllability results, Theorem 15.1 for system (15.1)–(15.3) and Corollary 15.1 for system (15.1)–(15.4), as well as their critical supporting result on small motions – Lemma 15.1. In Section 15.3 we invoke necessary auxiliary results, which are used to prove Lemma 15.1 in Section 15.6. Theorem 15.1 is proven in Section 15.4 and Corollary 15.1 in Section 15.5. The main results are further discussed in Section 15.7.

15.2 Main Results

The global controllability of highly nonlinear swimming models described by coupled fluid/ode systems of equations appears to be an open problem. It should be noted in this respect that, in its current state, the mathematical controllability theory is comprehensive and thorough only when dealing with linear pde's. The developed linear controllability methods were quite successfully extended to the semilinear pde's in the case when the nonlinear terms (as well as controls) are *additive* and are to be viewed and dealt with as though they are generic nonlinear disturbances (to be "overwhelmed").

In this regard, the methodological novelty of our approach in this chapter is the introduction of a qualitative technique, based on a careful study of asymptotically small motions of the swimmer at hand. It exploits the specifics of the concrete nonlinear mapping:

Multiplicative Controls + Geometric controls → *Position of the Swimmer.*

The *crux* of the issue here is the question on *how the geometric shape of a body in a fluid affects the projections of the forces (acting upon it) onto the fluid velocity space.* The respective results were obtained in Chapter 13, particularly, for narrow rectangles (see Lemmas 15.3 and 15.4 below). This is one of the reasons why our schematic swimmer consists of such rectangles, which, to achieve a desirable swimming motion, have to change their spatial orientation. Our intention here is to explore an empiric observation that *a body will move in a fluid in the direction where it meets the least resistance.*

We link the study of controllability of our swimmer to the study of the motion of the "head-rectangle" $S_1(z_1(t))$ on Figs. 1,2, 4–6, while, of course, making sure that its shape is preserved. Re-phrasing the controllability question from the above, we ask: *Can the point z_1 reach any point in the space domain Ω from any given initial position by making use of available controls*, assuming that the geometry of Ω permits to connect the aforementioned initial and target positions with a "corridor" which can accommodate the size our swimmer. The latter circumstance is formally addressed in the following assumption.

Assumption 2. *The initial and target positions $z_1(0)$ and z_{1T} for the center of the head rectangle $S_1(z_1)$ for the desirable motion of the swimmer in (15.1)–(15.3) can be connected by a straight line segment which lies inside Ω along with its neighborhood of radius σ, $\sigma > l$.*

The only purpose of this assumption is to avoid any dealing with the situation when the swimmer "collides" with the boundary $\partial\Omega$ during its motion. It can be relaxed at no extra cost. For example, the straight line can be replaced by a broken line. Our plan is to try to steer our swimmer from $z_1(0)$ to z_{1T} within the *"corridor of steering"* in Assumption 2.

Theorem 15.1 (Global controllability property). *Let $\varepsilon > 0, \sigma > l > 0, \delta \in (0, \sigma - l)$, $T > 0, y_0 \in H(\Omega)$ be given (the space $H(\Omega)$ is defined in Appendix A). Let the points $z_i(0), i = 1, 2, 3$ and $z_{1T} \in \Omega$ be given, $z_1(0)$ and z_{1T} satisfy Assumption 2, and*

$$\max\{|\,\| z_2(0) - z_1(0) \|_{R^2} - l\,|, |\,\| z_3(0) - z_1(0) \|_{R^2} - l\,|\} < \delta.$$

1. *There exist a $T_* \in [0, T]$, a positive integer n, the instants of time $0 < t_1 < \ldots < t_n < T_*$, piecewise constant control functions $v_i, i = 1, 2, 3, 4$ and orientations of sets $S_i(z_i(t))$'s that do not change between the aforementioned instants of time such that the respective solution to (15.1)–(15.3) satisfies the following inequality:*

$$\| z_1(T_*) - z_{1T} \|_{R^2} < \varepsilon, \tag{15.6}$$

provided that the parameters r and q (i.e., the dimensions of rectangles $S_i(z_i(t))$'s) are small enough, namely, for sufficiently small $h > 0$

$$\max\{q^a, q^{1-a}/r, r, \sqrt{rq}\} \leq C_1 h^b, \quad q \geq C_2 h^{1-2c}$$

for some $a \in (0, 1)$, $b > 0, c \in (0, 0.5)$ and $C_1, C_2 > 0$.
2. *The above steering can be achieved, while preserving the initial structure of the object at hand at all times, namely:*

$$\max\{|\,\| z_2(t) - z_1(t) \|_{R^2} - l\,|, |\,\| z_3(t) - z_1(t) \|_{R^2} - l\,|\} < \delta, \ t \in [0, T_*]. \tag{15.7}$$

Remark 15.1. The conditions linking q and h in Theorem 15.1 hold, for example, for $C_1 = C_2 = 1, r = h^{0.2}, q = h^{0.9}, a = 0.5, b = 0.2, c = 0.05$ as $h \to 0+$.

The main idea of the proof of Theorem 15.1 from the technical viewpoint is to obtain the required controlled swimming motion to the target state as a sequence of suitably small incremental motions shown on Figs. 2–6. In this respect, the following result, Lemma 15.1, is the most critical.

Denote by $\Pi_i b, i = 1, 2, 3$ the *vector projections* of vector $b \in R^2$ on the straight lines co-linear to the longer sides of the rectangles $S_i(z_i(t))$. Also set:

$$\alpha(t) = t^2 \sum_{i=1}^{4} |v_i|,$$

$$\beta(t) = \frac{3t}{\sqrt{rq}(1 - 3\alpha(t))} \| y(\cdot, 0) \|_{(L^2(\Omega))^2} + \frac{3\alpha(t)}{1 - 3\alpha(t)} \text{diam}(\Omega).$$

Throughout the rest of Chapter 15 we use the same generic symbol C to denote positive, possibly different, constants.

Lemma 15.1 (Swimming micro ("small") motions for model (15.1)–(15.3)).
Let $t > 0$ be given. Assume that $v_i, i = 1, 2, 3, 4$ and the orientation of $S_i(z_i(t))$'s do not change on $[0, t]$. Assume that the distances between $S_i(z_i(\tau))$'s, $\tau \in [0, t]$ and from them to the boundary $\partial\Omega$ exceed some value $d_* > 0$. Then, if $\alpha(t) < 1/3$,

$$z_1(t) = z_1(0) - \frac{t^2 v_1}{2} \Pi_1 \left[\begin{pmatrix} 0 & 1 \\ -1 & 0 \end{pmatrix} (z_2(0) - z_1(0)) \right]$$

$$- \frac{t^2 v_2}{2} \Pi_1 \left[\begin{pmatrix} 0 & 1 \\ -1 & 0 \end{pmatrix} (z_3(0) - z_1(0)) \right]$$

$$+ \frac{t^2 v_3}{2} \Pi_1 [z_2(0) - z_1(0)] + \frac{t^2 v_4}{2} \Pi_1 [z_3(0) - z_1(0)] + R_1(t), \quad (15.8)$$

$$z_2(t) = z_2(0) + \frac{t^2 v_1}{2} \Pi_2 \left[\begin{pmatrix} 0 & 1 \\ -1 & 0 \end{pmatrix} (z_2(0) - z_1(0)) \right]$$

$$+ \frac{t^2 v_3}{2} \Pi_2 [z_1(0) - z_2(0)] + R_2(t), \quad (15.9)$$

$$z_3(t) = z_3(0) + \frac{t^2 v_2}{2} \Pi_3 \left[\begin{pmatrix} 0 & 1 \\ -1 & 0 \end{pmatrix} (z_3(0) - z_1(0)) \right]$$

$$+ \frac{t^2 v_4}{2} \Pi_3^t [z_1(0) - z_3(0)] + R_3(t), \quad (15.10)$$

where for some constant $C > 0$ and function $\gamma_*(t, z(0)) \to 0+$ as $t \to 0+$ we have:

$$\| R_i(t) \|_{R^2} \le C \left\{ \max\{q^a, q^{1-a}/r, r\} + \frac{\sqrt{rq}}{d_*^4} \right\} \alpha(t)$$

$$+ \frac{C}{\sqrt{rq}} \alpha(t) \gamma_*(t, z(0))$$

$$+ \frac{Ct}{\sqrt{rq}} \| y(\cdot, 0) \|_{(L^2(\Omega))^2} + C\alpha(t) \beta(t)$$

$$+ \frac{C}{\sqrt{q}} \beta^{1/2}(t) \alpha(t), \quad i = 1, 2, 3, \quad (15.11)$$

provided that for some $a \in (0, 1)$ the value of $\max\{q^a, q^{1-a}/r, r\}$ is small enough.

We refer to the discussion on the existence of parameter $d_* > 0$ in Lemma 15.1 to Proposition 15.1 and Lemma 15.5 below in Section 15.6.

Corollary 15.1. *The result of Theorem 15.1 holds for system (15.1)–(15.4), that is, with a single multiplicative control v_1 and v_2 as in (15.4), and with the same geometric control – the orientation of rectangles $S_i(z_i), i = 1, 2, 3$.*

15.3 Preliminary Results

Recall that the unique solution to (15.1)–(15.3) lies in the space $J_o(\Omega)$ at all times and admits the following implicit Fourier series representation:

$$y(x,t) = \sum_{k=1}^{\infty} e^{-\lambda_k t} \left(\int_{\Omega} y_0^T \omega_k dx \right) \omega_k(x)$$

$$+ \sum_{k=1}^{\infty} \int_0^t e^{-\lambda_k(t-\tau)} \left(\int_{\Omega} (F(z,v))^T \omega_k dx d\tau \right) \omega_k(x). \quad (15.12)$$

Here the 2-D vector functions $\omega_k, k = 1, \ldots$ and the real numbers $-\lambda_k, k = 1, \ldots$ denote respectively the orthonormalized in $(L^2(\Omega))^2$ eigenfunctions and eigenvalues of the spectral problem associated with (15.1):

$$v\Delta\omega_k - \nabla p_k = -\lambda_k\omega_k \quad \text{in } \Omega, \quad \operatorname{div}\omega_k = 0 \text{ in } \Omega, \quad \omega_k = 0 \text{ in } \partial\Omega.$$

The functions $\{\omega_k\}_{k=1}^{\infty}$ also form a basis in $J_o(\Omega)$.

Denote, as in Chapter 13, the orthogonal projection operator from the space $(L^2(\Omega))^2$ onto $J_o(\Omega)$ by P. Let us recall now the result of Lemma 13.1 from Chapter 13. For our specific model in this section we can rewrite it, more explicitly, as follows:

Lemma 15.2. *Let on some time-interval $[t_0, T], t_0 \geq 0$ the sets $S_i(z_i(t)), i = 1, \ldots, n$ do not change their orientation and $v_i, i = 1, 2, 3, 4$ remain constant. Then we have the following formula for solutions to systems (15.1)–(15.3):*

$$\frac{dz_i(t)}{dt} = \frac{1}{rq} \int_{S_i(z_i(t_0))} y(x,t_0)dx + \frac{t - t_0}{rq} \int_{S_i(z_i(t_0))} (PF(z,v))(x,t_0)dx$$

$$+ u(t - t_0) + (t - t_0)v(t - t_0), \quad i = 1, 2, 3, \quad (15.13)$$

where

$$\| u(t-t_0) \|_{R^2} \leq \frac{1}{\sqrt{rq}}\mu(t,y(\cdot,t_0))$$

$$+ \frac{2\max^{1/2}\{\mathrm{mes}\,\{S_i(z_i(t_0))\backslash S_i(z_i(t)),\mathrm{mes}\,S_i(z_i(t))\backslash S_i(z_i(t_0))\}}{rq}$$

$$\times \| y(\cdot,t_0) \|_{(L^2(\Omega))^2} \tag{15.14}$$

and $\mu(t,y(\cdot,t_0)) \to 0$, $v(t-t_0) \to 0$ *as* $t \to t_0+$, $0 \leq \mu(t,y(\cdot,t_0)) \leq \| y(\cdot,t_0) \|_{(L^2(\Omega))^2}$.

More precisely, it was shown in Chapter 13 that, say, for $t_0 = 0$:

$$\mu(t,y(\cdot,0)) = \left(\sum_{k=1}^{\infty} \left(e^{-\lambda_k t}-1\right)^2 \left(\int_{\Omega} y(\cdot,0)^T \omega_k ds\, dx \right)^2 \right)^{1/2}, \tag{15.15}$$

$$0 \leq v(t) \leq Cw_*(\cdot) + \frac{C}{\sqrt{rq}}\gamma(t)$$

$$+ \frac{C}{\sqrt{rq}}\max_{i=1,2,3}\left(\mathrm{mes}^{1/2}\{S_i(z_i(t))\backslash S_i(z_i(0))\}\right.$$

$$\left. + \mathrm{mes}^{1/2}\{S_i(z_i(0))\backslash S_i(z_i(t))\}\right) w(t). \tag{15.16}$$

In (15.16) $C > 0$ is, as usual, a generically denoted positive constant,

$$w(t) = max_{j=1,\dots,8}\| f_j(\cdot) \|_{(C[0,t])^2}, \tag{15.17}$$

$$w_*(t) = max_{j=1,\dots,8}\| f_j(\cdot) - f_j(0) \|_{(C[0,t])^2}, \tag{15.18}$$

$f_j(t), j = 1,\dots,8$ denote the eight coefficients at the characteristic functions $\xi_i(x,t)$, $i = 1,2,3$ in (15.3) (respectively repeated twice or four times) in the order of appearance, and

$$\gamma(t) = max_{j=1,\dots,8}\left(\sum_{k=1}^{\infty} \left[\frac{1-e^{-\lambda_k t}}{\lambda_k t} - 1 \right]^2 \left(\int_{\Omega} f_j^T(0)\xi_j(x,0)\omega_k dx \right)^2 \right)^{1/2}. \tag{15.19}$$

Note:

$$\left[\frac{1-e^{-\lambda_k t}}{\lambda_k t} - 1 \right]^2 \leq 1.$$

The formula (15.13) implies that, if sufficiently large control forces v_i's are applied, the directions at which the points $z_i(t)$'s will try to move from their current positions are primarily determined by the projections of the object's internal *control* forces on the fluid velocity space *at this moment*, averaged over their corresponding supports $S_i(z_i(t))$'s.

Below we intend to make use of the following results from Chapter 13, describing an asymptotic representation of the averaged forces acting *in the fluid* on each of the rectangles $S_i(z_i)$'s.

Consider any point $x_0 = (x_{10}, x_{20})$ and denote

$$S_0 = \{x = (x_1, x_2) \mid -r/2 < x_1 - x_{10} < r/2, \ -q/2 < x_2 - x_{20} < q/2\}. \quad (15.20)$$

For this set, associated with our specific model in this section, we can rewrite Theorem 13.1 and Lemma 13.2 from Chapter 13, as follows:

Lemma 15.3. *Let $b = (b_1, b_2)$ be a given 2-D vector and S_0 lie strictly inside of Ω. Let $q, r, q^{1-a}/r \to 0+$ for some $a \in (0, 1)$. Then*

$$\frac{1}{rq} \int_{S_0} (Pb\xi)(x)dx = (b_1, 0) + \left(O(q^a) + O(q^{1-a}/r) + O(r)\right) \| b \|_{R^2} \quad (15.21)$$

as $q, r, q^{1-a}/r \to 0+$, where $\xi(x)$ is the characteristic function of S_0.

In the above and below, the notation $O(s)$ means that $\| O(s) \|_{R^2} \leq Cs$ as $s \to 0+$ for some positive constant C.

This result can be interpreted as that the average of the projection on the fluid velocity space of the constant force (i.e., b in (15.21)), acting upon a narrow rectangle in a fluid, will try to move this rectangle in the direction of the least resistance – the narrow side. This observation is taken into account in the swimming strategy shown on Figs. 2–6.

Lemma 15.4. *Let $b = (b_1, b_2)$ be a given 2-D vector. Assume that $S_0 \subset \Omega$ is strictly separated from $\partial\Omega$ and has the diameter of size s. Then for any subset Q of Ω of positive measure of diameter (not to exceed) s, which lies strictly outside of S_0 and is strictly separated from $\partial\Omega$, we have:*

$$\left\| \frac{1}{\text{mes}\,\{Q\}} \int_Q (Pb\xi)(x)dx \right\|_{R^2} \leq C \frac{\| b \|_{R^2}}{d_*^4} \text{mes}^{1/2}\{S_0\} \quad (15.22)$$

as $s \to 0+$, where $C > 0$ is a (generic) constant and d_* is the smallest out of the distances from Q to S_0 and to $\partial\Omega$.

We can interpret Lemma 15.4 as that the effect of the force $b\xi(x)$ on similarly sized sets outside of its support $S(0)$ is "negligible" as this size decreases.

15.4 Proof of Theorem 15.1: Global Controllability

The result of Theorem 15.1 is achieved (that is, in our specific proof below) as a combination of two motions:

- The *turning motion* (as on Fig. 6) with the goal to *aim the swimming object at the desirable target position* z_{1T} along the bisector of the angle $\angle z_2 z_1 z_3$.
- When the above is achieved, we apply the *forward* motion (Figs. 2–5) to reach this position.

To make this plan work we need sufficient room within Ω, which is provided by Assumption 2. We begin the proof with the case of the forward motion shown on Figs. 2–5.

15.4.1 The Forward Motion: Plan of Proof

Let us assume that we need to move our swimmer from its original position $z(0) = (z_1(0), z_2(0), z_3(0))$ as on Fig. 2 forward, i.e., to the left by P units.

The respective target position for the center of the head-rectangle $S_1(z_1)$ denote by z_{1T}. Assume that Assumption 2 holds for $z_1(0)$ and z_{1T} as required by Theorem 15.1. Our plan is as follows:

1. We intend to accomplish the required motion in an N incremental steps of length of order h, where h will be selected below.
2. These incremental motions are to be performed during suitably selected time-intervals $(0,t_1], (t_1,t_2], \ldots, (t_{N-1},t_N]$.
3. We assume that on each of the above-mentioned time-intervals we apply

- either the constant rotation forces (see Fig. 2) only, in which case:

$$
\begin{aligned}
(F(z,v))(x,t) = & -\xi_1(x,t)v_1(t)A\,(z_2(t)-z_1(t)) \\
& +\xi_2(x,t)v_1(t)A\,(z_2(t)-z_1(t)) \\
& -\xi_1(x,t)v_2(t)A\,(z_3(t)-z_1(t)) \\
& +\xi_3(x,t)v_2(t)A\,(z_3(t)-z_1(t))
\end{aligned}
\tag{15.23}
$$

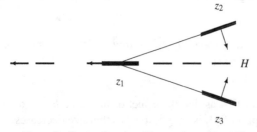

Figure 2 : Forward – propelling motion with rotation forces only

- or the constant structural forces (see Fig. 4) only, in which case:

$$(F(z,v))(x,t) = \xi_1(x,t)v_3(z_2(t) - z_1(t))$$
$$+\xi_2(x,t)v_3(z_1(t) - z_2(t))$$
$$+\xi_1(x,t)v_4(z_3(t) - z_1(t))$$
$$+\xi_3(x,t)v_4(z_1(t) - z_3(t)). \quad (15.24)$$

4. The orientation of sets $S_i(z_i(t)), i = 1, 2, 3$ remains unchanged on each of $(0, t_1]$, $(t_1, t_2], \ldots, (t_{N-1}, t_N]$.

The proof below deals with selection of suitable parameters N; the instants t_i's; the values for respective controls v_i's and the orientations for $S_i(z_i(t))$'s between the aforementioned moments with the goal to ensure that the object at hand swims forward for approximately P units. *The main idea* here is to chose the above parameters in such a way that they will make "errors" R_i's in (15.8)–(15.10)–(15.11) "small", namely, relative to the remaining "easily calculable" explicit terms in (15.8)–(15.10). This will allow us to use these terms to steer our swimmer to the desirable target position in a constructive manner.

15.4.2 Proof of Theorem 15.1: The Forward Motion

Without loss of generality (otherwise, we would apply the motion on Fig. 4 first), we can assume that at time $t = 0$

$$\| z_1(0) - z_2(0) \|_{R^2} = \| z_1(0) - z_3(0) \|_{R^2} = l$$

and that the swimmer is positioned symmetrically as on Fig. 1:

$$\theta_1(0) = \angle z_2(0)z_1(0)H = \theta_2(0) = \angle z_3(0)z_1(0)H.$$

In our further actions we intend to keep these angles between two preassigned values θ_* and θ^* as follows:

$$0 < \theta_* \leq \theta_1(t) = \angle z_2(t)z_1(t)H, \theta_2(t) = \angle z_3(t)z_1(t)H \leq \theta^* < \pi/4. \quad (15.25)$$

We assume that (15.25) also holds at $t = 0$ (otherwise, we would apply the motion on Fig. 5 first). Without loss of generality, *we can further assume* that r and q are small enough (recall that Theorem 15.1 deals with "small" r and q) to ensure that $S_2(z_2(t))$ and $S_3(z_3(t))$ do not overlap, regardless of their orientation, as long as (15.25) holds.

Step 1. Forward propelling motion and selection of parameters on $(0,t_1]$. We assume that on this time-interval only the the rotation forces (RF) are active, that is,

$$v_3 = v_4 = 0, \quad t \in (0,t_1].$$

We intend to apply the aforementioned forces to make the points z_2 and z_3 move toward each other:

$$v_1 > 0, \quad v_2 = -v_1.$$

Restriction on v_1 and v_2 in terms of t_1 and h. We assume that on $(0,t_1]$ constant controls v_1 and v_2 can be selected arbitrarily except for the assumption that, whatever t_1 is chosen, their values satisfy the following equalities:

$$| v_j | = \frac{h}{t_1^2}, \quad j = 1,2, \ t \in (0,t_1], \tag{15.26}$$

where $h > 0$ will be selected below. Condition (15.26) implies that

$$\alpha(t) = 2h, \quad t \in (0,t_1], \tag{15.27}$$

where $\alpha(t)$ is defined in Section 15.2 before Lemma 15.1.

1-st restriction on h. Assume from now on that

$$h \le \min\left\{ \frac{1}{24}, \sqrt{\mathrm{diam}\,(\Omega)}, \frac{1 - \sin 2\theta^*}{\sin^2 \theta^*} \right\}. \tag{15.28}$$

Restriction on p and q in terms of h. Assume that

$$\max\{q^a, q^{1-a}/r, r, \sqrt{rq}\} \le C_1 h^b, \quad q \ge C_2 h^{1-2c} \text{ or } \frac{\sqrt{h}}{\sqrt{q}} \le \frac{h^c}{\sqrt{C_2}} \tag{15.29}$$

for some $a \in (0,1)$, $b > 0, c \in (0,0.5)$ and $C_1, C_2 > 0$ (see an example in Remark 15.1).

1-st restriction on t_1 in terms of h. Given y_0, in addition to (15.26) and (15.29), assume that t_1 is so small that

$$\frac{t_1}{\sqrt{rq}} \max\{1, \| y(\cdot,0) \|_{(L^2(\Omega))^2}\} \le h^2. \tag{15.30}$$

(15.27)–(15.30) imply (see (15.47) below) that the motion of z_i's is restricted on $[0,t_1]$ as follows:

$$\sum_{i=1}^{3} \| z_i(\cdot) - z_i(0) \|_{(C[0,t])^2} \le M_0 h, \quad t \in [0,t_1] \tag{15.31}$$

for some $M_0 > 0$ independent of $z_i(0)$'s. Inequality (15.31) allows us to impose the following condition on h (if necessary).

___2-nd restriction on h.___ We will assume from now on that, in addition to the above, h is so small that: *(i)* the shape condition (15.7) holds on $[0, t_1]$ and *(ii)* the swimmer stays on $[0, t_1]$ within the corridor of steering outlined in Assumption 2 before Theorem 15.1. The latter means that Proposition 15.1 from subsection 6 below holds for some $d_* > \sigma - l > 0$.

___2-nd restriction on t_1 in terms of h.___ In addition to the above, we will also assume that t_1 is so small that, given $z(0)$,

$$\frac{1}{\sqrt{rq}} \gamma(t_1, z(0)) \leq h. \tag{15.32}$$

___Calculation of the incremental motion on $(0, t_1]$.___ Denote $K = \min\{b, c\}$. Applying (15.26)–(15.32) to (15.11) provides us with the following estimate:

$$\| R_i(t_1) \|_{R^2} \leq M(d_*) h^{1+K}, \quad i = 1, 2, 3, \tag{15.33}$$

where $M(d_*)$ is a positive-valued nondecreasing function of d_*, independent of $z_i(0)$'s and y_0, whenever we select t_1, v_1, v_2 and r, q to satisfy (15.26)–(15.32). In other words, the above choice of control parameters resulted in making errors $R_i(t_1)$'s *smaller* than the remaining explicit motion terms in (15.8)–(15.10) by the order of h^K as $h \to 0+$.

___Step 2.___ We will evaluate next *where these terms in (15.8)–(15.10) will take our swimmer at time* $t = t_1$. Assuming that the vector $(1, 0)$ points to the right and $(0, 1)$ points up, we obtain (see Fig. 2 for $t = 0$ and also Fig. 3) that:

$$-\frac{t^2 v_1}{2} \Pi_1 [A (z_2(0) - z_1(0))] - \frac{t^2 v_2}{2} \Pi_1 [A (z_3(0) - z_1(0))] = lh(-1, 0) \sin \theta_1(0).$$

Hence, due to (15.8)–(15.10):

$$\| z_1(t_1) - z_1(0) - lh(-1, 0) \sin \theta_1(0) \|_{R^2} \leq M(d_*) h^{1+K}, \tag{15.34}$$
$$\| z_j(t_1) - z_j(0) \|_{R^2} \leq M(d_*) h^{1+K}, \quad j = 2, 3. \tag{15.35}$$

Thus, for our above-selected "small" h, the lengths between $z_1(t_1)$ and $z_j(t_1), j = 2, 3$ are longer than the original value l and the horizontal shift of point $z_1(t_1)$ to the left from point $z_1(0)$ satisfies the following inequality:

$$| (1, 0)^T (z_1(t_1) - z_1(0)) + lh \sin \theta_1(0) | \leq M(d_*) h^{1+K}. \tag{15.36}$$

We now consider an *ideal "zero-error" situation* as follows.

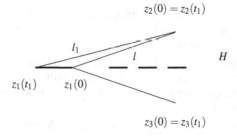

Figure 3 : The case when $R_i(t_1) = 0, i = 1, 2, 3$

Step 3. The "ideal" case $R_i(t_1) = 0, i = 1, 2, 3$, see **Fig. 3**. In *this case* denote by l_1 the distance from $z_j(0), j = 1, 2$ to the point $z_1(t_1)$, positioned $lh \sin \theta_1(0)$ units to the left of $z_1(0)$. Making use of the Cosine Law (see Fig. 3), we further obtain:

$$l_1^2 = l^2 + \| z_1(t_1) - z_1(0) \|_{R^2}^2 + 2l \| z_1(t_1) - z_1(0) \|_{R^2}^2 \cos \theta_1(0)$$
$$= l^2 + (lh \sin \theta_1(0))^2 + 2l^2 h \sin \theta_1(0) \cos \theta_1(0).$$

Therefore,

$$| l_1 - l | = \frac{l^2}{l_1 + l} \left((h \sin \theta_1(0))^2 + 2h \sin \theta_1(0) \cos \theta_1(0) \right)$$
$$\leq \frac{lh}{2} \left(h \sin^2 \theta^* + \sin 2\theta^* \right)$$
$$\leq \frac{lh}{2} \left(\frac{1 - \sin 2\theta^*}{\sin^2 \theta^*} \sin^2 \theta^* + \sin 2\theta^* \right)$$
$$\leq \frac{lh}{2}, \tag{15.37}$$

in view of (15.25) and (15.28).

Step 4. Note that if $R_i(t_1) = 0, i = 1, 2, 3$, then $\theta_i(t_1) < \theta_i(0), i = 1, 2$. This allows us to impose the following restrictions on h, making use of the fact that in the general case the smallness of errors $R_i(t_1)$'s is of higher order h^{1+K} as $h \to 0+$ compared to h.

Further restrictions on h. We further assume that, in addition to the above, h is so small that:

- $\theta_1(t_1), \theta_2(t_1) \leq \theta_1(0), \theta_2(0) \leq \theta^*$.
- $lh/2 > 2M(d_*)h^{1+K}$. This, in view of (15.37), means that the distances from $z_j(t_1), j = 2, 3$ to $z_1(t_1)$ increased, compared to the initial value l at time $t = 0$.

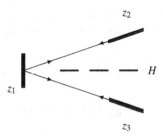

Figure 4 : Length – recovering motion only

Step 5. Length-recovering motion: Selection of parameters on $(t_1, t_2]$. We
assume that *on this time-interval only the structural forces (SF) are active* (i.e.,
$v_1 = v_2 = 0$) and that the sets $S_i(z_i)$'s are oriented as on Fig. 4:

Note also that, after the completion of swimming motion on $[0, t_1]$, the distances
between $z_1(t_1)$ and $z_j(t_1), j = 2,3$ may differ from each other. We established in
Step 1 that they exceed l by no more than $lh/2 + 2M(d_*)h^{1+K}$ units as given in
(15.34)–(15.37). (The factor 2 appeared here to reflect the fact that the motions of
the point z_1 and each of $z_j, j = 2,3$ may add an error of size $M(d_*)h^{1+K}$.)

Our goal on this time-interval is to shrink the distances between z_1 and $z_j, j =
2,3$ back to the original distance l, while staying within the same corridor of steer-
ing described in Assumption 2. To this end, we will select two, possibly different,
controls v_3 and v_4 of the form:

$$v_{j+1} = \frac{h_j}{(t_2 - t_1)^2}, \quad 0 < h_j \le h, \quad j = 2,3, \tag{15.38}$$

where h is *the same as on the time-interval* $(0, t_1]$ and the values for h_2 and h_3 will
be determined below.

Step 6. Motion on $(t_1, t_2]$. If the values of $(t_2 - t_1)$, r and q are selected to satisfy
(15.29)–(15.32) with $(t_2 - t_1)$ in place of t_1 and with $y(\cdot, t_1)$ in place of y_0, once
again we will have

$$\| R_i(t_2) \|_{R^2} \le M(d_*)h^{1+K}, \quad i = 1,2,3 \tag{15.39}$$

for the same $M(d_*)$ as in (15.33), where, without loss of generality, we can use
the same d_* as before. Indeed, the latter follows from the initial configuration of
our swimming object at time $t = t_1$ as on Fig. 2 inside of the corridor of steering,
outlined in Assumption 2 (recall also the angle condition (15.25)), and the fact that
the "size" of the motions of z_i's on $(t_1, t_2]$, implied by (15.47) for this time-interval,
is of order h or less. This also ensures that the swimming motion occurs within the
corridor of steering.

This, similar to (15.34)–(15.35), will result in the following estimates (see (15.8)–(15.10), applied now on $(t_1, t_2]$ in place of $(0, t_1]$):

$$\| z_1(t_2) - z_1(t_1) - \frac{h_2}{2} \| z_2(t_1) - z_1(t_1) \|_{R^2} \sin \theta_1(t_1)(0, 1)$$

$$- \frac{h_3}{2} \| z_3(t_1) - z_1(t_1) \|_{R^2} \sin \theta_2(t_1)(0, -1) \|_{R^2} \leq M(d_*) h^{1+K}, \quad (15.40)$$

$$\| z_j(t_2) - z_j(t_1) - \frac{h_j}{2}(z_1(t_1) - z_j(t_1)) \|_{R^2} \leq M(d_*) h^2, \quad j = 2, 3. \quad (15.41)$$

Recalling that $\| z_j(t_1) - z_1(t_1) \|_{R^2} > l$, we have:

$$\| \frac{h_j}{2}(z_1(t_1) - z_j(t_1)) \|_{R^2} > \frac{h_j l}{2}, \quad j = 2, 3. \quad (15.42)$$

We will now consider two cases.

Step 7. The "ideal" case $R_i(t_2) = 0, i = 1, 2, 3$. In this case the points $z_2(t_1)$ and $z_3(t_1)$ lie at equal distance from the point $z_1(t_1)$ and are symmetrically placed about the horizontal line passing through it (as on Fig. 3). Hence, the points $z_j, j = 2, 3$ will move toward the point $z_1(t_1)$ and each will cover the distance greater than $h_j l/2$, which, due to (15.37), in case of $h_j = h$ will exceed the value of $l_1 - l$. This means that in such case, given t_1 and t_2, we can select the equal values for h_2 and h_3 in (15.38) so that

$$\| z_1(t_2) - z_j(t_2) \|_{R^2} = l, \quad j = 2, 3. \quad (15.43)$$

Step 8. The general case: Errors $R_i(t_2), i = 1, 2, 3$ are present. In the general case, the above calculations will be distorted by errors of order h^{1+K} (see (15.33) and (15.39)) and one can use the Cosine Law to show it). Therefore, we still can achieve (15.43) if h is sufficiently *small*, since we may just need to make correction with respect to the terms of order h^{1+K} for each of positions z_i's, compared to the allowed incremental motion of order h.

One more restriction on h. In addition to the above restrictions on the smallness of h, we will also assume that h is small enough to ensure the just described property. The respective upper bound for h will depend on $M(d_*), l, \theta_*, \theta^*$ and K.

Thus, we can write:

$$| z_1(t_2) - z_1(t_1) | \leq \bar{M}(d_*) h^{1+K} \quad (15.44)$$

for some $\bar{M}(d_*) > 0$, which is determined by the values of $M(d_*), l, \theta_*, \theta^*$, and K.

Furthermore, (15.41) implies that during the length-recovering motion as on Fig. 2 the points z_2 and z_3 will make a vertical motion towards each other *going further inside of the corridor of steering*, which can be evaluated as follows (see (15.37) and (15.41)):

$$| (0,1)^T (z_j(t_2) - z_j(t_1)) | \leq \frac{h_j}{2} \| z_j(t_1) - z_1(t_1) \|_{R^2} \sin \theta^* + M(d_*)h^{1+K}$$

$$\leq h(l_1 + 2M(d_*)h^{1+K}) \sin \theta^* + M(d_*)h^{1+K}$$

$$\leq h \left(l + \frac{lh}{2} + 2M(d_*)h^{1+K} \right) \sin \theta^*$$

$$+ M(d_*)h^{1+K}, \quad j = 2,3. \tag{15.45}$$

Step 9. Let us summarize the results of our actions on $[0,t_2]$:

- The point $z_1(t_2)$ is positioned at least $lh \sin \theta_* - 2M(d_*)h^{1+K}$ units left of $z_1(0)$, see (15.36), (15.25) and (15.40). It is shifted by no more than $(\bar{M}(d_*)h^{1+K} + M(d_*)h^{1+K}$ units vertically from the level of $z_1(0)$, see (15.44), (15.33), (15.34).
- The distances between $z_1(t_2)$ and $z_j(t_2), j = 2,3$ once again are equal to l.
- However, the positions of points $z_j(t_2), j = 2,3$ are not necessarily symmetrically positioned relative to the horizontal line $z_1 H$.
- The angles $\theta_1(t_2)$ and $\theta_2(t_2)$ became smaller than the original $\bar{\theta}$ but still satisfy (15.25).

The above means that our swimmer at time $t = t_2$ is deeper inside of the corridor of steering than it was at time $t = 0$. Hence, we can repeat our actions on this time-interval to move the swimmer from its new position at time $t = t_2$ to the left in a similar way *without changing the upper bound for h* chosen for steering on $[0,t_2]$.

Step 10. Selection of parameters on $(t_2,t_3], (t_3,t_4],\ldots$. On the subsequent pairs of time-intervals $\{(t_2,t_3], (t_3,t_4]\},\ldots$ we intend to repeat our actions on $\{(0,t_1], (t_1,t_2]\}$ as long as the angle restriction (15.25) holds. Let us consider first the following "ideal" situation.

Case I: The points $z_j(t_2), z_j(t_4), z_j(t_6),\ldots, j = 2,3$ remain symmetric relative to the line $z_1 H$, see Figs. 2–4. In this case, we are in fact in the same situation as at the initial moment $t = 0$, except that the swimming object is now closer to the target position z_{1T}, lies deeper inside of the pre-assigned corridor of steering, and $z_1(t_2)$ is possibly shifted a "little bit" vertically relative to the initial position of $z_1(0)$.

Therefore, we can repeat the motions of Steps 1–9 (as on Figs. 2–4) on $\{(t_2,t_3], (t_3,t_4]\}$ and subsequent *pairs of time-intervals*, thus moving the point z_1 further and further to the left toward the target, each time at least by additional $lh \sin \theta_* - 2M(d_*)h^{1+K}$ units, while possibly also shifting it vertically by no more than $\bar{M}(d_*)h^{1+K} + M(d_*)h^{1+K}$ units each time.

Assume that, given $h = h_*$ (for which all the above assumptions in Steps 1–9 hold), we can accomplish N pairs of the above-described forward-propelling and length-recovering motions without violating the angle restriction (15.25). Then the total motion of z_1 to the left will be no less than

$$S(h_*, N) = Nh_* l \sin \theta_* - 2M(d_*)Nh_*^{1+K}.$$

In turn, its vertical motion will be bounded by $N(\bar{M}(d_*)h^{1+K} + M(d_*)h^{1+K})$, while the vertical motions of each of the points z_2 and z_3 will not exceed

$$Nh\left(l + \frac{lh}{2} + 2M(d_*)h^{1+K}\right) \sin \theta^* + 2M(d_*)Nh^{1+K}$$

(see (15.45), (15.33) and (15.35)).

Denote $Nh_* = D$. Then for any positive integer $N = N(h)$ and $h \in (0, h_*)$ such that

$$N(h)h = D,$$

we obtain that

$$S(h, N) \rightarrow Dl \sin \theta_* \quad \text{as } h \rightarrow 0+.$$

Simultaneously, the upper bounds for the vertical motions of each of the points z_2 and z_3 toward each other will converge to $Dl \sin \theta^*$, while the vertical shift of z_1 will tend to zero. In other words, we can move the point z_1 from its original position $z_1(0)$ to any position on the left of it (i.e., on the same horizontal line) within any distance under $Dl \sin \theta_*$ units as precisely as we want, while preserving the deviations of the distances between $z_1(t_1)$ and $z_j(t_1), j = 2, 3$ as close to l as we wish. This will give us the result of Theorem 15.1 in the case of forward motion only, provided that the target position for z_1 lies no further than $Dl \sin \theta_*$ to the left of $z_1(0)$.

Step 11. Recharging motion. Of course, it can happen that at least one of the inequalities in (15.25) will be about to be violated, but our swimmer still needs to "swim" forward to reach the desirable target position for z_1. In this case, we will need to apply the *"recharging"* motion as shown on Fig. 5, combined with

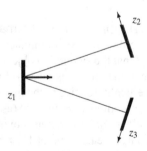

Figure 5 : Recharging motion with no length – recovering forces

necessary length-recovering motions (that is, *separated in time*), to "return" it to a configuration as on Fig. 2, for which the angle condition (15.25) holds. The methods and results of Steps 1–9 apply in this case as well.

This way, after finally many repetitions of the above-described actions, the center of the head-rectangle $S_1(z_1)$ will be able to approach any point on the left of $z_1(0)$ (i.e., on the same horizontal line) as close as we wish. This ends the proof of Theorem 15.1 in the case of forward motion when the points $z_j(t_2), z_j(t_4), \ldots, j = 2, 3$ remain symmetric relative to the horizontal line $z_1 H$.   □

Case II: The points $z_j(t_2), z_j(t_4), z_j(t_6), \ldots, j = 2, 3$ are NOT necessarily symmetrically positioned relative to the line $z_1 H$. In this case, applying the same scheme of actions as described in **Case I** in the above will yield the same steering result. Indeed, in **Case I** we did not deal with the actual incremental motions of points z_i's. Instead, we used in our calculations the *"worst possible" estimates for them*, which, in particular, include all the non-symmetrical shifts of the Case II. Namely, all incremental motions were evaluated in terms of the limiting angles θ_* and θ^*, which are the same in **Case I** and **Case II**.

4.3. Proof of Theorem 15.1: The turning motion. The turning motion needed to finish the proof of Theorem 15.1 (see Fig. 6) can be analyzed and performed in the same fashion as it is described in the above for the case of forward-propelling, length-recovering and recharging motions (i.e., incrementally, separating them in time).

For example, one may want to turn the swimmer *clockwise*. This can be achieved by subsequent rotation of its "legs" $z_1 z_2$ and $z_1 z_3$. Fig. 6 illustrates how one can rotate first the leg $z_1 z_3$, while not changing "much" the positions of z_1 and z_2:

In a similar way, we can further rotate the leg $z_1 z_2$, changing the orientation of rectangles $S_i(z_i)$'s respectively. After that, we can re-orient these rectangles into a configuration similar to that on Fig. 2 *now aimed at the new desirable direction* for further forward straight line motion. This ends the proof of Theorem 15.1.   □

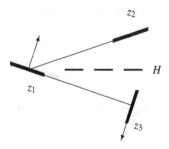

Figure 6 : Turning the leg $z_1 z_3$ clockwise

15.5 Proof of Corollary 15.1: The Case of Explicitly Conserved Angular Momentum

In this case of condition (15.4) the expression (15.3) is to be replaced with the following:

$$
\begin{aligned}
(F(z,v))(x,t) = {} & \xi_1(x,t)v_3(z_2(t) - z_1(t)) + \xi_2(x,t)v_3(z_1(t) - z_2(t)) \\
& + \xi_1(x,t)v_4(z_3(t) - z_1(t)) + \xi_3(x,t)v_4(z_1(t) - z_3(t)) \\
& - \xi_1(x,t)v_1(t) \begin{pmatrix} 0 & 1 \\ -1 & 0 \end{pmatrix} (z_2(t) - z_1(t)) \\
& + \xi_2(x,t)v_1(t) \begin{pmatrix} 0 & 1 \\ -1 & 0 \end{pmatrix} (z_2(t) - z_1(t)) \\
& - \xi_1(x,t)(-v_1) \frac{\| z_2(t) - z_1(t) \|_{R^2}^2}{\| z_3(t) - z_1(t) \|_{R^2}^2} \begin{pmatrix} 0 & 1 \\ -1 & 0 \end{pmatrix} (z_3(t) - z_1(t)) \\
& + \xi_3(x,t)(-v_1) \frac{\| z_2(t) - z_1(t) \|_{R^2}^2}{\| z_3(t) - z_1(t) \|_{R^2}^2} \begin{pmatrix} 0 & 1 \\ -1 & 0 \end{pmatrix} (z_3(t) - z_1(t)).
\end{aligned}
$$

$$(15.46)$$

Let us outline how Corollary 15.1 can be proved.

In Steps 5–8 in the above, when we deal with the length recovering motion, the rotation controls are inactive, which makes systems (15.1)–(15.3) and (15.1)–(15.4)) identical. Hence, this part of the proof applies to the latter system with no change.

This also allows us to assume that we can start the argument of Steps 1–4 for the forward motion of system (15.1)–(15.4) at the same position as given on Fig. 2 (otherwise, we would apply Steps 5–8 first) with the same *initial* constant controls $v_1 = -v_2$ as in (15.26).

Lemma 15.1 was extended along the lines of proof in Section 15.6 below (and in the same form with the exception that formulas (15.8)–(15.10) need to be re-written as given in (15.3)–(15.4) or (15.46) in place of (15.3)) in [80], [81] to the case of a system more general than (15.1)–(15.4), while Lemma 15.2 in the above remains valid for the expression (15.46) in place of (15.3).

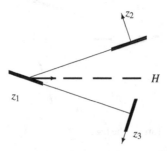

Figure 7 : Turning the leg $z_1 z_3$ clockwise for system $(1.1) - - (1.4)$

These two observations allow us to apply the argument of Steps 1–4 to system (15.1)–(15.4) as well, with the difference that errors R_i's will be evaluated with possibly different constants. The same reasoning applies to the recharging and the turning motions. For the former Fig. 5 remains the same, while for the latter we will deal with the changes in force distribution due to (15.4) as shown on Fig. 7.

This ends the proof of Corollary 15.1. □

15.6 Derivation of the Formula for Micromotions

We begin with the following result providing the foundation for the existence of parameter $d_* > 0$ in Lemma 15.1.

Lemma 15.5. *Let $t > 0$ be given. Assume that $v_i, i = 1,2,3,4$, and the orientation of $S_i(z_i(t))$'s do not change on $[0,t]$. Then:*

$$\sum_{i=1}^{3} \| z_i(\cdot) - z_i(0) \|_{(C[0,t])^2} \leq \beta(t) = \frac{3t}{\sqrt{rq}(1 - 3\alpha(t))} \| y(\cdot,0) \|_{(L^2(\Omega))^2}$$

$$+ \frac{3\alpha(t)}{1 - 3\alpha(t)} \mathrm{diam}\,(\Omega), \qquad (15.47)$$

provided that $\alpha(t) = t^2 \sum_{i=1}^{4} | v_i | < \frac{1}{3}$ ($\alpha(t)$ is defined in Section 15.2 before Lemma 15.1).

Proof of Lemma 15.5. Let us evaluate the following values:

$$\gamma_i(0) = \max_{\tau \in [0,t]} \| z_i(\tau) - z_i(0) \|_{R^2}, \quad i = 1,2,3.$$

It follows from (15.2) and the solution formula (15.12) that for $\tau \in [0,t]$ we have:

$$\| z_i(\tau) - z_i(0) \|_{R^2} \leq \frac{\tau}{\sqrt{rq}} \| y(\cdot,0) \|_{(L^2(\Omega))^2}$$

$$+ \tau^2 \{ (| v_3 | + | v_1 |) \| z_1 - z_2 \|_{(C[0,\tau])^2}$$

$$+ (| v_4 | + | v_2 |) \| z_1 - z_3 \|_{(C[0,\tau])}^2 \}, \quad i = 1,2,3. \quad (15.48)$$

Making use of the following inequalities:

$$\| z_j - z_1(\tau) \|_{(C[0,\tau])^2} \leq \gamma_1(0) + \gamma_j(0) + \| z_j(0) - z_1(0) \|_{R^2}, \quad j = 2,3,$$

we further obtain from (15.48) that:

$$\gamma_i(0) \leq \frac{t}{\sqrt{rq}} \parallel y(\cdot,0) \parallel_{(L^2(\Omega))^2} + t^2 \sum_{i=1}^{4} |v_i| \left(\gamma_1(0) + \gamma_2(0) + \gamma_3(0) + \text{diam}(\Omega)\right).$$

This yields (15.47) and ends the proof of Lemma 15.5.   □

Lemma 15.5 implies the following proposition.

Proposition 15.1. *Let on some interval $[0,\bar{t}]$ we apply constant controls $v_i, i = 1,2,3,4$, and the orientation of rectangles $S_i(z_i(t)), i = 1,2,3$ does not change in (15.1)–(15.3). Suppose that at the initial moment $t = 0$ the distances between these rectangles and from them to the boundary $\partial\Omega$ exceed some value $d_* > 0$. Then there is a $t_* \in (0,\bar{t}]$ such that the aforementioned distances will exceed this d_* for any $t \in [0,t_*]$.*

Proof of Lemma 15.1. We further assume that $t \in [0,t_*]$ as in Proposition 15.1.

Formula for $z_i(t) - z_i(0)$. The formal integration of (15.13) with $t_0 = 0$ yields
for $t \in [0,t_*]$:

$$z_i(t) - z_i(0) = \frac{t}{rq} \int_{S_i(z_i(0))} y(x,0)dx + \frac{t^2}{2rq} \int_{S_i(z_i(0))} (PF(z,v))(x,0)dx$$

$$+ \int_0^t u(\tau)d\tau + \int_0^t \tau v(\tau)d\tau, \quad i = 1,2,3. \tag{15.49}$$

Let the expressions $F_i(z,v), i = 1,2,3$ represent the terms in (15.3) supported respectively on $S_i(z_i(0)), i = 1,2,3$. Making use of (15.3) and Lemma 15.4, we further obtain, for example for $i = 1$:

$$\frac{1}{rq} \int_{S_1(z_1(0))} (PF(z,v))(x,0)dx = \frac{1}{rq} \int_{S_1(z_1(0))} (PF_1(z,v))(x,0)dx + \sqrt{rq}\zeta_1(0),$$

$$\tag{15.50}$$

where, in view of (15.22),

$$\parallel \zeta_1(0) \parallel_{R^2} \leq \frac{C}{d_*^4} \zeta(0), \tag{15.51}$$

$$\zeta(0) = |v_3| \parallel z_2(0) - z_1(0) \parallel_{R^2}$$
$$+ |v_4| \parallel z_3(0) - z_1(0) \parallel_{R^2}$$
$$+ |v_1| \parallel z_1(0) - z_2(0) \parallel_{R^2}$$
$$+ |v_2| \parallel z_1(0) - z_3(0) \parallel_{R^2}, \tag{15.52}$$

and

$$(F_1(z,v))(x,0) = \xi_1(x,0)[v_3(z_2(0) - z_1(0))$$
$$+v_4(z_3(0) - z_1(0)) - v_1 A(z_2(0) - z_1(0))$$
$$-v_2 A(z_3(0) - z_1(0))].$$
(15.53)

Evaluation of the 3-rd and the 4-th terms in (15.49). Making use of (15.14)–(15.15), we obtain:

$$\| \int_0^t u(\tau)d\tau \|_{R^2} \leq \frac{t}{\sqrt{rq}} \| y(\cdot,0) \|_{(L^2(\Omega))^2} + \frac{1}{rq} \| y(\cdot,0) \|_{(L^2(\Omega))^2}$$

$$\times 2t \max_{\tau \in [0,t]}$$

$$\times [\max\{\text{mes}\, \{S_i(z_i(0)) \backslash S_i(z_i(\tau)), \text{mes}\, S_i(z_i(\tau)) \backslash S_i(z_i(0))\}]^{1/2}.$$
(15.54)

Furthermore, making use of (15.16)–(15.19),

$$\| \int_0^t \tau v(\tau)d\tau \|_{R^2} \leq \frac{Ct^2}{2} \kappa(t),$$
(15.55)

where, in view of (15.5),

$$\kappa(t) = w_*(t) + \frac{1}{\sqrt{rq}} \| \gamma \|_{C[0,t]} + \frac{1}{\sqrt{rq}} \max_{\tau \in [0,t]} \max_{i=1,2,3} (\text{mes}^{1/2} \{S_i(z_i(\tau)) \backslash S_i(z_i(0))\}$$

$$+\text{mes}^{1/2} \{S_i(z_i(0)) \backslash S_i(z_i(\tau))\}) w(t)$$

$$\leq w_*(t) + \frac{1}{\sqrt{rq}} \| \gamma \|_{C[0,t]}$$

$$+\frac{1}{\sqrt{q}} \max_{i=1,2,3} \| z_i(\cdot) - z_i(0) \|_{(C[0,t])^2}^{1/2} w(t)$$
(15.56)

In turn, since in $f_j(0)$'s in (15.19) parameters $v_i, i = 1,2,3,4$ enter as factors, and

$$\frac{1 - e^{-\lambda_k t}}{\lambda_k t} \to 1 \quad \text{as } t \to 0,$$

we obtain from (15.19) that

$$\| \gamma \|_{C[0,t]} \leq \gamma_*(t,z(0)) \sum_{i=1}^4 | v_i |,$$
(15.57)

where $\gamma_*(t, z(0)) \to 0$ as $t \to 0$ for any $z(0)$ and $\gamma_*(t, z(0))$ does not depend on v_1, v_2, v_3 and v_4.

Evaluation of the 2-nd term in (15.49). This can be done making use of (15.21) in Lemma 15.3 and above estimates, recall that $\Pi_i b, i = 1, 2, 3$ stands for the *vector projections* of a vector $b \in R^2$ on the straight lines co-linear to the longer sides of the rectangles $S_i(z_i(t))$.

Combining (15.49)–(15.57) produces the following result:

$$\| R_i(t) \|_{R^2} \leq Ct^2 \left\{ \max\{q^a, q^{1-a}/r, r\} + \frac{\sqrt{rq}}{d_*^4} \right\} \zeta(0)$$

$$+ \frac{Ct^2}{\sqrt{rq}} \gamma_*(t, z(0)) \sum_{i=1}^{4} | v_i | + \frac{Ct}{\sqrt{rq}} \| y(\cdot, 0) \|_{(L^2(\Omega))^2}$$

$$+ Ct^2 w_*(t)$$

$$+ \frac{Ct^2}{\sqrt{q}} \max_{i=1,2,3} \| z_i(\cdot) - z_i(0) \|_{(C[0,t])^2}^{1/2} w(t), \quad i = 1, 2, 3, \quad (15.58)$$

where $C > 0$ is a (generic) constant.

Finally, noticing that in (15.58) (see (15.17)–(15.18)):

$$\| w(t) \|_{R^2} \leq \mathrm{diam}(\Omega) \sum_{i=1}^{4} | v_i |,$$

and, due to (15.47),

$$\| w_*(t) \|_{R^2} \leq 2\beta(t) \sum_{i=1}^{4} | v_i | = \frac{2\beta(t)\alpha(t)}{t^2},$$

we establish (15.11). This ends the proof of Lemma 15.1.   □

Remark 15.2. The result of Lemma 15.1 was extended in [80] to the case of micro motions of a swimming model similar to (15.1)–(15.3), but in the case of fish-like motion with the body containing finitely many parts and with the elastic structural forces that satisfy Hooke's Law.

15.7 Discussion of Main Results and Concluding Remarks

- Theorem 15.1 and Corollary 15.1 state that our swimmer can swim between any two points in Ω arbitrarily fast, regardless of the initial datum for the fluid. It might appear to be surprising at the first glance. However, the reader should recall here that we assume that the swimmer's recourses are *unlimited* in our models. If so, the results of Theorem 15.1 and Corollary 15.1 are quite natural and similar in this respect to the known global controllability results dealing with the processes of instantaneous speed of propagation (such as, e.g., the parabolic equation). Of course, if we change the problem and assume that the magnitudes of controls v_i's are a priori bounded and the duration of swimming is fixed, one should not expect Theorem 15.1 and Corollary 15.1 to hold. Furthermore, the position where we will be able to steer our swimmer will also depend on the initial fluid velocity y_0.

- We assume in this chapter that the orientation of rectangles $S_i(z_i(t))$'s can be changed instantaneously, i.e., without any rotating motion. It is definitely a substantial simplification of the problem. However, we can point out at a few "real life" situations (out of many available) when such type of "geometric control" already exists. Let us give just two examples:

  - Modern airplanes employ variable wing geometry, achieved, e.g., by making use of retractable parts. Respectively, the rectangles on Figs. 1 and 2 can also be viewed as, e.g., "retractable fins", which can be employed when necessary (for different spatial orientations).
  - Duck's feet make use of a similar "technology". A duck can either spread the membranes (web) in its feet or hide them, turn the feet and then spread membranes again to swim in a different direction.

Models that include rotation of rectangles $S_i(z_i)$'s around their centers z_i' pose an interesting problem for future research.

- To prove Theorem 15.1, we used a specific strategy to move the swimmer from one position in the space domain to another. Obviously, one can pick a different swimming strategy for a desirable motion as it is allowed by the (universal) incremental formulas (15.8)–(15.11). The results of Theorem 15.1 and Corollary 15.1 establish a principal possibility of controlled motion anywhere in the space domain for respective swimmers (where their size permits). This in turn provides a foundation for setup of suitable optimal control problems, that is, optimizing some criterion over all possible swimming motions between given initial and desirable target positions of the swimmer at hand.

- Although in the model (15.1)–(15.3) we deal with rectangles $S_i(z_i(t))$'s of equal size, our results can be extended to the case when they are different in size and/or shape, or when the swimming object consists of more than just three pieces (for example, it can be a model of a robotic eel or fish with multiple joints).

- The global controllability result of Theorem 15.1 is achieved for asymptotically small h, which in turn implies that the dimensions r and $q, q \ll r$ of rectangles $S_i(z_i(t))$'s are respectively small as well. More precisely, the given

  - initial datum $\{z(0), y_0\}$ for (15.1)–(15.3),
  - target position z_{1T} and
  - the desirable precision of steering

will determine the *necessary smallness of r and q*, for which a needed swimming motion can be achieved by means of our asymptotic technique described in the proof of Theorem 15.1. It seems plausible that this technique can be extended to swimmers that are not "asymptotically small."

- The fact that we deal with a fluid governed in a bounded domain by the nonstationary Stokes equation implies that our models (15.1)–(15.3)/(15.4) are infinite dimensional pde models, which we study in the *same original infinite dimensional framework* of controllability theory for pde's, without reducing them to systems of ODE's. From this viewpoint, it seems plausible that the methods we used in Chapter 15 can be extended to the case of the Navier-Stokes equation (making use of the methods developed to investigate its wellposedness in the 2-*D* case).

Part IV
Multiplicative Controllability Properties of the Schrödinger Equation

Chapter 16
Multiplicative Controllability for the Schrödinger Equation

Abstract In this chapter we discuss some recent results obtained for multiplicative controllability of the Schrödinger equation.

In recent years a substantial progress has been made in investigating the controllability properties of the Schrödinger equation governed by multiplicative control. In this chapter we will discuss some of these results due to Beauchard [11, 12, 14], Beauchard and Coron [16], Beauchard and Mirrahimi [18], Chambrion, Mason, Sigalotti and Boscain [26], Nersesyan [124] and others.

It should be noted that the Schrödinger equation that we study below has certain property which sets it apart from the other partial differential equations considered in this monograph. Namely, the L^2-norms of its solutions are conserved, regardless of the value of real-valued multiplicative control applied. Therefore, *all the results below deal with controllability properties on the unit L^2-sphere S,*

$$S = \{\varphi \mid \varphi \in L^2(\Omega, C), \ \int_\Omega \mid \varphi(x) \mid^2 dx = 1\} \subset L^2(\Omega, C),$$

where Ω is the system's space domain.

The Schrödinger equation with real-valued control is also *complex-conjugate time-reversible* in the sense that if control $u(t), t \in [0, T]$ steers it from u_0 to u_1 at time $t = T$, then control $u_*(t) = u(T - t)$ steers this equation from $\bar{u}_1$ to $\bar{u}_0$.

16.1 Exact Controllability Properties of the Schrödinger Equation

16.1.1 Local Controllability of Quantum Particle in the Case of the Fixed Potential Well

K. Beauchard studied in [11] the one dimensional nonrelativistic motion of a single charged particle of mass 1 with potential V in a uniform electrical field $u(t), t \geq 0$. Assuming that

$$V(x) = 0 \text{ for } x \in (-0.5, 0.5) = \Omega \text{ and } V(x) = \infty \ x \notin \Omega,$$

A.Y. Khapalov, *Controllability of Partial Differential Equations Governed by Multiplicative Controls*, Lecture Notes in Mathematics 1995,
DOI 10.1007/978-3-642-12413-6_16, © Springer-Verlag Berlin Heidelberg 2010

it can be described by the following initial and boundary value problem:

$$i\frac{\partial \psi}{\partial t}(t,x) = -\frac{1}{2}\frac{\partial^2 \psi}{\partial x^2}(t,x) - u(t)x\psi(t,x) \quad x \in \Omega, \ t \geq 0, \tag{16.1}$$

$$\psi(0,x) = \psi_0(x) \text{ in } \Omega, \quad \psi(t,-0.5) = \psi(t,0.5) = 0, \ t \geq 0.$$

The state of system (16.1) at time $t \geq 0$ is given by $\psi(\cdot,t)$ with $\int_\Omega |\psi(t,x)|^2 \, dx = 1$ and the control parameter is the *real-valued* electrical field $u(t), t \geq 0$. The solution of (16.1) is defined as follows.

Let T_1 and T_2 be two real numbers satisfying $T_1 \leq T_2$, $u : [T_1,T_2] \rightarrow R$ be a continuous function, and $\psi_0 \in H^2(\Omega,C) \cap H_0^1(\Omega,C)$ be such that $\| \psi_0 \|_{L^2(\Omega,C)} = 1$. Then a function $\psi : [T_1,T_2] \times \Omega \rightarrow C$ is a solution to (16.1) if:

- $\psi \in C^0([T_1,T_2]; H^2(\Omega,C) \cap H_0^1(\Omega,C)) \cap C^1([T_1,T_2]; L^2(\Omega,C))$,
- the equation (16.1) holds in $L^2(\Omega,C)$ for every $t \in [T_1,T_2]$,
- the initial condition holds in $L^2(\Omega,C)$.

We further refer to the pair (ψ,u) as a trajectory of (16.1).

It was pointed out by Turinici in [144], Chapter 4, that Theorem 9.2 (due to [8]) in Chapter 9 implies that system (16.1) is not controllable in the space $H^2(\Omega,C) \cap H_0^1(\Omega,C)$, where its solution naturally lies. However, as we discussed it Section 9.5 in Chapter 9, one can try to study controllability of (16.1) in more "narrow spaces".

In this respect the main result in [11], Theorem 16.1 below, gives an example of a specific trajectory $(\psi_1,0)$ of system (16.1) such that this system can *locally* be steered anywhere near it within the set $S \cap H_{(0)}^7(\Omega,C)$,

$$H_{(0)}^7(\Omega,C) = \{\varphi \mid \varphi \in H^7(\Omega,C), \ \varphi^{(2n)}(0.5) = \varphi^{(2n)}(-0.5) = 0, \ n = 0,1,2,3\}.$$

Namely,

$$\psi_1(t,x) = \sqrt{2}\cos(\pi x)e^{-i\frac{\pi^2}{2}t} = \varphi_1(x)e^{-i\frac{\pi^2}{2}t}, \tag{16.2}$$

where

$$\varphi_1(x) = \sqrt{2}\cos \pi x \text{ and } \pi^2/2 \tag{16.3}$$

are respectively the 1-st eigenfunction and eigenvalue of the associated uncontrolled spectral problem for (16.1), i.e., with $u = 0$.

Theorem 16.1. *Let $\alpha,\beta \in R$. There exist $T > 0$ and $\eta > 0$ such that for every ψ_0 and ψ_f in $S \cap H_{(0)}^7(\Omega,C)$ satisfying*

$$\| \psi_0 - \varphi_1 e^{i\alpha} \|_{H_{(0)}^7(\Omega,C)} < \eta, \quad \| \psi_f - \varphi_1 e^{i\beta} \|_{H_{(0)}^7(\Omega,C)} < \eta$$

there exists a trajectory (ψ,u) of controlled system (16.1) such that $\psi(0,\cdot) = \psi_0$ and $\psi(T,\cdot) = \psi_f$.

The proof of Theorem 16.1 (we refer for all details to [11]) explores the classical idea of using the inverse or implicit function theorem near the drifting trajectory of (16.1), complemented by the moment theory technique (the proof of Theorem 9.4 in Chapter 9 follows the same approach), and further combined with the return method and quasi-static transformations technique due to M. Coron [31, 32]. The proof employs the following strategy:

- It is possible to find a family of trajectories $(\tilde{\psi}, \tilde{u}), t \geq 0$ of system (16.1) such that its linearized systems around them are *controllable at some finite time $T > 0$.* Such trajectories can be selected as the explicit solutions to (16.1) similar to (16.2), (16.3), calculated for constant *non-zero* controls $u \equiv \gamma$ and for the initial conditions to be the 1-st associated eigenfunctions (in place of φ_1 in (16.2)). The moment theory can be used to establish the aforementioned (linear) controllability, which will require that $\gamma \neq 0$ (see [131] for the lack of controllability otherwise).
- Using the Nash-Moser implicit function theorem (in place of the classical one, due to some lack of regularity), it is then possible to show that system (16.1) is locally controllable as follows: *For any $(\tilde{\psi}, \tilde{u})$, one can steer (16.1) from any initial state within some neighborhood U_0 of $\tilde{\psi} \mid_{t=0}$ to any state within some neighborhood U_T of $\tilde{\psi} \mid_{t=T}$.*
- Then to steer (16.1) from ψ_0 to ψ_f:

  - one needs to select them "close enough" respectively to $\varphi_1 e^{i\alpha}$ and $\varphi_1 e^{i\beta}$,
  - in which case it is possible to steer (16.1) from ψ_0 to some ψ_2 as close as we wish to some $(\tilde{\psi}, \tilde{u}) \mid_{t=0}$, making use of quasi-static transformations (see Theorem 11 in [11]),
  - and then to move it from ψ_2 to a state $\psi_3 \in U_T$ near $(\tilde{\psi}, \tilde{u}) \mid_{t=T}$, making use of the above-mentioned local controllability of (16.1),
  - and, finally, move it from ψ_3 to ψ_f, again, making use of quasi-static transformations.

In [14] the author proposed a different version of this proof.   □

16.1.2 The Case of Potential Well of Variable Length

In [12] a different version of the Schrödinger equation was considered:

$$i\frac{\partial \psi}{\partial t}(t,x) = -\frac{\partial^2 \psi}{\partial x^2}(t,x)$$

$$+[\dot{u}(t) - 4u^2(t)]x^2\psi(t,x) \quad x \in \Omega = (0,1), \ t \geq 0, \quad (16.4)$$

$$\psi(0,x) = \psi_0(x) \text{ in } (0,1), \quad \psi(t,0) = \psi(t,1) = 0, \ t \geq 0,$$

where control u satisfies $u \in H_0^2((0,t_f),R)$, $\int_0^{t_f} u dt = 0$ and t_f is the duration of steering. The mixed problem (16.4) describes the motion of the particle in a potential of well of variable length.

The result of [12] is similar to Theorem 16.1 except for it establishes the local controllability from *some neighborhood of any eigenfunction* φ_{n_0} *to some neighborhood of any other eigenfunction* φ_{n_f} within the set $S \cap H^{5+\varepsilon}(\Omega,C), \varepsilon > 0$. Its physical interpretation is that the wave function can be moved from one eigenstate to another by changing the length of potential well.

The proof in [12] modifies the above-described proof of Theorem 16.1 to include a compactness argument that makes use of local controllability results around solutions which are (finite) linear combinations of trajectories like in (16.2) (in place of $(\tilde{\psi},\tilde{u})$ in the above). This argument is necessary to show that (16.4) can be moved between these trajectories, that is, "from one eigenfunction to the next one", and so forth. We refer for all details of this proof to the original paper. The sketch of the proof of Theorem 16.2 below employs a similar approach.   □

16.1.3 The Case of Moving Well

Consider a quantum particle of mass $m = 1$ with potential V,

$$V(x) = 0 \text{ for } x \in \Omega = (-0.5,0.5) \text{ and } V(x) = \infty \ x \notin \Omega,$$

in a non-Galilean frame of absolute position $D(t)$ in R^1. One can describe the state of this system as a triplet (ψ,S,D), satisfying the following system of equations:

$$i\frac{\partial \psi}{\partial t}(t,x) = -\frac{1}{2}\frac{\partial^2 \psi}{\partial x^2}(t,x) - u(t)x\psi(t,x)\ x \in \Omega = (-0.5,0.5),\ t \geq 0, \quad (16.5)$$

$$\psi(0,x) = \psi_0(x) \text{ in } \Omega,\ \psi(t,-0.5) = \psi(t,0.5) = 0,\ t \geq 0,$$

$$\dot{S}(t) = u(t),\ \dot{D}(t) = S(t),\ t \geq 0,\ S(0) = S_0,\ D(0) = D_0. \quad (16.6)$$

The goal of controllability problem is to control (ψ,S,D) by changing the acceleration of the well (box) $u(t), t \geq 0$. One can notice that system (16.1) is a part of (16.5)–(16.6). Respectfully, the following result due to Beauchard and Coron [16] can be viewed as an extension of Theorem 16.1.

Theorem 16.2. *For every natural number n there exists $\eta_n > 0$ such that for every natural numbers n_0, n_f, every (ψ_0,S_0,D_0), $(\psi_f,S_f,D_f) \in S \cap H_{(0)}^7(\Omega,C) \times R \times R$ with*

$$\| \psi_0 - \varphi_{n_0} \|_{H_{(0)}^7(\Omega,C)} + | S_0 | + | D_0 | < \eta_{n_0},$$

$$\| \psi_0 - \varphi_{n_f} \|_{H_{(0)}^7(\Omega,C)} + | S_f | + | D_f | < \eta_{n_f},$$

there exists a time $T > 0$ and a trajectory (ψ, S, D, u) of controlled system (16.5)–(16.6) on $[0, T]$ which satisfies

$$(\psi(0, \cdot), S(0), D(0)) = (\psi_0, S_0, D_0), \quad (\psi(T, \cdot), S(T), D(T)) = (\psi_f, S_f, D_f)$$

and $u \in H_0^1((0, T), R)$.

In other words, *given ψ_0 close enough to an eigenfunction φ_{n_0} of uncontrolled system (16.5) ((or (16.1)) and ψ_f close enough to another such eigenfunction φ_{n_f}, the wave function can be moved from ψ_0 to ψ_f in finite time*, whose duration depends on the choice of n_0 and n_f.

Strategy of the proof. Here we provide the plan of the proof, following the original paper [16].

It is enough to consider the case $n_0 = 1, n_f = n_0 + 1$. Respectively, the plan outlined below shows how one can move system (16.5)–(16.6) from a neighborhood of the fist eigenstate to a neighborhood of the second eigenstate.

It is possible to prove the local controllability of (16.5)–(16.6) around the following uncontrolled trajectories with $u = 0$ for every $\theta \in [0, 1]$:

$$Y^{\theta,0,0}(t) = (\psi_\theta(t), S(t) = 0, D(t) = 0),$$
$$\psi_\theta = \sqrt{1 - \theta} \psi_1(t) + \sqrt{\theta} \psi_2(t), \quad \theta \in (0, 1),$$
$$Y^{k,0,0}(t) = (\psi_{k+1}(t), S(t) = 0, D(t) = 0), \quad k = 0, 1,$$

where ψ_2 is calculated as in (16.2), $\varphi_n = \sqrt{2} \sin(n\pi x)$, $n = 2, 4, \ldots$, $\varphi_n = \sqrt{2} \cos(n\pi x)$, $n = 1, 3, \ldots$ and $\lambda_n = (n\pi)^2/2, n = 1, \ldots$.

Note that the classical approach to prove the aforementioned local controllability, based on establishing the controllability of linearized system and the use of implicit function (or inverse mapping) theorem, does not apply to the trajectories $Y^{\theta,0,0}(t)$, because the linearized system is not controllable in this case. Instead, the authors of [16] first established the local controllability of (16.5)–(16.6) up to codimension one (this is due to the fact that the linearized controllability result "misses" exactly two directions), making use of the argument as in Theorem 16.1. Namely, because of the loss of regularity, one has to apply the Nash-Moser theorem. Then it can be shown that the nonlinear term in (16.5)–(16.6) allows the system to move in the two directions which are not covered by the linearized system.

Then:

- there exists a nonempty open ball V_0 (resp. V_1) centered at $Y^{0,0,0}(0)$ (resp. $Y^{1,0,0}(0)$) such that (16.5)–(16.6) can be moved in finite time between any two points in V_0 (resp. V_1),
- for every $\theta \in (0, 1)$, there exists a nonempty open ball V_θ with center at $Y^{\theta,0,0}(0)$ such that this system can be moved in finite time between any two points in V_θ.
- Making use of a compactness argument, it can be shown that the following "curve" with respect to the parameter λ from a closed interval:

$$[Y^{0,0,0}(0), Y^{1,0,0}(0)] = \{\sqrt{\lambda} Y^{0,0,0}(0) + \sqrt{1 - \theta} Y^{1,0,0}(0) \mid \lambda \in [0, 1]\}$$

is compact in $L^2(I,R) \times R \times R$ and covered by $\bigcup_{0 \le \theta \le 1} V_\theta$. Therefore there exists an increasing finite family $\theta_n, n = 1, \ldots, N$ such that $[Y^{0,0,0}(0), Y^{1,0,0}(0)]$ is covered by $\bigcup_{1 \le n \le N} V_{\theta_n}$. We can assume that $V_n \cap V_{n+1} \ne \emptyset, n = 1, \ldots, N-1$. Given $Y_0 \in V_1, Y_f \in V_N$, we can move (16.5)–(16.6) from point Y_0 to a point $Y_1 \in Y_{\theta_1} \cap Y_{\theta_2}$, then from Y_1 to a point $Y_2 \in Y_{\theta_2} \cap Y_{\theta_2}$, and so forth till we reach some point $Y_f \in V_N$.

We refer for all details of the proof to [16].   □

16.2 Geometric Control Theory Approach to Approximate Controllability

In [26] T. Chambrion, P. Mason, M. Sigalotti and U. Boscain applied the geometric control approach to study the *approximate* multiplicative controllability of the Schrödinger equation in the case of several variables. Their methods follow the techniques and terminology developed for the finite dimensional systems governed by multiplicative controls. These methods were first applied to Galerkin approximations and then extended to the original equation by a density argument, making use of the fact that the dynamics preserve the Hilbert sphere.

A key ingredient of this approach in [26] is a time re-parametrization that inverts the roles of the drift and control operator, which allowed the authors to apply the techniques, developed in [1] for finite-dimensional systems on compact semisimple Lie groups, to the Galerkin approximations.

It should be noted that, in the traditions of the geometric control theory, the abstract results of [26] deal with solutions to the abstract evolution system like

$$\frac{d\psi}{dt}(t) = A\psi(t) + u(t)B\psi(t), \ t \ge 0,$$

which are *only* defined as a result of subsequent actions of sequences of constant controls:

$$\psi(t) = e^{(t - \sum_{l=1}^{j-1} t_l(A + u_j B)} \circ e^{t_{j-1})(A + u_{j-1}B)} \circ \ldots \circ e^{t_1(A + u_1 B)}(\psi(0)),$$

where

$$u(t) = u_j \text{ if } \sum_{l=1}^{j-1} t_l \le t \le \sum_{l=1}^{j} t_l.$$

Results of [26] apply to both bounded and unbounded domains. Below we will focus only on the former.

The authors of [26] considered the following form of the Schrödinger equation:

$$i\frac{\partial \psi}{\partial t}(t,x) = -\Delta \psi(t,x) + (V + u(t)W)\,\psi(t,x), \quad x \in \Omega, \; t \geq 0, \quad (16.7)$$

$$\psi\,|_{\partial \Omega} = 0, t \geq 0, \quad \psi(0,\cdot) = \psi_0 \in L^2(\Omega,C), \quad (16.8)$$

assuming that Ω is a bounded domain in R^n, control $u(\cdot)$ is a piecewise constant function, $u(t) \in U \subset R, t \geq 0$ and $V, W \in L^\infty(\Omega, R)$.

Denote by $\{\lambda_k, \phi_k, k = 1, \ldots\}$ the eigenvalues and orthonormal (in the L^2-norm) eigenfunctions solving the spectral problem

$$-\Delta \phi + V\phi = \lambda \phi, \quad \phi\,|_{\partial \Omega} = 0.$$

The main controllability result in [26] is as follows (Theorem 3.4 in [26]).

Theorem 16.3. *Let $(0,\delta) \subset U$ for some $\delta > 0$. Assume that $(\lambda_{k+1} - \lambda_k), k = 1, \ldots$ are Q-linear independent and*

$$\int_\Omega W\phi_k\phi_{k+1}dx \neq 0, \quad k = 1,\ldots.$$

Then system (16.7), (16.8) is approximately controllable in $L^2(\Omega,C)\cap S$ (that is, from any initial state in $L^2(\Omega,C)\cap S$ to a dense subset on this sphere).

In the above, the elements of the sequence $\{\lambda_{k+1} - \lambda_k\}_{k=1}^\infty$ are said to be Q-linearly independent if for every positive integer n and $(q_1, \ldots, q_n) \in Q^n\setminus\{0\}$ one has $\sum_{k=1}^{K} q_n(\lambda_{k+1} - \lambda_k) \neq 0$.

It should be noted that the space for controllability in Theorem 16.3 is $L^2(\Omega,C)$ (intersected with S), which is less regular than the spaces in Section 16.1 for the (exact) controllability of the Schrödinger equation. Below we discuss a similar controllability due to Nersesyan [124], see Theorem 16.4.

Applications of Theorem 16.3 in [26] include two examples: the harmonic oscillator with unbounded space domain and the 3-D well of potential, both controlled by suitable potentials. In the latter case, the governing equation is as follows:

$$i\frac{\partial \psi}{\partial t}(t,x) = -\Delta \psi(t,x) + u(t)W\psi(t,x) \quad x \in \Omega = (0,l_1) \times (0,l_2) \times (0,l_3), \; t \geq 0.$$

$$(16.9)$$

Theorem 16.3 yields that if $(l_1 l_2)^2, (l_1 l_3)^2, (l_2 l_3)^2$ are Q-linear independent,

$$W(x_1,x_2,x_3) = e^{\alpha_1 x_1 + \alpha_2 x_2 + \alpha_3 x_3}, \quad \alpha_i \neq 0, \; \alpha_i \in R, i = 1,2,3,$$

and $(\pi/\alpha_i l_i)^2$'s are algebraically independent, then (16.9), (16.8) is approximately reachable in $L^2(\Omega,C)\cap S$.

16.3 Approximate Controllability via Stabilization

16.3.1 Stabilization Near Eigenstates

In [18] the authors considered the system (16.1) with control constructed as a feedback $u = u(\psi) \in R, t \geq 0$:

$$i\frac{\partial \psi}{\partial t}(t,x) = -\frac{1}{2}\frac{\partial^2 \psi}{\partial x^2}(t,x) - u(\psi)x\psi(t,x), \quad x \in \Omega, \ t \geq 0, \qquad (16.10)$$

$$\psi(0,x) = \psi_0(x) \text{ in } \Omega, \quad \psi(t,-0.5) = \psi(t,0.5) = 0, \ t \geq 0.$$

They showed that solutions to (16.10) can be steered as close as one wishes to the eigenstates $\varphi_{n,\sigma}$ that solve the following spectral problem:

$$A_\sigma \varphi_{n,\sigma} = \lambda_{n,\sigma}\varphi_{n,\sigma}, \quad \varphi_{n,\sigma}(-0.5) = \varphi_{n,\sigma}(0.5) = 0,$$

where $\sigma \in R, A_\sigma \varphi = -\frac{1}{2}\frac{\partial^2 \varphi}{\partial x^2} - \sigma x\varphi, D(A_\sigma) = H^2(\Omega,C)\bigcap H_0^1(\Omega,C)$.

More precisely, given any $\varepsilon > 0, s > 0, \Gamma > 0, \gamma \in (0,1), n$ there exists a $\sigma_* > 0$ such that for every $\sigma \in (-\sigma_*, \sigma_*)$ there is a suitable feedback law $u(\psi)$ such that for every $\psi_0 \in H^2(\Omega,C)\bigcap H_0^1(\Omega,C)\bigcap S\bigcap D(A_\sigma^{s/2})$,

$$\| \psi_0 \|_{D(A_\sigma^{s/2})} \leq \Gamma, \quad |< \psi_0, \varphi_{n,\sigma} >| > \gamma,$$

where $< \cdot, \cdot >$ is the inner product in $L^2(\Omega,C)$, the respective solution to (16.10) satisfies

$$\limsup_{t \to \infty} \text{dist}_{L^2(\Omega,C)}(\psi(t,\cdot), \mathscr{C}_{n,\sigma}) < \varepsilon,$$

where $\mathscr{C}_{n,\sigma} = \{\varphi_{n,\sigma}e^{i\theta} \mid \theta \in [0,2\pi)\}$.

16.3.2 Approximate Controllability via Stabilization

In [124] V. Nersesyan studied approximate controllability of system (16.7)–(16.8) under the following conditions (of the nature similar to conditions in Theorem 16.3):

Assumption 1.

- Boundary $\partial\Omega$ of the domain Ω is smooth and $V,W \in C^\infty(\bar{\Omega},R)$,
- $< W\phi_1, \phi_j > \neq 0, j = 2,\ldots,$
- $\lambda_1 - \lambda_j \neq \lambda_k - \lambda_l$ for all $j,k,l \geq 1$ such that $\{1,j\} \neq \{k,l\}, j \neq 1$.

Under these conditions $\psi \in C([0,\infty), L^2(\Omega,C)\bigcap S)$ (recall that $S \subset L^2(\Omega,C)$) for $u \in L^1_{loc}([0,\infty),R)$. The main controllability result in [124] is as follows.

Theorem 16.4 (Approximate controllability in S). *Let Assumption 1 hold. Then for any $\varepsilon > 0$ $\psi_0, \psi_1 \in S$ there is $T > 0$ and a suitable control u such that the respective solution to (16.7), (16.8) satisfies*

$$\| \psi(T, \cdot) - \psi_1 \|_{L^2(\Omega, C)} < \varepsilon.$$

Sketch of the proof.

Step 1. The proof of Theorem 16.4 is heavily based on the following stabilization result for a version of system (16.7), (16.8) with feedback:

$$i\frac{\partial z}{\partial t}(t,x) = -\Delta z(t,x) + (V(x) + u(z)W(x)) z(t,x) \ \ x \in \Omega, \ t \geq 0, \quad (16.11)$$

$$z\mid_{\partial\Omega} = 0, \ z(0, \cdot) = z_0 \in L^2(\Omega, C),$$

where

$$u(z) = -\delta \operatorname{Im}\left(< \alpha(-\Delta + V)P_1(Wz), (-\Delta + V)P_1(z) > - < Wz, \phi_1 >< \phi_1, z >\right),$$
$$(16.12)$$

t$\delta, \alpha > 0$ and $P_1 z = z - < z, \phi_1 > \phi_1$ is the orthogonal projection of z in $L^2(\Omega, C)$ onto the closure of the vector span of $\{\phi_k\}_{k \geq 2}$.

Introduce the following Lyapunov function:

$$\mathcal{V}(z) = \alpha \| (-\Delta + V)P_1 z \|^2_{L^2(\Omega, C)} + 1 - |< z, \phi_1 >|^2.$$

Theorem 16.5. *Under Assumption 1, there is a finite or countable set J of positive numbers such that for any $\alpha \notin J$ and $z_0 \in S \cap H^2(\Omega, C) \cap H_0^1(\Omega, C)$ with $< z_0, \phi_1 > \neq 0$ and $0 < \mathcal{V}(z_0) < 1$ there is a sequence $\{k_n \geq 1\}_{n=1}^{\infty}$ verifying*

$$\| z(k_n, \cdot) - c\phi_1 \|_{H^2(\Omega, C)} \to 0$$

as n increases, where $c \in C, |c| = 1$.

Step 2. Making use of the fact that the space $H^2(\Omega, C) \cap H_0^1(\Omega, C)$ is dense in $L^2(\Omega, C)$ and the fact that the distance between two solutions to the Schrödinger equation with the same control remains constant in L^2-norm, without loss of generality we can assume that $\psi_0, \psi_1 \in H^2(\Omega, C) \cap H_0^1(\Omega, C)$ and conditions of Theorem 16.5 hold for these two functions treated as z_0.

Step 3. Theorem 16.5 yields that we can steer system (16.7), (16.8) from ψ_0 to some $\hat{\psi}$ in the $\varepsilon/4$-neighborhood of ϕ_1. It remains to show that we can steer this system from $\hat{\psi}$ to the ε-neighborhood of ψ_1.

Step 4. Similarly to Step 3, we can steer system (16.7), (16.8) from $\bar{\psi}_1$ to some ψ_* in the $\varepsilon/4$-neighborhood of ϕ_1. The complex-conjugate time-reversibility of equation (16.7) implies that we can also steer system (16.7), (16.8) from $\bar{\psi}_*$ to ψ_1 applying some control u_*.

Step 5. Since the distance between two solutions to the Schrödinger equation with the same control remains constant in L^2-norm, we can apply control u_* to steer (16.7), (16.8) from $\hat{\psi}$ to the ε-neighborhood of ψ_1. This ends the sketch of the proof of Theorem 16.4. $\square$

References

1. A. Agrachev and T. Chambrion, An estimation of the controllability time for single-input systems on compact Lie groups, *ESAIM Control Optim. Calc. Var.*, **12** (2006), pp. 409–441.
2. F. Alouges, A. DeSimone, and A. Lefebvre, Optimal Strokes for Low Reynolds Number Swimmers: An Example, *J. Nonlinear Sci.*, (2008) 18: 277–302.
3. W. Alt and D.A. Lauffenberger, Transient behavior of a chemotaxis system modeling certain types of tissue information, *J. Math. Biol.*, **24** (1987), pp. 691–722.
4. S. Anita and V. Barbu, Null controllability of nonlinear convective heat equations, *ESAIM Control Optim. Calc. Var.*, **5**, (2000), pp. 157–173.
5. A. Baciotti, *Local Stabilizability of Nonlinear Control Systems*, Series on Advances in Mathematics and Applied Sciences, vol. 8, World Scientific, Singapore, 1992.
6. J.M. Ball and M. Slemrod, Feedback stabilization of semilinear control systems, *Appl. Math. Opt.*, **5** (1979), pp. 169–179.
7. J.M. Ball and M. Slemrod, Nonharmonic Fourier series and the stabilization of distributed semi-linear control systems, *Comm. Pure. Appl. Math.*, **32** (1979), pp. 555–587.
8. J.M. Ball, J.E. Mardsen, and M. Slemrod, Controllability for distributed bilinear systems, *SIAM J. Contr. Opt.*, 1982, pp. 575–597.
9. V. Barbu, Exact controllability of the superlinear heat equation, *Appl. Math. Opt.*, 42 (2000), pp. 73–89.
10. C. Bardos and L. Tartar, Sur l'unicité rétrograde des équations paraboliques et quelques questions voisines, *Arch. Rational Mech. Anal.*, **50** (1973), pp. 10–25.
11. K. Beauchard, Local controllability of a 1-D Schrödinger equation, *L. Math. Pures Appl*, **84** (2005), pp. 851–956.
12. K. Beauchard, Controllability of a quantum partucle in a 1-D variable domain, *ESAIM: COCV*, **14** (2008), pp. 105–147.
13. K. Beauchard, Local controllability of a 1-D beam equation, *SIAM J. Cont. Optim.*, **47** (2008), pp. 1219–1273.
14. K. Beauchard, Local controllability of a 1-D bilinear Schrödinger equation: a simpler proof, preprint.
15. K. Beauchard, Local controllability and noncontrollability of a 1-D wave equation, preprint.
16. K. Beauchard and J.-M. Coron, Controllability of a quantum partucle in a moving potential well, *J. Finc. Anal.*, **232** (2006), pp. 328–389.
17. K. Beauchard, J.-M. Coron and P. Rouchon, Controllability issues for continuous-spectrum systems and ensemble controllability of Bloch equations, *Comm. Math. Phys.*, to appear.
18. K. Beauchard and M. Mirrahimi, Practical stabilization of a quantum particle in a one-dimensional infinite square potential well, *SIAM J. Cont. Optim.*, **48** (2009), pp. 1179–1205.
19. L. E. Becker, S. A. Koehler and H. A. Stone, On self-propulsion of micro-machines at low Reynolds number: Purcell's three-link swimmer, *J. Fluid Mech.*, (2003), 490:15–35.
20. C. Bardos, G. Lebeau, J. Rauch, Sharp sufficient conditions for the observation, control and stabilization of waves from the boundary, *SIAM J. Control and Opt.*, **30** (1992), pp. 1024–1065.

21. M.E. Bradley, S. Lenhart, and J. Yong, Bilinear optimal control of the velocity term in a Kirchhoff plate equation, *J. Math. Anal. Appl.*, **238** (1999), 451–467.
22. A.G. Butkovskii, *Structural Theory of Distributed Systems*, Ellis Horwood, Chichester, 1983.
23. P. Cannarsa and A.Y. Khapalov, Multiplicative controllability for reaction-diffusion equations with target states admitting finitely many changes of sign, Special issue of Discr. Cont. Dyn. Systems, Ser. B in honor of D.L. Russell, in press.
24. P. Cannarsa, V. Komornik, and P. Loreti, Well-posedness and control of semilinear evolution equations with iterated logarithms, *ESAIM: COCV*, **4** (1999), pp. 37–56.
25. A. Chambolle and F. Santosa, Control of the wave equation by time-dependent coefficient, *ESAIM: Contrôle, Optimisation et Calcul des Variations*, **8** (2002), pp. 375–392.
26. T. Chambrion, P. Mason, M. Sigalotti and U. Boscain, Controllability of the discrete-spectrum Schrödinger equation driven by an external field, *Ann. Inst. H. Poincaré Anal. Non Linéaire* **26** (2009), pp. 329–349.
27. S. Childress, *Mechanics of swimming and flying*, Cambridge University Press, 1981.
28. J.E. Colgate and K.M. Lynch, Mechanics and control of swimming: A Review, *IEEE J. Oceanic Engr.*, **29** (2004), pp. 660–673.
29. C. Conca, J. San Martin, M. Tucsnak, Existence of solutions for the equations modelling the motion of a rigid body in a viscous fluid, *Comm. Partial Differential Equations*, **25** (2000), pp. 1019–1042.
30. R. Cortez, The method of regularized stokeslets, *SIAM. J. Sci. Comp.*, **23** (2001), pp. 1204–1225.
31. J.-M. Coron, Global asymptotic stabilization for controllable systems without drift, *Math. Control Signals Systems*, **5** (1992), pp. 295–312.
32. J.-M. Coron, Local controllability of 1-D tank containing fluid modeled by the shallow water equations, *ESAIM COCV*, **8** (2002), pp. 513–554.
33. R. Dáger and E. Zuazua, *Wave propagation, observation and control in 1-d flexible multi-structures* , New York : Springer, 2006.
34. S. Dolecki and D.L. Russel, A general theory of observatiion and control, *SIAM. J. Contr.*, **15** (1977), pp. 185–219.
35. S. Ervedoza and J.-P. Puel, Approximate controllability for a system of Schrödinger equations modeling a single trapped ion, *Ann. Inst. H. Poincaré Anal. Non Linéaire*, **26** (2009), pp. 2111–2136.
36. C. Fabre, J.-P. Puel and E. Zuazua, Approximate controllability for the semilinear heat equations, *Proc. Royal Soc. Edinburg*, **125A** (1995), pp. 31–61.
37. T. Fakuda et al, Steering mechanism and swimming experiment of micro mobile robot in water, . *Proc. Micro Electro Mechanical Systems (MEMS'95)* pp. 300–305, 1995.
38. H.O. Fattorini and D.L. Russell, *Uniform bounds on biorthogonal functions for real exponentials with an application to the control theory of parabolic equations*, Quarterly of Appl. Mathematics, April, 1974, pp. 45–69.
39. H.O. Fattorini, Local controllability of a nonlinear wave equation, *Math. Systems Theory,* **9** (1975), pp.30–45. pp. 45–69.
40. L.J. Fauci and C.S. Peskin, A computational model of aquatic animal locomotion, *J. Comp. Physics*, **77** (1988), pp. 85–108.
41. L.J. Fauci, Computational modeling of the swimming of biflagellated algal cells, *Contemporary Mathematics*, **141** (1993), pp. 91–102.
42. E. Feireisl, On the motion of rigid bodies in a viscous compressible fluid, *Arch. Rational Mech. Anal.*, **167** (2003), pp. 281–308.
43. E. Fernández-Cara, Null controllability of the semilinear heat equation, *ESAIM: COCV,* (1997), pp. 87–103.
44. L. A. Fernándes and E. Zuazua, Approximate controllability of the semilinear heat equation via optimal control, *J. Optim. Theory Appl.*, 101 (1999), no. 2, 307–328.
45. E. Fernández-Cara and E. Zuazua, The cost of approximate controllability for heat equations: the linear case, *Adv. Differential Equations*, **5** (2000), pp. 465–514.

46. E. Fernández-Cara and E. Zuazua, Null and approximate controllability for weakly blowing-up semilinear heat equations, *Annales de l'Institut Henri Poincare, Analyse non linéaire*, 17 (5) (2000), pp. 583–616.
47. L.A. Fernández, Controllability of some semilnear parabolic problems with multiplicativee control, a talk presented at the Fifth SIAM Conference on Control and its applications, held in San Diego, July 11–14, 2001.
48. L.A. Fernández and E. Zuazua, Approximate controllability for the semilinear heat equation involving gradient terms, *J. Optim. Theory Appl.*, **101** (1999), pp. 307–328.
49. A. Friedman, *Partial Differential equations of Parabolic Type*, Prentice-Hall, 1964.
50. A. Fursikov and O. Imanuvilov, *Controllability of evolution equations*, Lect. Note Series 34, Res. Inst. Math., GARC, Seoul National University, 1996.
51. G.P. Galdi, On the steady self-propelled motion of a body in a viscous incompressible fluid, *Arch. Ration. Mech. Anal.*, **148** (1999), no. 1, 53–88.
52. I.C. Gohberg and M.G. Krein, *Introduction to the theory of linear non-self-adjoint operators*, AMS Transl., vol 18, American Mathematical Society, Providence, RI, 1969.
53. J. Gray, Study in animal locomotion IV - the propulsive power of the dolphin, *J. Exp. Biology*, **10** (1032), pp. 192–199.
54. J. Gray and G.J. Hancock, The propulsion of sea-urchin spermatozoa, *J. Exp. Biol.*, **32**, 802, 1955.
55. S. Guo et al., Afin type of micro-robot in pipe, *Proc. of the 2002 Int. Symp. on micromechatronics and human science (MHS 2002)*, pp. 93–98, 2002.
56. M.F. Hawthorne, J.I. Zink, J.M. Skelton, M.J. Bayer, Ch. Liu, E. Livshits, R. Baer and D. Neuhauser, Electrical or photocontrol of rotary motion of a metallacarborane, *Science*, f303, 1849, 2004.
57. J. Henry, Étude de la contrôlabilité de certaines équations paraboliques non linéaires, *Thèse détat, Université Paris VI*, 1978.
58. S. Hirose, *Biologically inspired robots: Snake-like locomotors and manipulators*, Oxford University Press, Oxford, 1993.
59. E. Kanso, J.E. Marsden, C.W. Rowley, and J. Melli-Huber Locomotion of Articulated Bodies in a Perfect Fluid, *J. Nonlinear Science*, 15(4) pp. 255–289, 2005.
60. J. Keener and J. Sneyd, *Mathematical Physiology*, Springer.
61. A.Y. Khapalov, Some aspects of the asymptotic behavior of the solutions of the semilinear heat equation and approximate controllability, *J. Math. Anal. Appl.*, **194** (1995), pp. 858–882.
62. A.Y. Khapalov, Exact controllability of second-order hyperbolic equations under impulse controls, *Applicable Analysis*, **63** (1996), pp. 223–238.
63. A.Y. Khapalov, Approximate controllability properties of the semilinear heat equation with lumped controls, *Int. J. Applied Math. & Comp. Sci.*, **9** (1999), pp. 751–765.
64. A.Y. Khapalov, Global approximate controllability properties for the semilinear heat equation with superlinear nonlinear term, *Revista Matematica Complutense*, **12** (1999), pp. 511–535.
65. A.Y. Khapalov, Approximate controllability and its well-posedness for the semilinear reaction-diffusion equation with internal lumped controls, *ESAIM: Contrôle, Optimisation et Calcul des Variations*, **4** (1999), pp. 83–98.
66. A.Y. Khapalov, A class of globally controllable semilinear heat equations with superlinear terms, *J. Math. Anal. Appl.*, **242** (2000), pp. 271–283.
67. A.Y. Khapalov, Bilinear control for global controllability of the semilinear parabolic equations with superlinear terms, *the Special volume "Control of Nonlinear Distributed Parameter Systems,"* dedicated to David Russell, G. Chen/I. Lasiecka/J. Zhou Eds., 2001, Marcel Dekker, pp. 139–155.
68. A.Y. Khapalov, Exact null-controllability for the semilinear heat equation with superlinear nonlinear term and mobile internal controls, *Nonlinear Analysis: TMA*, **43** (2001), pp. 785–801.
69. A.Y. Khapalov, Mobile point controls versus locally distributed ones for the controllability of the semilinear parabolic equation, *SIAM J. Contr. Opt.*, **40** (2001), pp. 231–252.

70. A.Y. Khapalov, Global non-negative controllability of the semilinear parabolic equation governed by bilinear control, *ESAIM: Contrôle, Optimisation et Calcul des Variations*, **7** (2002), pp. 269–283.
71. A.Y. Khapalov, On bilinear controllability of the parabolic equation with the reaction-diffusion term satisfying Newton's Law, the special issue of the *Journal of Computational and Applied Mathematics*, dedicated to the memory of J.-L. Lions, **21** (2002), pp. 1–23.
72. A.Y. Khapalov, *Controllability of the semilinear parabolic equation governed by a multiplicative control in the reaction term: A qualitative approach*, SIAM J. Control. Optim., **41** (2003), pp. 1886–1900.
73. A.Y. Khapalov, Bilinear controllability properties of a vibrating string with variable axial load and damping gain, *Dynamics of Continuous, Discrete, and Impulsive systems*, **10** (2003), pp. 721–743.
74. A.Y. Khapalov, Controllability properties of a vibrating string with variable axial load, *Discrete and Cont. Dynamical Systems*, **11** (2004), pp. 311–324.
75. A.Y. Khapalov, Reachability of nonnegative equilibrium states for the semilinear vibrating string by varying its axial load and the gain of damping, ESAIM: Contr., Optimisation et Calcul des Variations, 12 (2006), pp. 231–252.
76. A.Y. Khapalov, The well-posedness of a model of an apparatus swimming in the 2-*D* Stokes fluid, Techn. Rep. 2005-5, *Washington State University, Department of Mathematics, Tech. Rep. Ser.*, http://www.math.wsu.edu/TRS/2005-5.pdf.
77. A.Y. Khapalov, Local controllability for a swimming model, SIAM J. Control. Optim., **46** (2007), pp. 655–682.
78. A.Y. Khapalov, Geometric aspects of force controllability for a swimming model, *Appl. Math. Opt.*, **57** (2008), pp. 98–124.
79. A.Y. Khapalov, Global controllability for a swimming model, submitted, also available as Techn. Rep. 2007-11, *Washington State University, Department of Mathematics, Tech. Rep. Ser.*, http://www.math.wsu.edu/TRS/2007-11.pdf.
80. A.Y. Khapalov, Micro motions of a 2-*D* swimming model governed by multiplicative controls, *Nonlinear Analysis: TMA,* **71** (2009), pp. 1970–1979.
81. A.Y. Khapalov, Swimming models and controllability, *Proc. Int. Conf. (Fes, Morocco) Sys. Theory: Modeling, Analysis and Control*, pp. 241–248, 2009.
82. A.Y. Khapalov and Sh. Eubanks, The well-posedness of a 2-D swimming model governed in the nonstationary Stokes fluid by multiplicative controls, *Applicable Analysis*, **88** (2009) pp. 1763–1783.
83. A.Y. Khapalov, W. Kolodziej, D. Kosterev, and R.R. Mohler, Controllability and placement of FACTS devices for power modulation and transient stability control, Proc. *1995 Stockholm Power Tech, Intern. Symp. Electr. Power Engineering*, paper SPT IS-06-3, Stockholm, Sweden, June 18–22, 1995.
84. A.Y. Khapalov, R.R. Mohler, R. Vedam, and R. Zakrzewski, On bilinear control design methodology, an invited session, IFAC'96, San Francisco, Pergamon Press.
85. A.Y. Khapalov and R.R. Mohler, Global controllability of the power transmission network, *Proc. 35th IEEE Conf. on Decision and Control,* Kyoto, Japan, Pergamon Press, 1996.
86. A.Y. Khapalov and R.R. Mohler, Reachable sets and controllability of bilinear time-invariant systems: A qualitative approach, *IEEE Trans. on Autom. Control,* 41 (1996), pp. 1342–1346.
87. A.Y. Khapalov and R.R. Mohler, Global asymptotic stabilizability of the bilinear time-invariant system via piecewise constant feedback, *Systems & Control Letters,* **33** (1997), pp. 47–54.
88. A.Y. Khapalov and R.R. Mohler, On global controllability of time-invariant nonhomogeneous bilinear system, In *Dynamics and control (Sopron, 1995), 71–79, Stability Control Theory Methods Appl.,* **9** (1999), Gordon and Breach, Amsterdam.
89. A.Y. Khapalov and R.R. Mohler, Bilinear control and application to Flexible a.c. Transmission Systems, *J. Optim. Th. Appl.,* **105** (2000), pp. 621–637 (Special Issue honoring D.G. Luenberger).
90. K. Kime, Simultaneous control of a rod equation and a simple Schrödinger equation, *Systems Control Lett.*, 24 (1995), pp. 301–306.

91. J. Koiller, F. Ehlers, and R. Montgomery, Problems and progress in microswimming, *J. Nonlinear Sci.*, (1996) 6: 507–541.
92. A.N. Kolmogorov and S.V. Fomin, *Elements of the theory of functions and functional analysis*, Imprint Rochester, N.Y., Graylock Press, 1957.
93. V. Komornik, *Exact controllability and stabilization. The multiplier method*, RAM: Research in Applied Mathematics. Masson, Paris; John Wiley & Sons, Ltd., Chichester, 1994. 156 pp.
94. V. Komornik, Vilmos and P. Loreti, *Fourier series in control theory*, Springer Monographs in Mathematics. Springer-Verlag, New York, 2005, 226 pp.
95. Jan, Kucera, Solution in large of control problem $\dot{x} = (Au + Bv)x$, *Czechoslovak Mathematical Journal*, **17** (1967), pp. 91–96.
96. O.H. Ladyzhenskaya, *The mathematical theory of viscous incompressible flow*, Cordon and Breach, New-York, 1963.
97. O.H. Ladyzhenskaya, V.A. Solonikov and N.N. Ural'ceva, "Linear and Quasi-linear Equations of Parabolic Type," AMS, Providence, Rhode Island, 1968.
98. J.E. Lagnese, Control of wave processes with distributed controls supported on subregion, *SIAM J. Control Opt.*, **21** (1983), pp. 68–85.
99. J. Lagnese and J.-L. Lions, *Modeling, analysis and control of thin plates,* Masson, Paris, 1988.
100. J. Lagnese, G. Leugering, and G., E.J.P.G. Schmidt, *Modelling, Analysis and Control of Multi-Link Flexible Structures*, Basel, Birkhäuser, 1994, 398 p.
101. I. Lasiecka, *Mathematical Control Theory of Coupled PDE's*, CBMS-NSF Series, vol 75, SIAM, 2002.
102. I. Lasiecka and R. Triggiani, *Exact controllability of semilinear abstract systems with application to waves and plates boundary control problems,* Appl. Math. Optim.,23 (1991), pp. 109–154.
103. I. Lasiecka and R. Triggiani, *Differential and Algebraic Riccati Equations with applications to boundary/point control problems: Continuous theory and approximation theory*, Vol. 164, Springer-Verlag Lecture Notes Series in Control and Information Sciences, 1991, 160 pp.
104. I. Lasiecka and R. Triggiani, *Research monographs: Control Theory for Partial Differential Equations: Continuous and Approximation Theories*, Vol. 1 (680 pp.), and Vol. 2 (422 pp.), Encyclopedia of Mathematics and Its Applications Series, Cambridge University Press, January 2000.
105. S. Lenhart, Optimal control of convective-diffusive fluid problem, *Math. Models and Methods in Appl. Sci.,* **5** (1995), pp. 225–237.
106. S. Lenhart and M. Liang, Bilinear optimal control for a wave equation with viscous damping, *Houston J. Math.*, **26** (2000), 575–595.
107. M. Liang, Bilinear optimal control for a wave equation, *Math. Models Methods Appl. Sci.*, **9** (1999), 45–68.
108. M.J. Lighthill, *Mathematics of biofluiddynamics*, Philadelphia, Society for Industrial and Applied Mathematics, 1975.
109. P. Lin, H. Gao and Xu Liu, Some results on controllability of a nonlinear degenerate parabolic system by bilinear control, *J. Math. Anal.Appl.*, **326** (2007), pp. 1149–1160.
110. P. Lin, , P. Leid, and H. Gao, Bilinear control system with the reaction-diffusion term satisfying Newton's law, *ZAMM - J. Appl. Math. Mech.*, **87**, (2006), pp. 14–23.
111. J.L. Lions, Contrôlabilité exacte des systèmes distribués, *C.R. Acad. Sci. Paris, Sér. I Math.*, **302** (1986), pp. 471–475.
112. J.L. Lions, Exact controllaiblity, stabilizability and perturbations for distributed systems, *SIAM Rev.*, **30** (1988), pp. 1–68.
113. J.L. Lions, Remarques sur la contrôlabilité approchée, In *Jornadas Hispano-Francesas sobre Control de Sistemas Distribuídos*, pp. 77–87, University of Malaga (Spain), 1991.
114. D.G. Luenberger, *Optimization by vector space methods*, John Wiley, New-York, 1969.
115. W.A.J. Luxemburg, and J. Korevaar, *Entire functions and Müntz-Szász type approximation*, Trans. of the AMS,157 (1971), pp. 23–37.
116. R. Mason and J.W. Burdick, Experiments in carangiform robotic fish locomotion, *Proc. IEEE Int. Conf. Robotics and Automation*, pp. 428–435, 2000.

117. K.A. McIsaac and J.P. Ostrowski, Motion planning for dynamic eel-like robots, *Proc. IEEE Int. Conf. Robotics and Automation*, San Francisco, 2000, pp. 1695–1700.
118. S. Martinez and J. Cortés, Geometric control of robotic locomotion systems, *Proc. X Fall Workshop on Geometry and Physics, Madrid, 2001, Publ. de la RSME*, Vol. 4 (2001), pp. 183–198.
119. V.P. Mikhailov, *Partial Differential Equations*, Mir, Moscow, 1978.
120. R. Mohler, *Nonlinear Systems, Vol. I Dynamics and Control and Vol. II Applications to Bilinear Control*, Prentice-Hall, Englewood Cliffs, New Jersey, 1991.
121. K.A. Morgansen, V. Duindam, R.J. Mason, J.W. Burdick and R.M. Murray, Nonlinear control methods for planar carangiform robot fish locomotion, *Proc. IEEE Int. Conf. Robotics and Automation*, pp. 427–434, 2001.
122. S. Müller, Strong convergence and arbitrarily slow decay of energy for a class of bilinear control problems, *J. Differential Equations*, **81** (1989), pp. 50–67.
123. A. Najafi and R. Golestanian, Simple swimmer at low Reynolds numbers: Three linkes spheres, *Phys. rev. E* **69**, 062901-1-2 (2004).
124. V. Nersesyan, Growth of Sobolev norms and controllability of the Schrödinger equation, *Comm. Math. Phys.*, **290** (2009), pp. 371–387.
125. C.S. Peskin, Numerical analysis of blood flow in the heart, *J. Comp. Physics*, **25** (1977), pp. 220–252.
126. C.S. Peskin and D.M. McQueen, A general method for the computer simulation of biological systems interacting with fluids, *SEB Symposium on biological fluid dynamics*, Leeds, England, July 5–8, 1994.
127. E.M. Purcell, Life at Low Reynolds Number, American Journal of Physics vol 45, pages 3–11, 1977.
128. M. Reed and B. Simon, *Methods of modern mathematical physics, I: Functional analysis*, Academic Press, 1972, London, 325p.
129. M. Reed and B. Simon, *Methods of modern mathematical physics, IV: Analysis of operators*, Academic Press, 1978, London, 396p.
130. R. Rink and R. Mohler, Completely Controllable Bilinear Systems, *SIAM J. Contr. Opt.*, **6** (1968), pp. 477–486.
131. P. Rouchon, Control of a quantum particule in a moving potential well, *Second IFAC Workshop on Lagrangian and Hamiltonian Methods for Nonlinear Control*, Seville, 2003.
132. D. Russell, Nonharmonic Fourier series in the control theory of distributed parameter systems, *J. Math. Anal. Appl.*, **18** (1967), pp. 542–560.
133. J. San Martin, T. Takashi and M. Tucsnak, A control theoretic approach to the swimming of microscopic organisms , *Quart. Appl. Math.*, **65** (2007), pp. 405–424.
134. J. San Martin, J.-F. Scheid, T. Takashi and M. Tucsnak, An initial and boundary value problem modeling of fish-like swimming, *Arch. Ration. Mech. Anal.*, **188** (2008), 429–455.
135. M. Sigalotti and J.-C. Vivalda, Controllability properties of a class of systems modeling swimming microscopic organisms, *ESAIM: COCV*, DOI: 10.1051/cocv/2009034, Published online August 11, 2009.
136. W.I. Smirnow, *Lehrgang der h"heren Mathematik. Teil IV/1*, (Translated from the Russian. Hochschulbucher fur Mathematik, 5a. VEB Deutscher Verlag der Wissenschaften, Berlin, 1988. 300 pp.
137. K.R. Symon, *Mechanics*, Addison-Wesley series in Physics, 1971.
138. D. Tataru, *Carleman estimates, unique continuation and controllability for anizotropic PDE's*, Contemporary Mathematics, 209 (1997), pp. 267–279.
139. G.I. Taylor 1951 Analysis of the swimming of microscopic organisms. Proc. R. Soc. Lond. A 209, 447–461, 1951.
140. G.I. Taylor, Analysis of the swimming of long and narrow animals, *Proc. R. Soc. Lond. A* **214**, 1952.
141. R. Temam, *Navier-Stokes equations*, North-Holland, 1984.
142. A.N. Tichonov and A.A. Samarski, "Partial differential equations of mathematical physics", Vol. 1, Holden-Day, 1964.

143. M.S. Trintafyllou, G.S. Trintafyllou and D.K.P. Yue, Hydrodynamics of fishlike swimming, *Ann. Rev. Fluid Mech.*, **32** (2000), pp. 33–53.

144. M. Defranceschi, C. Le Bris, *Mathematical Models and Methods for Ab Initio Quantum Chemistry*, Springer-Verlag, Berlin/New York, 2000.

145. T.Y. Wu, Hydrodynamics of swimming fish and cetaceans, *Adv. Appl. Math.*, **11** (1971), pp. 1–63.

146. E. Zuazua, Exact controllability for the semilinear wave equation, *J. Math. pures et appl.,* **69** (1990), p. 1–31.

147. E. Zuazua, Exact controllability for the semilinear wave equations in one space dimension, *Annales de l'Institut Henri Poincare, Analyse non linéaire,* **10** (1993), pp. 109–129.

148. E. Zuazua, Recent results on the large time behavior for scalar parabolic conservation laws. *2nd European Conference on elliptic and parabolic problems. "Elliptic and parabolic problems"*, Pont à Mousson, 1994, C. Bandle, J. Bemelmans, M. Chipot, J. Saint Jean Paulin & I. Shafrir eds., *Pitman Research Notes in Mathematics Series 325*, Longman Scientific & Technical, 1995.

149. E. Zuazua, Finite dimensional null controllability for the semilinear heat equation, *J. Math. Pures et Appl.*, **76** (1997), pp. 237–264.

150. E. Zuazua, Controllability of partial differential equations and its semi-discrete approximation, *Discrete and Continuous Dynamical Systems*, **8** (2002), pp. 469–513.

151. E. Zuazua, Remarks on the controllability of the Schrdinger equation. A. Bandrauk, M.C. Delfour, and C. Le Bris, eds. *Quantum Control: mathematical and numerical challenges, CRM Proc. Lect. Notes Ser.*, **33**, AMS Publications, Providence, R.I., 2003, pp. 181–199.

152. E. Zuazua, Controllability and observability of partial differential equations: Some results and open problems, in *Handbook of Differential Equations: Evolutionary Equations, vol. 3*, C. M. Dafermos and E. Feireisl eds., Elsevier Science, 2006, pp. 527–621.

Index

Lecture Notes in Mathematics

For information about earlier volumes
please contact your bookseller or Springer
LNM Online archive: springerlink.com

Recent Reprints and New Editions

LECTURE NOTES IN MATHEMATICS  Springer

Edited by J.-M. Morel, F. Takens, B. Teissier, P.K. Maini

Editorial Policy (for the publication of monographs)

1. Lecture Notes aim to report new developments in all areas of mathematics and their applications - quickly, informally and at a high level. Mathematical texts analysing new developments in modelling and numerical simulation are welcome.

   Monograph manuscripts should be reasonably self-contained and rounded off. Thus they may, and often will, present not only results of the author but also related work by other people. They may be based on specialised lecture courses. Furthermore, the manuscripts should provide sufficient motivation, examples and applications. This clearly distinguishes Lecture Notes from journal articles or technical reports which normally are very concise. Articles intended for a journal but too long to be accepted by most journals, usually do not have this "lecture notes" character. For similar reasons it is unusual for doctoral theses to be accepted for the Lecture Notes series, though habilitation theses may be appropriate.

2. Manuscripts should be submitted either online at www.editorialmanager.com/lnm to Springer's mathematics editorial in Heidelberg, or to one of the series editors. In general, manuscripts will be sent out to 2 external referees for evaluation. If a decision cannot yet be reached on the basis of the first 2 reports, further referees may be contacted: The author will be informed of this. A final decision to publish can be made only on the basis of the complete manuscript, however a refereeing process leading to a preliminary decision can be based on a pre-final or incomplete manuscript. The strict minimum amount of material that will be considered should include a detailed outline describing the planned contents of each chapter, a bibliography and several sample chapters.

   Authors should be aware that incomplete or insufficiently close to final manuscripts almost always result in longer refereeing times and nevertheless unclear referees' recommendations, making further refereeing of a final draft necessary.

   Authors should also be aware that parallel submission of their manuscript to another publisher while under consideration for LNM will in general lead to immediate rejection.

3. Manuscripts should in general be submitted in English. Final manuscripts should contain at least 100 pages of mathematical text and should always include

   - a table of contents;
   - an informative introduction, with adequate motivation and perhaps some historical remarks: it should be accessible to a reader not intimately familiar with the topic treated;
   - a subject index: as a rule this is genuinely helpful for the reader.

   For evaluation purposes, manuscripts may be submitted in print or electronic form (print form is still preferred by most referees), in the latter case preferably as pdf- or zipped ps-files. Lecture Notes volumes are, as a rule, printed digitally from the authors' files. To ensure best results, authors are asked to use the LaTeX2e style files available from Springer's web-server at:

   ftp://ftp.springer.de/pub/tex/latex/svmonot1/ (for monographs) and
   ftp://ftp.springer.de/pub/tex/latex/svmultt1/ (for summer schools/tutorials).

Additional technical instructions, if necessary, are available on request from: lnm@springer.com.

4. Careful preparation of the manuscripts will help keep production time short besides ensuring satisfactory appearance of the finished book in print and online. After acceptance of the manuscript authors will be asked to prepare the final LaTeX source files and also the corresponding dvi-, pdf- or zipped ps-file. The LaTeX source files are essential for producing the full-text online version of the book (see http://www.springerlink.com/openurl.asp?genre=journal&issn=0075-8434 for the existing online volumes of LNM).

The actual production of a Lecture Notes volume takes approximately 12 weeks.

5. Authors receive a total of 50 free copies of their volume, but no royalties. They are entitled to a discount of 33.3% on the price of Springer books purchased for their personal use, if ordering directly from Springer.

6. Commitment to publish is made by letter of intent rather than by signing a formal contract. Springer-Verlag secures the copyright for each volume. Authors are free to reuse material contained in their LNM volumes in later publications: a brief written (or e-mail) request for formal permission is sufficient.

Addresses:
Professor J.-M. Morel, CMLA,
École Normale Supérieure de Cachan,
61 Avenue du Président Wilson, 94235 Cachan Cedex, France
E-mail: Jean-Michel.Morel@cmla.ens-cachan.fr

Professor F. Takens, Mathematisch Instituut,
Rijksuniversiteit Groningen, Postbus 800,
9700 AV Groningen, The Netherlands
E-mail: F.Takens@rug.nl

Professor B. Teissier, Institut Mathématique de Jussieu,
UMR 7586 du CNRS, Équipe "Géométrie et Dynamique",
175 rue du Chevaleret,
75013 Paris, France
E-mail: teissier@math.jussieu.fr

For the "Mathematical Biosciences Subseries" of LNM:

Professor P.K. Maini, Center for Mathematical Biology,
Mathematical Institute, 24-29 St Giles,
Oxford OX1 3LP, UK
E-mail: maini@maths.ox.ac.uk

Springer, Mathematics Editorial, Tiergartenstr. 17,
69121 Heidelberg, Germany,
Tel.: +49 (6221) 487-259
Fax: +49 (6221) 4876-8259
E-mail: lnm@springer.com